Forest-Tree Gene Regulation in Response to Abiotic and Biotic Stress

Forest-Tree Gene Regulation in Response to Abiotic and Biotic Stress

Editor

Yuepeng Song

MDPI • Basel • Beijing • Wuhan • Barcelona • Belgrade • Manchester • Tokyo • Cluj • Tianjin

Editor
Yuepeng Song
Beijing Forestry University
China

Editorial Office
MDPI
St. Alban-Anlage 66
4052 Basel, Switzerland

This is a reprint of articles from the Special Issue published online in the open access journal *Forests* (ISSN 1999-4907) (available at: https://www.mdpi.com/journal/forests/special_issues/gene_regulation_stress).

For citation purposes, cite each article independently as indicated on the article page online and as indicated below:

LastName, A.A.; LastName, B.B.; LastName, C.C. Article Title. *Journal Name* **Year**, *Volume Number*, Page Range.

ISBN 978-3-0365-5947-6 (Hbk)
ISBN 978-3-0365-5948-3 (PDF)

Contents

About the Editor vii

Yue Xiao, Menglei Wang and Yuepeng Song
Abiotic and Biotic Stress Cascades in the Era of Climate Change Pose a Challenge to Genetic Improvements in Plants
Reprinted from: *Forests* **2022**, *13*, 780, doi:10.3390/f13050780 1

Jiaojiao Wu, Jingyan Wang, Wenkai Hui, Feiyan Zhao, Peiyun Wang, Chengyi Su and Wei Gong
Physiology of Plant Responses to Water Stress and Related Genes: A Review
Reprinted from: *Forests* **2022**, *13*, 324, doi:10.3390/f13020324 5

Xiaojuan Liu, Yingying Yang, Huihui Xu, Dan Yu, Quanxin Bi and Libing Wang
Transcriptome Analysis of Apricot Kernel Pistils Reveals the Mechanisms Underlying ROS-Mediated Freezing Resistance
Reprinted from: *Forests* **2022**, *13*, 1655, doi:10.3390/f13101655 21

Jiahui Dong, Shance Niu, Ji Qian, Juan Zhou, Mengnan Zhao, Yu Meng and Bao Di
Identification of Aquaporin Gene Family in Response to Natural Cold Stress in *Ligustrum* × *vicaryi* Rehd.
Reprinted from: *Forests* **2022**, *13*, 182, doi:10.3390/f13020182 35

Rongxue Li, Baseer Ahmad, Delight Hwarari, Dong'ao Li, Ye Lu, Min Gao, Jinhui Chen and Liming Yang
Genomic Survey and Cold-Induced Expression Patterns of *bHLH* Transcription Factors in *Liriodendron chinense* (Hemsl) Sarg.
Reprinted from: *Forests* **2022**, *13*, 518, doi:10.3390/f13040518 55

Shuang Sun, Xingxing Liang, Hu Chen, La Hu and Zhangqi Yang
Identification of AP2/ERF Transcription Factor Family Genes and Expression Patterns in Response to Drought Stress in *Pinus massoniana*
Reprinted from: *Forests* **2022**, *13*, 1430, doi:10.3390/f13091430 75

Miaomiao Ma, Lu Li, Xuhui Wang, Chunyan Zhang, Solme Pak and Chenghao Li
Comprehensive Analysis of GRAS Gene Family and Their Expression under GA3, Drought Stress and ABA Treatment in *Larix kaempferi*
Reprinted from: *Forests* **2022**, *13*, 1424, doi:10.3390/f13091424 95

Ting Wang, Ying Hu, Hu Chen, Jianhui Tan, Huilan Xu, Peng Li, Dongshan Wu, Jie Jia and Zhangqi Yang
Transcriptome Analysis of Response to Aluminum Stress in *Pinus massoniana*
Reprinted from: *Forests* **2022**, *13*, 837, doi:10.3390/f13060837 113

Qiaoli Chen, Feng Wang and Danlei Li
Coexpression Network Analysis Based Characterisation of the R2R3-MYB Family Genes in Tolerant Poplar Infected with *Melampsora larici-populina*
Reprinted from: *Forests* **2022**, *13*, 1255, doi:10.3390/f13081255 127

Xuri Hao, Lei Liu, Peng Liu, Menglei Wang and Yuepeng Song
Genome-Wide Identification of miRNAs and Its Downstream Transcriptional Regulatory Network during Seed Maturation in *Tilia tuan*
Reprinted from: *Forests* **2022**, *13*, 1750, doi:10.3390/f13111750 149

About the Editor

Yuepeng Song

Dr. Yuepeng Song is currently a Professor at the College of Biological Sciences and Technology, Beijing Forestry University. He researches molecular mechanisms of DNA methylation and noncoding RNAs on gene regulation for abiotic stress tolerance in perennial woody plants. He has published 35 peer-reviewed publications, owned seven patents and received two of the ministry of education's natural science awards. Presently, he is undertaking two research projects supported by the National Natural Science Foundation of China (NSFC).

forests

MDPI

Editorial

Abiotic and Biotic Stress Cascades in the Era of Climate Change Pose a Challenge to Genetic Improvements in Plants

Yue Xiao [1,2], Menglei Wang [1,2] and Yuepeng Song [1,2,*]

1 National Engineering Research Center of Tree Breeding and Ecological Restoration, College of Biological Sciences and Technology, Beijing Forestry University, No. 35 Qinghua East Road, Beijing 100083, China; yuexiao33@bjfu.edu.cn (Y.X.); lea2021@bjfu.edu.cn (M.W.)

2 Key Laboratory of Genetics and Breeding in Forest Trees and Ornamental Plants, Ministry of Education, College of Biological Sciences and Technology, Beijing Forestry University, No. 35 Qinghua East Road, Beijing 100083, China

* Correspondence: yuepengsong@bjfu.edu.cn

Citation: Xiao, Y.; Wang, M.; Song, Y. Abiotic and Biotic Stress Cascades in the Era of Climate Change Pose a Challenge to Genetic Improvements in Plants. *Forests* **2022**, *13*, 780. https://doi.org/10.3390/f13050780

Received: 30 April 2022
Accepted: 14 May 2022
Published: 18 May 2022

Forest ecosystems are vast, second in expanse only to marine ecosystems. Prior to widespread deforestation, forests covered two-thirds of the Earth's surface. Forest ecosystems have high biological productivity and have the potential to maintain carbon and oxygen balances and mitigate the temperature increases associated with global climate change. However, climate change, which is primarily driven by anthropogenic greenhouse gas emissions, poses a severe challenge to such ecosystems [1]. The Intergovernmental Panel on Climate Change's (IPCC) Sixth Assessment Report (AR6) indicated that CO_2 concentrations have risen from approximately 280 ppm prior to the industrial revolution to approximately 410 ppm in 2019, and average temperatures between 2011 and 2020 were 1.09 °C higher than those of the preindustrial period (1850–1900). The frequency and intensity of extreme thermal events will significantly increase with each 0.5 °C increase in temperature. In addition, each 0.5 °C of warming will significantly alter precipitation regimes, increasing agricultural and ecological drought in some regions [1]. As climate risks continue to increase, forest ecosystems will reach their limits.

The effects of climate change are most apparent in terms of CO_2 concentrations, temperature, rainfall intensity, and the probability of extreme weather events. In particular, extreme heat, extreme drought and intense rainfall will become more frequent and widespread. In addition, a growing population will also contribute to the need to reduce the damage of low temperatures and saline–alkali to trees by using high latitude and saline–alkali lands. Of course, researchers have investigated the means by which trees resist such abiotic stresses. Stress alters numerous physiological processes, including photosynthesis and transpiration, as well as chlorophyll content, and plants transmit signals through ABA-dependent and ABA-independent pathways to synthesize transcription factors and promote the expression of stress-resistance genes. For example, MYB transcription factors are induced, and DREB transcription factors are produced under drought, salt, and low-temperature stress, thus enhancing resistance to these stresses. The analysis of molecular mechanisms in plants under stress has great practical value for improving stress resistance in trees. Nevertheless, studies need to consider the compound effects of such stresses in conjunction with climate change, as abiotic stresses operate at longer and more intense time scales. Although CO_2 policies have been introduced, they have not been effective. However, from the perspective of improving plant photosynthesis, higher CO_2 concentrations are not entirely negative, and the effects of changes in CO_2 concentrations may be particularly complex in perennial woody plants. Hence, it is important to better understand the regulatory mechanisms forest trees employ in response to multiple concurrent stresses. In addition, low-carbon energy policies have spurred the development of new energy sources, which may produce heavy metals (HMs), plastics, and radionuclides (Figure 1). Researchers have proposed using plants to absorb HMs from soil and have pointed to genetic improvements in perennial woody plants that can enhance such remediation efforts. Food safety is currently the subject of much attention, as plastic waste is released and degrades into microplastics and nanoplastics that accumulate in plants. One study explored the uptake and accumulation of nanoplastics in Arabidopsis thaliana and demonstrated that these contaminants inhibit plant growth [2]. How the accumulation of microplastics and nanoplastics may affect forests and the underlying physiological and molecular mechanisms involved have yet to be studied. Moreover, the effects of radionuclides on forest growth and development can be severe and lead to die offs. This is a topic that requires further research.

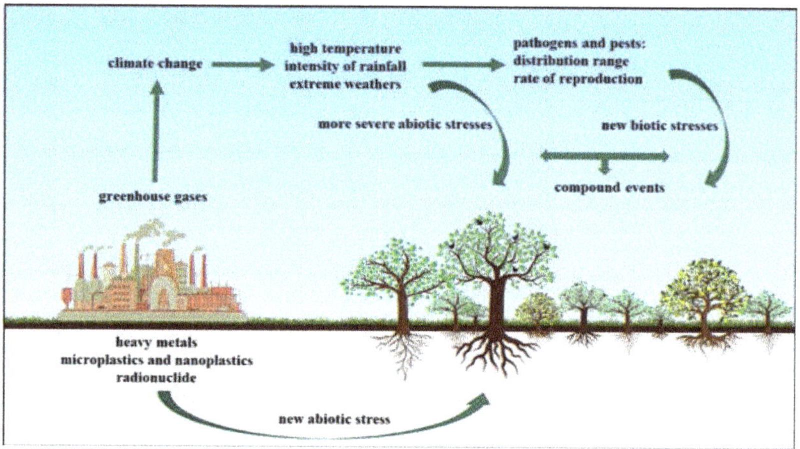

Figure 1. This picture shows the impact of human activities on climate change and new stresses. Climate change will affect temperature, intensify of rainfall and extreme weather, and environmental change will further affect the distribution range and reproduction rate of pathogens and pests. There may also be multiple abiotic stresses simultaneous compound events or biotic and abiotic stress simultaneous compound events.

Biotic stresses, such as pathogen and pest infestations, are also affected by changes in temperature and humidity. Warmer, more humid conditions tend to accelerate the development of pests and facilitate range expansions. Range expansions of beneficial rhizosphere microorganisms can benefit plants by improving their stress resistance and remediating polluted soils. However, the spread of pests is concerning. Adverse climatic conditions have preceded spruce budworm outbreak episodes, leading to tree mortality [3]. This implies that new pathogens and pests could emerge in areas where they have not previously occurred, increasing damage to trees. In addition, extreme weather events can trigger outbreaks of pathogens and pests. Models developed to predict future disease outbreaks [4,5] can be used to proactively plan for such events. When pathogenic bacteria infect plants, they release effectors into host cells that inhibit defense responses. However, plants can recognize these effectors and initiate an immune response, but the process by which the effectors target the host is not fully understood. The combined effects of abiotic and biotic stresses could cause massive mortality events. Essentially, interactions among multiple phenomena are expected to be more harmful to trees than a single phenomenon or event. Hence, it is important to understand the resistance mechanisms trees have to cope with concurrent biotic and abiotic stresses.

Global climate change will substantially impact forest ecosystems, which, as the largest carbon pool on Earth, plays a critical role in mitigating climate change. It is crucial to understand how climate change affects the growth and development of trees, how multiple concurrent stresses affect their regulatory mechanisms, and how trees regulate the effects of novel stresses. Exploring these issues requires ongoing work on forest genomes and the understanding of their complex regulatory network. Furthermore, new techniques can be applied to this end. For example, the bioaccumulation and transport of microplastics and nanoplastics have been studied in vegetables and other crops using europium chelate Eu-β-diketonate doped polystyrene particles with a diameter of 200 nm [6]; this tracer technique can also be used to study the effects of such plastics on perennial woody plants. In the future, such new technologies should be used to study tree stress to better understand how forests may respond to future challenges.

Author Contributions: Conceptualization, Y.S.; Investigation, Y.S.; writing—original draft preparation, Y.X.; Writing—Review and Editing, Y.S. and M.W.; Visualization, Y.X. and Y.S.; Supervision, Y.S.; All authors have read and agreed to the published version of the manuscript.

Conflicts of Interest: The author declares no conflict of interest.

References

1. IPCC. *Climate Change 2021: The Physical Science Basis. Contribution of Working Group I to the Sixth Assessment Report of the Intergovernmental Panel on Climate Change*; Cambridge University Press: Cambridge, UK; New York, NY, USA, 2021; pp. 113–119.
2. Sun, X.D.; Yuan, X.Z.; Jia, Y.; Feng, L.J.; Zhu, F.P.; Dong, S.S.; Liu, J.; Kong, X.; Tian, H.; Duan, J.L.; et al. Differentially charged nanoplastics demonstrate distinct accumulation in Arabidopsis thaliana. *Nat. Nanotechnol.* **2020**, *15*, 755–760. [CrossRef] [PubMed]
3. De Grandpré, L.; Kneeshaw, D.D.; Perigon, S.; Boucher, D.; Marchand, M.; Pureswaran, D.; Girardin, M.P. Adverse climatic periods precede and amplify defoliator-induced tree mortality in eastern boreal North America. *J. Ecol.* **2019**, *107*, 452–467. [CrossRef]
4. Matsuhashi, S.; Hirata, A.; Akiba, M.; Nakamura, K.; Oguro, M.; Takano, K.T.; Nakao, K.; Hijioka, Y.; Matsui, T. Developing a point process model for ecological risk assessment of pine wilt disease at multiple scales. *For. Ecol. Manag.* **2020**, *463*, 118010. [CrossRef]
5. Bosso, L.; Luchi, N.; Maresi, G.; Cristinzio, G.; Smeraldo, S.; Russo, D. Predicting current and future disease outbreaks of *Diplodia sapinea* shoot blight in Italy: Species distribution models as a tool for forest management planning. *For. Ecol. Manag.* **2017**, *400*, 655–664. [CrossRef]
6. Luo, Y.; Li, L.; Feng, Y.; Li, R.; Yang, J.; Peijnenburg, W.J.; Tu, C. Quantitative tracing of uptake and transport of submicrometre plastics in crop plants using lanthanide chelates as a dual-functional tracer. *Nat. Nanotechnol.* **2022**, *17*, 424–431. [CrossRef] [PubMed]

forests

Review

Physiology of Plant Responses to Water Stress and Related Genes: A Review

Jiaojiao Wu, Jingyan Wang, Wenkai Hui, Feiyan Zhao, Peiyun Wang, Chengyi Su and Wei Gong *

College of Forestry, Sichuan Agricultural University, Chengdu 611130, China; jwu0929@163.com (J.W.); wangjingyan@sicau.edu.cn (J.W.); wkxi@sicau.edu.cn (W.H.); zfywawzj1@163.com (F.Z.); wangpeiyunxj@163.com (P.W.); scy904525759@163.com (C.S.)
* Correspondence: gongwei@sicau.edu.cn

Abstract: Drought and waterlogging seriously affect the growth of plants and are considered severe constraints on agricultural and forestry productivity; their frequency and degree have increased over time due to global climate change. The morphology, photosynthetic activity, antioxidant enzyme system and hormone levels of plants could change in response to water stress. The mechanisms of these changes are introduced in this review, along with research on key transcription factors and genes. Both drought and waterlogging stress similarly impact leaf morphology (such as wilting and crimping) and inhibit photosynthesis. The former affects the absorption and transportation mechanisms of plants, and the lack of water and nutrients inhibits the formation of chlorophyll, which leads to reduced photosynthetic capacity. Constitutive overexpression of 9-cis-epoxydioxygenase (NCED) and acetaldehyde dehydrogenase (ALDH), key enzymes in abscisic acid (ABA) biosynthesis, increases drought resistance. The latter forces leaf stomata to close in response to chemical signals, which are produced by the roots and transferred aboveground, affecting the absorption capacity of CO_2, and reducing photosynthetic substrates. The root system produces adventitious roots and forms aerenchymal to adapt the stresses. Ethylene (ETH) is the main response hormone of plants to waterlogging stress, and is a member of the ERFVII subfamily, which includes response factors involved in hypoxia-induced gene expression, and responds to energy expenditure through anaerobic respiration. There are two potential adaptation mechanisms of plants ("static" or "escape") through ETH-mediated gibberellin (GA) dynamic equilibrium to waterlogging stress in the present studies. Plant signal transduction pathways, after receiving stress stimulus signals as well as the regulatory mechanism of the subsequent synthesis of pyruvate decarboxylase (PDC) and alcohol dehydrogenase (ADH) enzymes to produce ethanol under a hypoxic environment caused by waterlogging, should be considered. This review provides a theoretical basis for plants to improve water stress tolerance and water-resistant breeding.

Keywords: drought stress; waterlogging stress; plant morphology; physiology and biochemistry; transcription factor

Citation: Wu, J.; Wang, J.; Hui, W.; Zhao, F.; Wang, P.; Su, C.; Gong, W. Physiology of Plant Responses to Water Stress and Related Genes: A Review. *Forests* **2022**, *13*, 324. https://doi.org/10.3390/f13020324

Academic Editor: Yuepeng Song

Received: 5 January 2022
Accepted: 8 February 2022
Published: 16 February 2022

1. Introduction

In recent years, drought and waterlogging stress have seriously affected the growth of plants due to extreme climate change; these stresses are an important limiting factor for global agricultural and forestry productivity [1]. Over the past decade, the total area of the world's drylands has increased dramatically, with a clear upward trend in the scope, extent and frequency of drought, resulting in a total global loss of crop production of approximately $30 billion [2,3]. Waterlogging is the second most important climate disaster after drought. Since the 1990s, the scope of waterlogging disasters has been expanding year by year, and the frequency has also been increasing [4,5]. Due to the frequency and severity of drought and waterlogging, the global vegetation loss caused by these stresses is equivalent. The response and adaptation mechanisms of plants have been the focus of physiological and ecological research related to water stress (including drought stress and waterlogging stress), and are also very important for breeding water-tolerant varieties.

When plants are damaged by water stress, they will respond to adverse environments with changes to different morphological structures and physiological metabolisms, such as leaf and root morphology, photosynthesis, antioxidant enzyme systems and hormone levels [6,7]. A large number of stress response genes are activated through complex signal transduction networks and synthesize many functional proteins to improve the ability of plants to resist water stress [8,9]. To date, it is believed that drought stress mainly affects the absorption and transport of nutrients from roots to leaves [10–12], while waterlogging stress is an anaerobic respiratory metabolism caused by the environment around the roots [13–15]. Based on the research results, this review discusses and compares the changes to plant morphology, structure, physiology and molecular mechanisms under drought and waterlogging stress. These are important factors to understand plant regulatory mechanisms in response to drought and waterlogging stress, and to increase plant productivity in adverse environments.

2. Morphological Structure Responses to Water Stress in Plants

The response of plants to water stress is mainly reflected in leaves and roots, and their external morphological characteristics and internal anatomical structure can best reflect the adaptability to adverse environments [16–19] (Table 1). Leaves are the most variable organs in long-term adaptation to the environment. They react similarly under drought and waterlogging stress, showing signs of etiolation, atrophy, curling, senescence and even abscission [20,21]. In some cases, stress resulted in stunted leaf growth and reduced leaf number and area [22–24] (Figure 1).

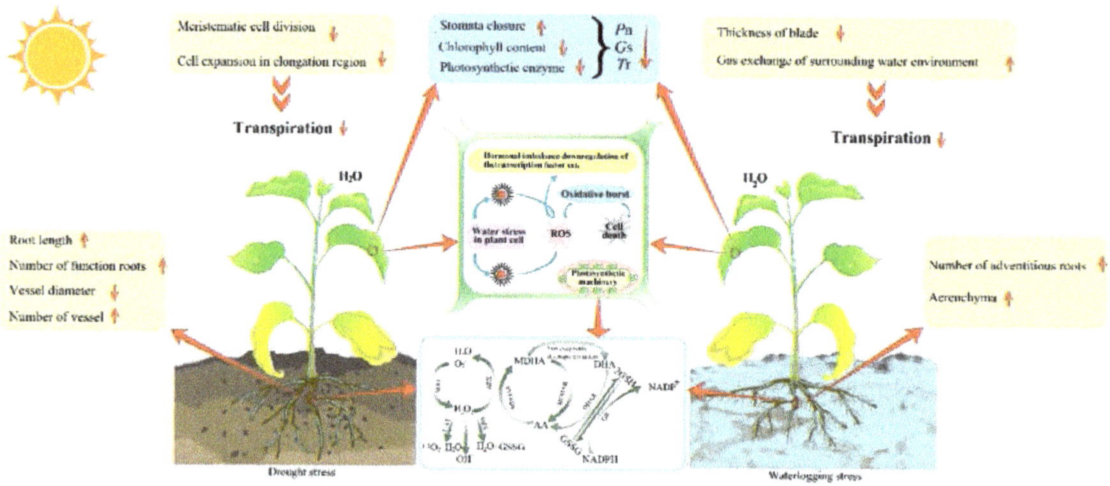

Figure 1. Changes to the morphological and anatomical structure of plant leaves and roots due to water stress. P_n: net photosynthetic rate; G_s: stomatal conductance; T_r: transpiration rate; ROS: reactive oxygen species; SOD: superoxide dismutase; CAT: catalase; APX: ascorbic peroxidase; GPX: peroxidase; GSSG: L-glutathione oxidized; MDHA: dehydroascorbic acid reductase; MDHAR: monodehydroascorbic acid reductase; DHAR: dehydroascorbate reductase glutathione; GR: glutathione reductase; GSH: glutathione peroxidase; AA: ascorbic acid.

2.1. Morphological Structure Responses to Drought Stress

Drought can limit plant growth by inhibiting the cell division of leaf meristematic tissue and cell expansion in elongation areas, as well as inducing complex changes in leaf thickness, palisade tissue and spongy tissue during adaptation [25–27]. Rueda et al. [28]

found that the conifers (water-holding capacity of plants) could be improved by increasing the thickness of leaves and decreasing the thickness of palisade tissue and spongy tissue in drought environments. However, Zheng et al. [29] found that *Lycium barbarum* increased the thickness of palisade tissue and reduced the thickness of spongy tissue, inhibiting transpiration and preventing tissue from excessive dehydration. The above results presented that the internal structure of the leaf changes resulted in transpiration reduction, as well as photosynthetic rate.

The root is an important organ for plants to fix and absorb substances from the soil. Drought stress reduces the stele area, vessel diameter and secondary root cortex cells and increases the number of vessels in the stele to facilitate water flow [30–32]. To improve water retention and drought resistance, plants not only extend the root system by increasing the number of functional roots, but also increase the water-absorbing capacity of the root sheath [33,34]. Furthermore, plants improve resistance by changing the root structure (such as root hair and root density) to influence root spatial distribution, soil fixation and nutrient absorption [35–37]. Therefore, plants could improve water absorption capacity by changing root length and internal structure under drought stress conditions.

2.2. Morphological Structure Responses to Waterlogging Stress

The main response symptoms of leaves to waterlogging stress are curling, yellowing, wilting, falling off, rotting, etc. Leaves have two kinds of adaptation to waterlogging stress: one is to increase the thickness, while the other is to reduce the thickness. For the former, the water loss is reduced and the water holding capacity of plants is improved by increasing palisade tissue and spongy tissue, as well as the decrease in leaf and stomata size [38–40]. The latter takes place because leaves cannot complete morphogenesis normally due to lack of water and nutrition [41]. Thereby, some plants thin their leaves or form special leaves to promote the infiltration ability of CO_2 and inorganic nutrients into the leaves [42,43], and improve gas exchange to restore and maintain respiration under waterlogging stress [44,45]. Therefore, the internal anatomy variation of the leaf is to adjust the stomata and improve transpiration under waterlogging stress, but the reason is uncertain and further study is needed.

Aerenchyma forming in the adventitious roots are the most obvious adaptation features under waterlogging stress. Meanwhile, the epithelial cell wall keratinizes gradually under a waterlogged environment to promote oxygen capture by underwater tissue, and enhance waterlogging tolerance [46,47]. Yamauchi et al. [48] found that there are a lot of root hairs in the adventitious roots, the surface area is large, and the cuticle of the adventitious root is thin, but the aerenchyma is well developed, which can improve the oxygen content of waterlogging-tolerant plants. Moreover, lignified and embolized vascular bundle cortical cells contribute to long-distance oxygen diffusion to the root tips, and block the entry of soil toxins into plants effectively. For instance, Ranathunge et al. [49] found that rice promoted the early formation and increased lignin deposition in both the internal and external epidermis of roots, and prevented ion penetration more effectively under waterlogged conditions. Abiko et al. [50] found that waterlogging-tolerant *teosinte* formed adventitious roots and produced larger aerenchyma, a stronger lignified vascular bundle cell barrier, and the transport of oxygen from stem base to root tip was better than normal maize under a waterlogging environment. Therefore, the ways of producing adventitious roots are diverse in different types of plants under waterlogging stress, and strong waterlogging-tolerant plants are more likely to have the ability to form adventitious roots. It has been indicated that roots could improve adaptability by creating air cavities in the aerenchyma to expand storage space, and block the entry of soil toxins into plants.

Table 1. Characteristics of plant roots and leaves under water stress.

Treatment	Root	Reference	Leaf	Reference
Drought stress	Root system lengthens; functional root number increases; distribution breadth increases.	[2,51]	Wilting; crimping; stomatal closure.	[52,53]
	Area of the stele reduces; number of vascular bundles increases but their diameter reduces.	[2,54,55]	Thickness of spongy tissue decreases; vascular bundles increase.	[56,57]
Waterlogging stress	Number of roots decreases; root activity decreases; adventitious roots are generated.	[58–61]	Etiolation; wilting; abscission; stomatal closure.	[62–64]
	Aerenchyma is formed in adventitious roots; size of the stele reduces.	[65–67]	Blade thickness is reduced; number and area of leaves decreases.	[62,68,69]

3. Photosynthetic Characteristics of Plant Responses to Water Stress

3.1. Photosynthetic Characteristics of Plant Responses to Drought Stress

To maintain photosynthesis, plants form a series of defense mechanisms to protect their photosynthetic organs from damage in the process of adapting to water stress [70,71]. For most plants, light water stress can control stomata and transpiration, directly regulate leaf water potential, and self-repair after a return to a normal water supply; some plants even increase photosynthesis [72,73]. For example, light drought stress usually leads to a stomatal conductance and transpiration increase, while moderate and severe drought stress results in a net photosynthetic rate (P_n), stomatal conductance (G_s) and transpiration rate (T_r) decrease. However, the intercellular carbon dioxide concentration (C_i) shows a different trend. C_i increases or decreases with the deepening of stress, while the stomatal limit (L_s) first increases and then decreases. These results indicate that the decrease in P_n under drought stress is mainly caused by nonstomatal factors [74,75]. Most nonstomatal factors, including chlorophyll content, photosynthetic enzyme activity and active oxygen metabolism, are induced by moderate and severe drought stress. Drought not only inhibits the formation of chlorophyll directly [76,77], but also causes difficulty in absorbing mineral elements from the soil, causing leaf nutrient deficiency (for example, leaf etiolation) [78,79] (Figure 1). The regulation of photosynthetic enzymes is a very complicated process. Light drought stress may slightly affect the photosynthetic carboxylation efficiency, but it can inhibit the activity of RuBPCase, which may result in a decrease in the photosynthetic carboxylation efficiency under severe drought stress [80].

3.2. Photosynthetic Characteristics of Plant Responses to Waterlogging Stress

Under waterlogging stress, both stomatal and nonstomatal factors inhibit photosynthesis. For stomatal factors, the chemical signals from roots are transferred to the ground, forcing the stomata of leaves to close, and reducing the photosynthetic rate by decreasing the absorption capacity of the photosynthetic substrate CO_2 [81–83]; Another aspect of stomatal conductance increasing is the supply of CO_2, which increases the amount of assimilates to maintain growth under waterlogging. For non-stomatal factors, there is the anaerobic respiration of the plant under hypoxic surroundings. Lactic acid and ethanol are produced, which break the balance of active oxygen metabolism, degrade chlorophyll and damage the photosynthetic apparatus, producing excess excitation energy and causing photoinhibition [84,85]. For severe waterlogging-tolerant plants, the stomata closed quickly due to the stress reaction of plants at the initial stage. For poor waterlogging-tolerant plants, leaf carbohydrates may accumulate rapidly within a few days, because root anaerobic respiration restrains sugar transfer from the stem to the root by reducing sugar consumption in the root, and the accumulation of photoassimilated products in leaves can form a negative feedback inhibition to the photosynthetic rate.

4. Antioxidant System of Plant Responses to Water Stress

Under normal physiological activities, plants produce reactive oxygen species (ROS), such as superoxide anion radicals (O_2^-), singlet oxygen (O_2), hydroxyl radicals ($\cdot OH$) and hydrogen peroxide (H_2O_2), as signal transmitters to regulate gene and protein expression in plant cells, and the production and elimination of ROS are always in a state of dynamic equilibrium [86]. When the plant is stressed, the balance will be broken, the physiological and biochemical functions of the plant cell membrane will be disturbed, and the production of reactive oxygen species will increase [87]. Plants have similar responses to drought and waterlogging, and both stresses activate the antioxidant defense system of plants to avoid cell damage. The components of the antioxidant defense system are enzymatic and nonenzymatic antioxidants. The enzymatic antioxidants include superoxide dismutase (SOD), catalase (CAT), peroxidase (POD), ascorbate peroxidase (APX), glutathione reductase (GR), dehydroascorbate reductase glutathione (DHAR) and monodehydroascorbic acid reductase (MDHAR). The nonenzymatic antioxidants are glutathione (GSH), ascorbic acid (AA) (both water soluble), carotenoids and tocopherols (lipid soluble). Both components counteract the harm caused by reactive oxygen species [88–91].

The response of antioxidant enzymes in plants to water stress is mainly related to tolerance and the level of stress. The activity of SOD in leaves and roots of the same species increases with an increasing level of water stress. Furthermore, the disproportionation conversion of O_2^- to H_2O_2 increases and the content of O_2^- decreases. POD and CAT decompose H_2O_2 to H_2O, inhibit the accumulation of H_2O_2 effectively, protect plants from oxidative damage, and reduce the toxic effect on plants caused by water stress [92]. This mechanism has been demonstrated in mosses [93], trifoliate orange seedlings [94], and tobacco [95]. There are different antioxidant enzyme activities in different tolerant varieties under the same water stress. The adaptive mechanism of plants is a very complicated process, and there are no fixed rules to follow. For example, the SOD activity of *Poa pratensis* and *Festuca arundinacea* increased briefly and then decreased, while the CAT activity of *F. arundinacea* decreased with increasing drought stress [96]. The SOD activity of the drought-sensitive cultivar *Trifolium repens* was inhibited under stress, but there was no significant change in the drought-tolerant cultivar Debut, which may be related to its higher ability to mitigate oxidative damage [97]. These results showed that plants could increase the activity of antioxidant enzymes to cope with adverse environments, but the dynamic changes across individuals and stress degrees.

5. Phytohormones and Related Genes in Plant Responses to Drought Stress

Phytohormones play a vital role in plant growth and metabolism, as well as the transport and distribution of nutrients, as their synthesis and signal transduction pathways are interrelated. The physiological function is changed to a specific antistress mechanism through regulating hormone metabolism and signal transduction [98–100]. Drought stimulates abscisic acid (ABA) production in different plant organs, especially in the root, which can reach leaf guard cells and send signals through xylem transport and transpiration. ABA combines cytokinin (CTK) and jasmonic acid (JA) to regulate stomatal movement. They reduce the leaf transpiration rate and guard cell turgor pressure, which causes stomatal closure to adapt to external environments stress [101–104], and ABA accumulation also activates downstream signal components and enhances root antioxidant capacity to improve stress resistance [105]. These results indicated that ABA could play an important role in plant cells receiving drought signals. Therefore, it is of great significance to understand the involvement of ABA in regulating cell metabolism, energy supply, growth, and the expression of functional genes at the transcriptional level under drought stress.

To avoid drought, plants have evolved complex mechanisms to adapt (such as strictly controlling stomatal opening and closing), and endogenous ABA plays an important role in this process [106]. There are many ways to synthesize ABA under a drought environment. One is the involvement of key regulatory factors (such as 9-cis-epoxydioxygenase (*NCED*) and acetaldehyde dehydrogenase (*ALDH*)) in the last step of the ABA biosynthesis pathway, as the accumulation of ABA activates downstream signals and specifically binds to genes, which play an important role in drought environments [107] (Figure 2). We grouped them into drought adjustment (Table 2). Increased expression of the *TaNCED1* gene isolated from *Triticum aestivum*, significantly improved drought tolerance in tobacco transgenic plants [108]. Moreover, different levels of *OsALDH* expression were detected in rice seedlings under drought stress. Transgenic rice overexpressing *OsALDH* showed elevated stress tolerances and a down-regulation of *OsALDH* in the RNA interference (RNAi). Repression transgenic lines manifest a declined stress tolerance [109].

The second method plays an important role in the upstream enhancement of the expression of downstream genes to increase NCED enzyme activity, and promote ABA biosynthesis. The ABA-mediated signal transduction pathway leads to stomatal closure involved in ABA synthesis, including *NGA1*, *ATAF1*, *HAT1* and *ATX1* [110,111]. *NGA1* (a B3 transcription factor) binds directly to the *NCED3* promoter and activates its expression in vitro and in vivo under drought stress [112]. The regulatory target gene of *ATAF1* (a NAC protein) is *NCED3*, which binds specifically to the transcription factor NAC, regulates the ABA biosynthesis gene directly, and activates its expression. Drought-stimulated plants can enhance the expression of downstream genes by binding specific transcription factors (such as B3, NAC and MIKC) to cis-regulatory elements. Transcription factors such as MYB and WRKY bind specifically to cis-regulatory elements and induce the expression of drought-responsive genes to maintain osmotic balance [113–115]. Moreover, some genes can suppress ABA synthesis and signaling, such as *HAT1* (an HD-ZIP transcription factor) binding to their promoters and the ABA/drought-responsive genes *RD29A* and *RD22* directly, by down-regulating the expression of *ABA3* and *NCED3* [116]. *ATX1* not only upregulates *NCED3* transcription but also affects ABA production in response to drought stress directly [117].

The third method is changes in leaf stomatal density, leaf water loss rate and reactive oxygen species levels. *AGL16* (a MIKC transcription factor) plays an important role in the upstream of the *AAO3* gene (abscisic aldehyde oxidase 3, the gene encodes an aldehyde oxidase). *AGL16* binds to the CArG motif in the *AAO3* promoter, regulates transcription, and changes ABA levels and leaf stomatal density [118]. *GbMYB5* and *GhWRKY17* play an active role by regulating the expression of drought-related genes and the production of reactive oxygen species under drought stress [119,120].

In addition, ABA-independent signaling includes both the NAC and DREB2 pathways [121–123]. The former, *SlNAC4*, plays a role as a transcription factor in the positive regulation of stress tolerance. Zhu et al. [9] found that the chlorophyll content and leaf water content of transgenic tomato with *SlNAC4-RNAi* were lower than those of wild-type plants, and the leaf water loss rate was higher under drought stress. Drought also directly induces the binding of *HcDREB2* to the *DRE* cis-regulatory element and activates downstream gene expression to significantly improve the drought resistance of plants [124] (Figure 2). These results showed that genes can regulate signal transduction and induce the drought resistance gene expression under drought stress, and the functional genes can transcribe and synthesize proteins that play a direct role in stress tolerance. The activity of transcription factors was enhanced, and the interaction between transcription factors and cis-regulatory elements could further induce the expression of functional genes under drought stress.

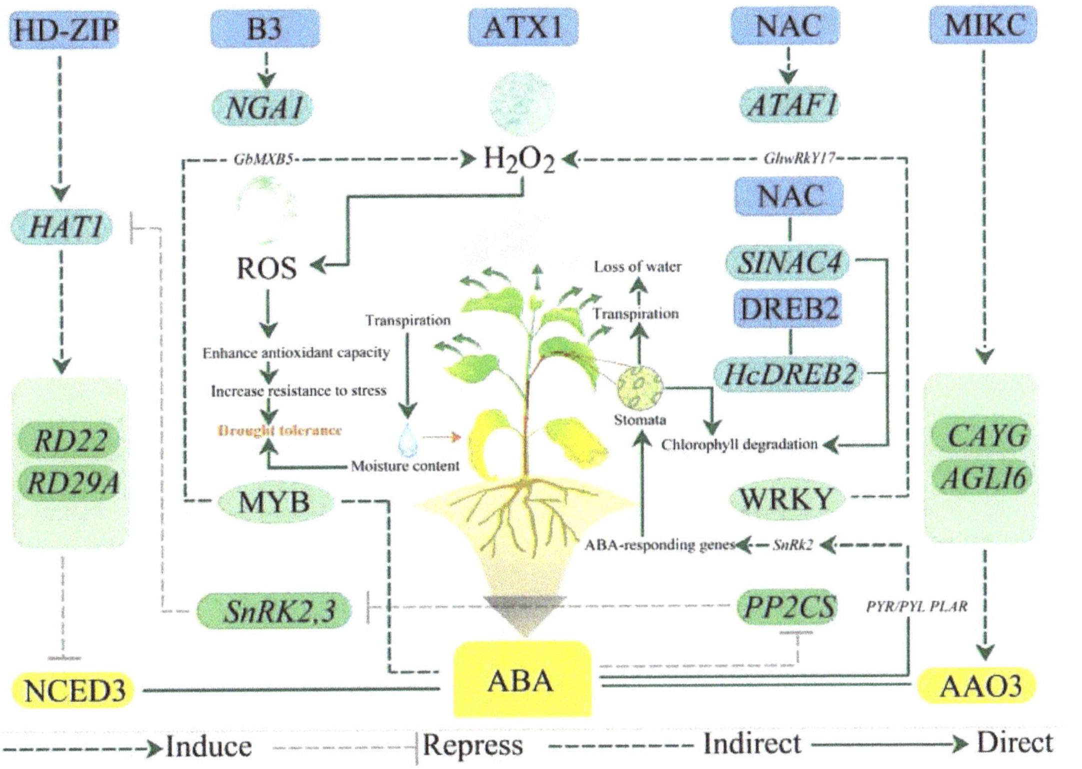

Figure 2. Regulatory mechanisms of abscisic acid (ABA) and related genes in response to drought stress in plants.

6. Phytohormones and Related Genes in Plant Responses to Waterlogging Stress

The root is the most sensitive and responsive organ, and its primary responsibility is to adapt to waterlogging by controlling growth [125,126]. Similar to drought stress, waterlogging stress induces ABA synthesis in the root system and adjusts stomatal movement to adapt to the external environment [127]. The difference is that ethylene (ETH) is one of the more sensitive hormones to waterlogging, and it is increased in an anoxic environment [128,129]. It has been reported that the regulatory mechanism of waterlogging in plants involves not only the production of ABA in the root system but also the regulation of stomatal opening and closing. First, plants respond to a lack of energy by increasing anaerobic respiration. Hypoxia stress caused by waterlogging leads to the inhibition of aerobic respiration to increase the ATP supply, and plants create energy through ethanol fermentation (mainly through pyruvate decarboxylase (PDC) and alcohol dehydrogenase (ADH)) [130,131]. Second, plants adapt to waterlogging through a "static" strategy [132,133]. ETH can regulate gibberellin (GA) synthesis, inhibit internode elongation and reduce energy consumption [134–136]. Third, plants adapt to long-term waterlogging through an "escape" strategy [137]. ETH maintains the stability of GA and ABA in plants to increase the contact between plants and the air, and promotes stem elongation to the water surface for photosynthesis and rapid aerobic absorption to maintain growth [138,139] (Figure 3).

Ethylene response factor (ERFVII) subfamily members are response factors involved in hypoxia-induced gene expression [140,141]. Plant hypoxia-responsive genes are involved in fermentation and glycometabolism pathways and affect gene expression related to ethylene biosynthesis [142]. When breathing is restricted, lactate dehydrogenase converts the pyruvate produced during glycolysis into lactic acid. The PDC and ADH genes can convert pyruvic acid into lactic acid and change it into ethanol; that is, PDC converts pyruvic acid into acetaldehyde, and ADH converts acetaldehyde into ethanol. Additionally, NAD^+ and a finite amount of ATP are produced [143,144]. At present, it has been shown that ADH and PDC activity are regulated by *SUB1*, *HRE1* and *HRE2* under waterlogging. We grouped them into waterlogging adjustment (Table 2), as waterlogging could increase the transcription level of *Sub1A* and *Sub1C* and affect PDC and ADH activity to inhibit the chlorophyll degradation and carbohydrate consumption of waterlogged plants [145]. *HRE1* overexpression increased the induction of anaerobic genes in a hypoxic environment. Compared with normal oxygen conditions, the overexpression of *HRE1* and *ATERF73/HRE1* has a positive regulatory role in the absence of oxygen, in which plants not only increase PDC enzyme activity, ADH enzyme activity, and ethanol content, but also induce elongated adventitious roots to adapt to waterlogging [146,147]. Moreover, amino-oxyacetic acid, an inhibitor of ethylene biosynthesis, can partially inhibit the anoxic induction of ADH, but this partial inhibition could be reversed by adding 1-aminocyclopropane-1-carboxylic acid, which is a direct precursor of ethylene [148,149]. *CgACO* (1-aminocyclopropane-1-carboxylate oxidase) expression in roots of the waterlogging-tolerant species (*Chrysanthemum zawadskii*) were higher than the sensitive species (*Chrysanthemum nankingense*) after 12 h waterlogging treatment. This indicated that higher *CgACO* expression possibly contributed to higher accumulation of ethylene in the waterlogging-tolerant species [150]. At present, research on this pathway mainly focuses on the enhancement of PDC and ADH enzyme activity after the overexpression of ERFVII subfamily members. The signal transduction mechanism of increased PDC and ADH activity in the synthesis of ethanol in an anoxic environment caused by waterlogging needs further study [151–154].

The waterlogging environment showed two opposite growth responses: "static" and "escape". Both were mainly regulated by *SK* and *Sub1* transcription factors induced by ETH [155,156]. *Sub1A* inhibits ETH production and the expression of the related downstream genes of ETH to promote the synthesis of brassinosteroids (BRs), and activates *Ga2oxidase7* expression to inhibit the synthesis of gibberellin (GA) while increasing the expression of the suppressor of the GA signaling pathway *SLR1* [157,158]. This process is a "static" strategy to adapt to short-term waterlogging by inhibiting internode elongation and reducing energy consumption until the stress is relieved [159]. Rice *SK1*, *SK2* and *Sub1* upregulate ABA-inactivating enzyme genes *OsCYP707A5* or *OsABA8ox1* and GA anabolism genes (*OsGA20ox* and *OsGA3ox*) under deep water, which induces a decline in ABA in rice internodes and increases the accumulation of GA in the subaqueous internodes, eventually upregulating growth-related genes to rapidly elongate stems to the water surface. This process is an "escape" strategy for the long-term submergence of plants [139,160]. The ERFVII transcripts downstream genes in a cascade amplification mode, which converts extracellular signals into intracellular, and then induces a series of adaptive mechanisms, such as accelerated glycolysis, elongated stem, formation of aerenchyma and increased oxygen transport rate, etc., to adapt to the waterlogging environment (Figure 3).

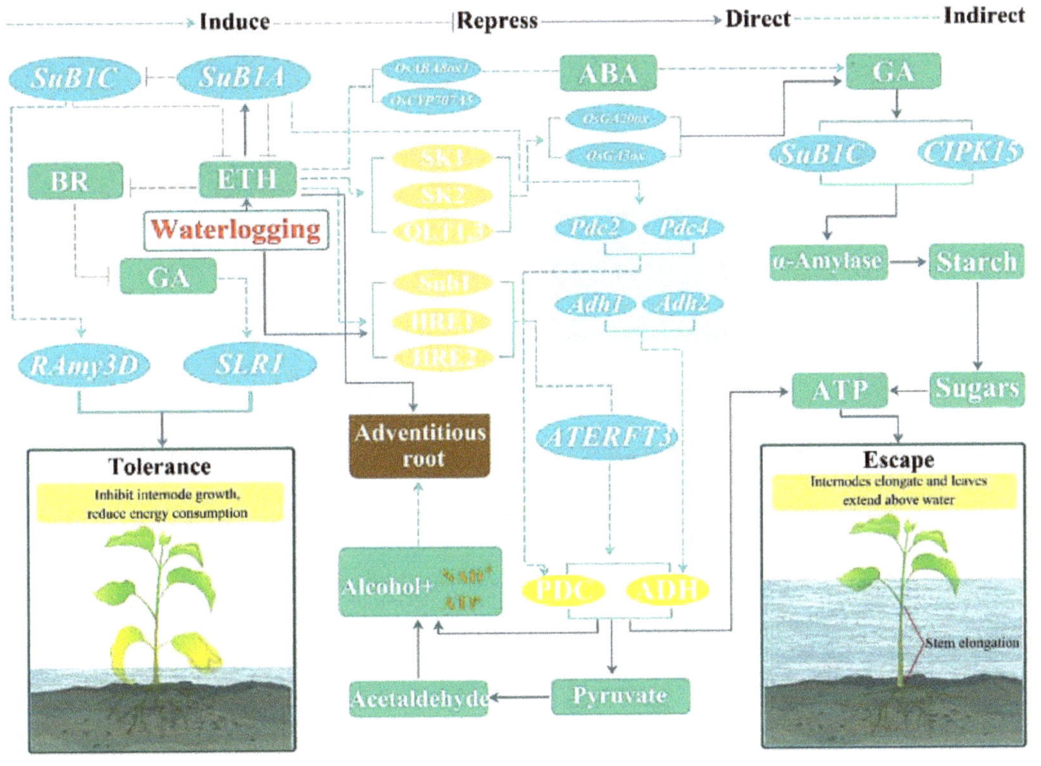

Figure 3. Regulatory mechanisms of phytohormones and related genes in response to waterlogging stress in plants.

Table 2. Genes involved in drought and waterlogging adjustment.

Type	Gene	Accession Nr.	Annotation	Function	References
Drought adjustment	*NCED1*	AT3G63520	9-cis-epoxycarotenoid dioxygenase	ABA biosynthesis	[109]
	ALDH	AT1G44170	Aldehyde dehydrogenase	ABA biosynthesis	[109]
	ATAF1	AT1G01720	Transcriptional activators with NAC domain	ABA signaling and synthesis	[110,111]
	NCED3	AT3G14440	9-cis-epoxycarotenoid dioxygenase	ABA biosynthesis	[112]
	NGA1	AT2G46870	DNA-binding proteins/contains B3 domain	Transcriptional regulation	[112]
	HAT1	AT4G17460	HD-ZIP transcription factor	Regulates meristematic	[116]
	RD29A	AT1G12610	DREB subfamily A-1 of ERF/AP2 transcription factor	Drought responsive gene	[116]
	RD22	AT5G25610	Dehydration 22 (RD22)-mediated	Responsive by ABA	[116]
	ABA3	AT1G16540	Molybdenum cofactor sulfurase	ABA biosynthesis	[116]
	ATX1	AT1G05830	Homolog of trithorax	Transcriptional regulation	[117]
	AGL16	AT3G57230	MIKC transcription factor	Transcriptional regulation	[118]
	AAO3	AT3G43600	Aldehyde oxidase	ABA biosynthesis	[118]
	WRKY17	AT2G24570	WRKY transcription factor	Transcriptional regulation	[119]
	MYB5	AT3G13540	MYB family of transcriptional regulators	Transcriptional regulation	[120]
	DREB2	AT1G75490	A-2 of ERF/AP2 transcription factor	Response to drought	[121–124]

Table 2. *Cont.*

Type	Gene	Accession Nr.	Annotation	Function	References
Waterlogging adjustment	*SLR1*	AT1G47440	S-locus related protein	Transcriptional regulation	[136]
	GA20	AT1G80330	Gibberellin 3-oxidase	Transcriptional regulation	[139]
	HRE1	AT1G72360	Ethylene response factor	Transcriptional regulation	[145,151]
	HRE2	AT2G47520	Ethylene response factor	Transcriptional regulation	[145,146]
	ATERF73	AT1G72360	Ethylene response factor	Transcriptional regulation	[146,147]
	ACO	AT1G12010	1-amino-cyclopropane-1-carboxylic acid oxidase	ETH biosynthesis	[150]
	SK	AT1G05180	RUB1-activating enzyme	Transcriptional regulation	[155]
	SUB1	AT4G08810	Calcium-binding protein	Transcriptional regulation	[156,158]
	SK2	AT3G62980	Auxin receptor	Transcriptional regulation	[160]
	SK1	AT1G06390	GSK3/shaggy-like protein kinase	Transcriptional regulation	[160]

7. A View to the Future

In recent years, more research has been devoted to the study of the harmful effects of extreme climate on plants, and some important progress has been made into the adaptability of different plants to drought and waterlogging. However, great differences were observed in the response mechanisms of different plants under water stress. To date, although scholars have proposed many mechanisms of plant tolerance, none of them have been universally accepted due to their complexity. Currently, gene cloning and genetic transformation are mainly focused on model plants and some crops, but these methods are still in their infancy in some species. On the one hand, the regulatory mechanism of plants under drought and waterlogging stress should be further compared to explore the gene expression regulation and functional identification of resistance genes. On the other hand, the response mechanism of roots and leaves to water stress and the generation and transformation of important regulatory factors should be further studied. In particular, the signal transduction pathway, after receiving a stimulus but before hormone production, should be focused on. In addition, the gene regulation mechanism of inducing PDC and ADH enzymes to create ethanol under an anoxic environment caused by waterlogging in order to improve the plant stress-resistance signaling network also needs further study.

Author Contributions: All authors contributed to the study conception and design. J.W. (Jiaojiao Wu) had the idea for the article; J.W. (Ji·aojiao Wu), P.W. and C.S. performed the literature search; F.Z. drew the pictures; J.W. (Jingyan Wang), W.H. and W.G. critically revised the work. All authors have read and agreed to the published version of the manuscript.

Funding: This work was carried out with the support of "The National Key Research and Development Program of China (Program No. 2018YFD1000605, 2020YFD1000700)", "The Central Government Forestry Science and Technology Demonstration Fund Project (Project No. Sichuan 2018-11)" and "The Forest and Bamboo Breeding Project of Sichuan Province for the Fifth Year Plan (Project No. 2016NYZ0035, 2021YFYZ0032)".

Conflicts of Interest: The authors declare no conflict of interest. We confirm that neither the manuscript nor any parts of its content are currently under consideration or published in another journal.

References

1. Teshome, D.T.; Zharare, G.E.; Naidoo, S. The threat of the combined effect of biotic and abiotic stress factors in forestry under a changing climate. *Front. Plant Sci.* **2020**, *11*, 601009. [CrossRef]
2. Gupta, A.; Rico-Medina, A.; Cano-Delgado, A.I. The physiology of plant responses to drought. *Science* **2020**, *368*, 266–269. [CrossRef]
3. Lesk, C.; Rowhani, P.; Ramankutty, N. Influence of extreme weather disasters on global crop production. *Nature* **2016**, *529*, 84–87. [CrossRef] [PubMed]
4. Du, W.; FitzGerald, G.J.; Clark, M.; Hou, X.Y. Health impacts of floods. *Prehosp. Disaster Med.* **2010**, *25*, 265–272. [CrossRef] [PubMed]

5. Shi, W.; Wang, M.; Liu, Y. Crop yield and production responses to climate disasters in China. *Sci. Total Environ.* **2021**, *750*, 141147. [CrossRef] [PubMed]
6. Khaleghi, A.; Naderi, R.; Brunetti, C.; Maserti, B.E.; Salami, S.A.; Babalar, M. Morphological, physiochemical and antioxidant responses of *Maclura pomifera* to drought stress. *Sci. Rep.* **2019**, *9*, 19250. [CrossRef] [PubMed]
7. Ren, B.; Zhang, J.; Dong, S.; Liu, P.; Zhao, B. Responses of carbon metabolism and antioxidant system of summer maize to waterlogging at different stages. *J. Agron. Crop Sci.* **2018**, *204*, 505–514. [CrossRef]
8. Wang, X.; Huang, M.; Zhou, Q.; Cai, J.; Dai, T.B.; Cao, W.X.; Jiang, D. Physiological and proteomic mechanisms of waterlogging priming improves tolerance to waterlogging stress in wheat (*Triticum aestivum* L.). *Environ. Exp. Bot.* **2016**, *132*, 175–182. [CrossRef]
9. Zhu, M.K.; Chen, G.P.; Zhang, J.L.; Zhang, Y.J.; Xie, Q.L.; Zhao, Z.P.; Pan, Y.; Hu, Z.L. The abiotic stress-responsive NAC-type transcription factor *SlNAC4* regulates salt and drought tolerance and stress-related genes in tomato (*Solanum lycopersicum*). *Plant Cell Rep.* **2014**, *33*, 1851–1863. [CrossRef]
10. Bista, D.R.; Heckathorn, S.A.; Jayawardena, D.M.; Boldt, J.K. Effect of drought and carbon dioxide on nutrient uptake and levels of nutrient-uptake proteins in roots of barley. *Am. J. Bot.* **2020**, *107*, 1401–1409. [CrossRef]
11. Jiao, P.P.; Wu, Z.H.; Wang, X.; Jiang, Z.B.; Wang, Y.Q.; Liu, H.; Qin, R.; Li, Z.J. Short-term transcriptomic responses of *Populus euphratica* roots and leaves to drought stress. *J. Forestry Res.* **2021**, *32*, 841–853. [CrossRef]
12. Zhao, Q.; Guo, J.; Shu, M.; Wang, P.; Hu, S. Impacts of drought and nitrogen enrichment on leaf nutrient resorption and root nutrient allocation in four Tibetan plant species. *Sci. Total Environ.* **2020**, *723*, 138106. [CrossRef]
13. Hartman, S.; Liu, Z.; van Veen, H.; Vicente, J.; Reinen, E.; Martopawiro, S.; Zhang, H.; van Dongen, N.; Bosman, F.; Bassel, G.W. Ethylene-mediated nitric oxide depletion pre-adapts plants to hypoxia stress. *Nat. Commun.* **2019**, *10*, 4020. [CrossRef] [PubMed]
14. Loreti, E.; van Veen, H.; Perata, P. Plant responses to flooding stress. *Curr. Opin. Plant Biol.* **2016**, *33*, 64–71. [CrossRef]
15. Coutinho, I.D.; Henning, L.M.; Dopp, S.A.; Nepomuceno, A.; Moraes, L.A.; Marcolino-Gomes, J.; Richter, C.; Schwalbe, H.; Colnago, L.A. Flooded soybean metabolomic analysis reveals important primary and secondary metabolites involved in the hypoxia stress response and tolerance. *Environ. Exp. Bot.* **2018**, *153*, 176–187. [CrossRef]
16. Cal, A.J.; Sanciangco, M.; Rebolledo, M.C.; Luquet, D.; Torres, R.O.; McNally, K.L.; Henry, A. Leaf morphology, rather than plant water status, underlies genetic variation of rice leaf rolling under drought. *Plant Cell Environ.* **2019**, *42*, 1532–1544. [CrossRef]
17. Liu, W.S.; Zheng, L.; Qi, D.H. Variation in leaf traits at different altitudes reflects the adaptive strategy of plants to environmental changes. *Ecol. Evol.* **2020**, *10*, 8166–8175. [CrossRef]
18. Lozano, Y.M.; Aguilar-Trigueros, C.A.; Flaig, I.C.; Rillig, M.C. Root trait responses to drought are more heterogeneous than leaf trait responses. *Funct. Ecol.* **2020**, *34*, 2224–2235. [CrossRef]
19. Pedersen, O.; Sauter, M.; Colmer, T.D.; Nakazono, M. Regulation of root adaptive anatomical and morphological traits during low soil oxygen. *New Phytol.* **2021**, *229*, 42–49. [CrossRef]
20. Patharkar, O.R.; Walker, J.C. Connections between abscission, dehiscence, pathogen defense, drought tolerance, and senescence. *Plant Sci.* **2019**, *284*, 25–29. [CrossRef]
21. Bhusal, N.; Kim, H.S.; Han, S.G.; Yoon, T.M. Photosynthetic traits and plant-water relations of two apple cultivars grown as bi-leader trees under long-term waterlogging conditions. *Environ. Exp. Bot.* **2020**, *176*, 104111. [CrossRef]
22. Wei, W.L.; Li, D.H.; Wang, L.H.; Ding, X.; Zhang, Y.X.; Gao, Y.; Zhang, X.R. Morpho-anatomical and physiological responses to waterlogging of sesame (*Sesamum indicum* L.). *Plant Sci.* **2013**, *208*, 102–111. [CrossRef] [PubMed]
23. Fang, Y.J.; Xiong, L.Z. General mechanisms of drought response and their application in drought resistance improvement in plants. *Cell Mol. Life Sci.* **2015**, *72*, 673–689. [CrossRef]
24. Nadal, M.; Roig-Oliver, M.; Bota, J.; Flexas, J. Leaf age-dependent elastic adjustment and photosynthetic performance under drought stress in Arbutus unedo seedlings. *Flora* **2020**, *271*, 151662. [CrossRef]
25. Nelissen, H.; Sun, X.H.; Rymen, B.; Jikumaru, Y.; Kojima, M.; Takebayashi, Y.; Abbeloos, R.; Demuynck, K.; Storme, V.; Vuylsteke, M. The reduction in maize leaf growth under mild drought affects the transition between cell division and cell expansion and cannot be restored by elevated gibberellic acid levels. *Plant Biotechnol. J.* **2018**, *16*, 615–627. [CrossRef]
26. Binks, O.; Meir, P.; Rowland, L.; Costa, A.C.; Vasconcelos, S.S.; Oliveira, A.A.R.; Ferreira, L.; Mencuccini, M. Limited acclimation in leaf anatomy to experimental drought in tropical rainforest trees. *Tree Physiol.* **2016**, *36*, 1550–1561. [CrossRef] [PubMed]
27. Meng, D.; Dong, B.; Niu, L.; Song, Z.; Wang, L.; Amin, R.; Cao, H.; Li, H.; Yang, Q.; Fu, Y. The pigeon pea CcCIPK14-CcCBL1 pair positively modulates drought tolerance by enhancing flavonoid biosynthesis. *Plant J.* **2021**, *106*, 1278–1297. [CrossRef]
28. Rueda, M.; Godoy, O.; Hawkins, B.A. Spatial and evolutionary parallelism between shade and drought tolerance explains the distributions of conifers in the conterminous United States. *Global Ecol. Biogeogr.* **2017**, *26*, 31–42. [CrossRef]
29. Zheng, G.Q.; Zhang, L.; Zheng, G.B.; Zhang, Y.P.; Wang, J.; Hu, Z.H. Effects of irrigation amount on leaf structure, photosynthetic physiology, and fruit yield of *Lycium barbarum* in arid area. *Ying Yong Sheng Tai Xue Bao* **2010**, *21*, 2806–2813.
30. De Bauw, P.; Vandamme, E.; Lupembe, A.; Mwakasege, L.; Senthilkumar, K.; Drame, K.N.; Merckx, R. Anatomical root responses of rice to combined phosphorus and water stress-relations to tolerance and breeding opportunities. *Funct. Plant Biol.* **2019**, *46*, 1009–1022. [CrossRef]
31. Thangthong, N.; Jogloy, S.; Punjansing, T.; Kvien, C.K.; Kesmala, T.; Vorasoot, N. Changes in root anatomy of peanut (*Arachis hypogaea* L.) under different durations of early season drought. *Agronomy* **2019**, *9*, 215. [CrossRef]
32. Hazman, M.; Brown, K.M. Progressive drought alters architectural and anatomical traits of rice roots. *Rice* **2018**, *11*, 1–16. [CrossRef] [PubMed]

33. Lee, D.K.; Jung, H.; Jang, G.; Jeong, J.S.; Kim, Y.S.; Ha, S.H.; Do Choi, Y.; Kim, J.K. Overexpression of the *OsERF71* transcription factor alters rice root structure and drought resistance. *Plant Physiol.* **2016**, *172*, 575–588. [CrossRef]
34. Pierret, A.; Maeght, J.L.; Clement, C.; Montoroi, J.P.; Hartmann, C.; Gonkhamdee, S. Understanding deep roots and their functions in ecosystems: An advocacy for more unconventional research. *Ann. Bot.* **2016**, *118*, 621–635. [CrossRef]
35. Henry, A.; Gowda, V.R.P.; Torres, R.O.; McNally, K.L.; Serraj, R. Variation in root system architecture and drought response in rice (*Oryza sativa*): Phenotyping of the *OryzaSNP* panel in rainfed lowland fields. *Field Crop Res.* **2011**, *120*, 205–214. [CrossRef]
36. Tanaka, N.; Kato, M.; Tomioka, R.; Kurata, R.; Fukao, Y.; Aoyama, T.; Maeshima, M. Characteristics of a root hair-less line of *Arabidopsis* thaliana under physiological stresses. *J. Exp. Bot.* **2014**, *65*, 1497–1512. [CrossRef]
37. Strock, C.F.; Burridge, J.D.; Niemiec, M.D.; Brown, K.M.; Lynch, J.P. Root metaxylem and architecture phenotypes integrate to regulate water use under drought stress. *Plant Cell Environ.* **2021**, *44*, 49–67. [CrossRef]
38. Yin, D.M.; Luo, H.L. Anatomical responses to waterlogging in *Chrysanthemum zawadskii*. *Sci. Hortic.* **2012**, *146*, 86–91. [CrossRef]
39. Tahira, M.H.; Muhammad, A.; Muhammad, S.A.A.; Riffat, B.; Sana, F. Anatomical and physiological adaptations in aquatic ecotypes of *Cyperus alopecuroides* Rottb. under saline and waterlogged conditions. *Aquat. Bot.* **2014**, *116*, 60–68.
40. Zuniga, F.A.; Bustos, S.A.; Alves, F.; Martinez, V.; Smith, R.C. Physiological and morphological responses to permanent and intermittent waterlogging in seedlings of four evergreen trees of temperate swamp forests. *Tree Physiol.* **2017**, *37*, 779–789. [CrossRef]
41. Fan, C.F.; Yang, Y.F. Water affects morphogenesis of growing aquatic plant leaves. *Phys. Rev. Lett.* **2020**, *124*, 038003.
42. Colmer, T.D.; Pedersen, O. Underwater photosynthesis and respiration in leaves of submerged wetland plants: Gas films improve CO_2 and O_2 exchange. *New Phytol.* **2008**, *177*, 918–926. [CrossRef] [PubMed]
43. Brodersen, K.E.; Hammer, K.J.; Schrameyer, V.; Floytrup, A.; Rasheed, M.A.; Ralph, P.J.; Kuhl, M.; Pedersen, O. Sediment resuspension and deposition on seagrass leaves impedes internal plant aeration and promotes phytotoxic H_2S intrusion. *Front. Plant Sci.* **2017**, *8*, 657. [CrossRef] [PubMed]
44. Lawson, J.R.; Fryirs, K.A.; Leishman, M.R. Interactive effects of waterlogging and atmospheric CO_2 concentration on gas exchange, growth and functional traits of Australian riparian tree seedlings. *Ecohydrology* **2017**, *10*, e1803. [CrossRef]
45. Mommer, L.; Visser, E.J. Underwater photosynthesis in flooded terrestrial plants: A matter of leaf plasticity. *Ann Bot.* **2005**, *96*, 581–589. [CrossRef]
46. Ayi, Q.L.; Zeng, B.; Liu, J.H.; Li, S.Q.; Bodegom, P.M.; Cornelissen, J.H.C. Oxygen absorption by adventitious roots promotes the survival of completely submerged terrestrial plants. *Ann. Bot.* **2016**, *118*, 675–683. [CrossRef]
47. Pedersen, O.; Nakayama, Y.; Yasue, H.; Kurokawa, Y.; Takahashi, H.; Floytrup, A.H.; Omori, F.; Mano, Y.; Colmer, T.D.; Nakazono, M. Lateral roots, in addition to adventitious roots, form a barrier to radial oxygen loss in *Zea nicaraguensis* and a chromosome segment introgression line in maize. *New Phytol.* **2021**, *229*, 94–105. [CrossRef]
48. Yamauchi, T.; Abe, F.; Tsutsumi, N.; Nakazono, M. Root cortex provides a venue for gas-space formation and is essential for plant adaptation to waterlogging. *Front. Plant Sci.* **2019**, *10*, 259. [CrossRef]
49. Ranathunge, K.; Lin, J.; Steudle, E.; Schreiber, L. Stagnant deoxygenated growth enhances root suberization and lignifications, but differentially affects water and NaCl permeabilities in rice (*Oryza sativa* L.) roots. *Plant Cell Environ.* **2011**, *34*, 1223–1240. [CrossRef]
50. Abiko, T.; Kotula, L.; Shiono, K.; Malik, A.I.; Colmer, T.D.; Nakazono, M. Enhanced formation of aerenchyma and induction of a barrier to radial oxygen loss in adventitious roots of *Zea nicaraguensis* contribute to its waterlogging tolerance as compared with maize (*Zea mays* ssp. *mays*). *Plant Cell Environ.* **2012**, *35*, 1618–1630. [CrossRef]
51. Djanaguiraman, M.; Prasad, P.V.V.; Kumari, J.; Rengel, Z. Root length and root lipid composition contribute to drought tolerance of winter and spring wheat. *Plant Soil* **2019**, *439*, 57–73. [CrossRef]
52. Song, H.; Li, Y.B.; Zhou, L.; Xu, Z.Z.; Zhou, G.S. Maize leaf functional responses to drought episode and rewatering. *Agr. For. Meteorol.* **2018**, *249*, 57–70. [CrossRef]
53. Chen, M.J.; Zhu, X.F.; Zhang, Y.; Du, Z.H.; Chen, X.B.; Kong, X.R.; Sun, W.J.; Chen, C.S. Drought stress modify cuticle of tender tea leaf and mature leaf for transpiration barrier enhancement through common and distinct modes. *Sci. Rep.* **2020**, *10*, 6696. [CrossRef] [PubMed]
54. Canales, F.J.; Rispail, N.; Garcia, T.O.; Arbona, V.; Perez, L.A.; Prats, E. Drought resistance in oat involves ABA-mediated modulation of transpiration and root hydraulic conductivity. *Environ. Exp. Bot.* **2021**, *182*, 104333. [CrossRef]
55. Xiao, S.; Liu, L.T.; Zhang, Y.J.; Sun, H.C.; Zhang, K.; Bai, Z.Y.; Dong, H.Z.; Li, C.D. Fine root and root hair morphology of cotton under drought stress revealed with RhizoPot. *J. Agron. Crop Sci.* **2020**, *206*, 679–693. [CrossRef]
56. Zhang, J.S.; Zhang, H.; Srivastava, A.K.; Pan, Y.J.; Bai, J.J.; Fang, J.J.; Shi, H.Z.; Zhu, J.K. Knockdown of rice MicroRNA166 Confers drought resistance by causing leaf rolling and altering stem xylem development. *Plant Physiol.* **2018**, *176*, 2082–2094. [CrossRef]
57. Ouyang, W.J.; Struik, P.C.; Yin, X.Y.; Yang, J.C. Stomatal conductance, mesophyll conductance, and transpiration efficiency in relation to leaf anatomy in rice and wheat genotypes under drought. *J. Exp. Bot.* **2017**, *68*, 5191–5205. [CrossRef]
58. Palta, J.A.; Ganjeali, A.; Turner, N.C.; Siddique, K.H.M. Effects of transient subsurface waterlogging on root growth, plant biomass and yield of chickpea. *Agr. Water Manag.* **2010**, *97*, 1469–1476. [CrossRef]
59. Domisch, T.; Qian, J.; Sondej, I.; Martz, F.; Lehto, T.; Piirainen, S.; Finer, L.; Silvennoinen, R.; Repo, T. Here comes the flood! Stress effects of continuous and interval waterlogging periods during the growing season on Scots pine saplings. *Tree Physiol.* **2020**, *40*, 869–885. [CrossRef]

60. Dresboll, D.B.; Thorup, K.K. Spatial variation in root system activity of tomato (*Solanum lycopersicum* L.) in response to short and long-term waterlogging as determined by N-15 uptake. *Plant Soil* **2012**, *357*, 161–172. [CrossRef]
61. Pan, D.L.; Wang, G.; Wang, T.; Jia, Z.H.; Guo, Z.R.; Zhang, J.Y. *AdRAP2.3*, a novel ethylene response factor vii from *Actinidia deliciosa*, enhances waterlogging resistance in transgenic tobacco through improving expression levels of PDC and ADH genes. *Int. J. Mol. Sci.* **2019**, *20*, 1189. [CrossRef]
62. Yin, D.M.; Chen, S.M.; Chen, F.D.; Guan, Z.Y.; Fang, W.M. Morpho-anatomical and physiological responses of two *Dendranthema* species to waterlogging. *Environ. Exp. Bot.* **2010**, *68*, 122–130. [CrossRef]
63. Ploschuk, R.A.; Miralles, D.J.; Colmer, T.D.; Ploschuk, E.L.; Striker, G.G. Waterlogging of winter crops at early and late stages: Impacts on leaf physiology, growth and yield. *Front. Plant Sci.* **2018**, *9*, 1863. [CrossRef] [PubMed]
64. Fukao, T.; Xu, K.; Ronald, P.C.; Bailey-Serres, J. A variable cluster of ethylene response factor-like genes regulates metabolic and developmental acclimation responses to submergence in rice. *Plant Cell* **2006**, *18*, 2021–2034. [CrossRef] [PubMed]
65. Celedonio, R.P.D.; Abeledo, L.G.; Mantese, A.I.; Miralles, D.J. Differential root and shoot biomass recovery in wheat and barley with transient waterlogging during preflowering. *Plant Soil* **2017**, *417*, 481–498. [CrossRef]
66. Zhang, X.C.; Shabala, S.; Koutoulis, A.; Shabala, L.; Johnson, P.; Hayes, D.; Nichols, D.S.; Zhou, M.X. Waterlogging tolerance in barley is associated with faster aerenchyma formation in adventitious roots. *Plant Soil* **2015**, *394*, 355–372. [CrossRef]
67. Sundgren, T.K.; Uhlen, A.K.; Lillemo, M.; Briese, C.; Wojciechowski, T. Rapid seedling establishment and a narrow root stele promotes waterlogging tolerance in spring wheat. *J. Plant Physiol.* **2018**, *227*, 45–55. [CrossRef]
68. Mommer, L.; Pons, T.L.; Wolters-Arts, M.; Venema, J.H.; Visser, E.J. Submergence-induced morphological, anatomical, and biochemical responses in a terrestrial species affect gas diffusion resistance and photosynthetic performance. *Plant Physiol.* **2005**, *139*, 497–508. [CrossRef]
69. Challabathula, D.; Zhang, Q.; Bartels, D. Protection of photosynthesis in desiccation-tolerant resurrection plants. *J. Plant Physiol.* **2018**, *227*, 84–92. [CrossRef]
70. Yang, X.; Li, Y.; Chen, H.; Huang, J.; Zhang, Y.; Qi, M.; Liu, Y.; Li, T. Photosynthetic response mechanism of soil salinity-induced cross-tolerance to subsequent drought stress in tomato plants. *Plants* **2020**, *9*, 363. [CrossRef]
71. Sharma, A.; Kumar, V.; Shahzad, B.; Ramakrishnan, M.; Sidhu, G.P.S.; Bali, A.S.; Handa, N.; Kapoor, D.; Yadav, P.; Khanna, K. Photosynthetic response of plants under different abiotic stresses: A review. *J. Plant Growth Regul.* **2020**, *39*, 509–531. [CrossRef]
72. Ding, L.; Lu, Z.; Gao, L.; Guo, S.; Shen, Q. Is Nitrogen a key determinant of water transport and photosynthesis in higher plants upon drought stress? *Front. Plant Sci.* **2018**, *9*, 1143. [CrossRef] [PubMed]
73. Killi, D.; Haworth, M. Diffusive and metabolic constraints to photosynthesis in quinoa during drought and salt stress. *Plants* **2017**, *6*, 49. [CrossRef]
74. Li, P.D.; Zhu, Y.F.; Song, X.L.; Song, F.P. Negative effects of long-term moderate salinity and short-term drought stress on the photosynthetic performance of *Hybrid Pennisetum*. *Plant Physiol. Biochem.* **2020**, *155*, 93–104. [CrossRef]
75. Tian, L.X.; Li, J.; Bi, W.S.; Zuo, S.Y.; Li, L.J.; Li, W.L.; Sun, L. Effects of waterlogging stress at different growth stages on the photosynthetic characteristics and grain yield of spring maize (*Zea mays* L.) under field conditions. *Agr. Water Manag.* **2019**, *218*, 250–258. [CrossRef]
76. Shivakrishna, P.; Reddy, K.A.; Rao, D.M. Effect of PEG-6000 imposed drought stress on RNA content, relative water content (RWC), and chlorophyll content in peanut leaves and roots. *Saudi. J. Biol. Sci.* **2018**, *25*, 285–289.
77. Chen, Y.E.; Liu, W.J.; Su, Y.Q.; Cui, J.M.; Zhang, Z.W.; Yuan, M.; Zhang, H.Y.; Yuan, S. Different response of photosystem II to short and long-term drought stress in *Arabidopsis thaliana*. *Physiol. Plant.* **2016**, *158*, 225–235. [CrossRef]
78. Smirnoff, N.; Colombe, S.V. Drought influences the activity of enzymes of the chloroplast hydrogen peroxide scavenging system. *J. Exp. Bot.* **1988**, *39*, 1097–1108. [CrossRef]
79. Bondada, B.R.; Oosterhuis, D.M. Canopy photosynthesis, specific leaf weight, and yield components of cotton under varying nitrogen supply. *J. Plant Nutr.* **2001**, *24*, 469–477. [CrossRef]
80. Parry, M.A.; Andralojc, P.J.; Khan, S.; Lea, P.J.; Keys, A.J. Rubisco activity: Effects of drought stress. *Ann Bot.* **2002**, *89*, 833–839. [CrossRef]
81. Pereira, T.S.; Lobato, A.K.S.; Alves, G.A.R.; Ferreira, R.N.; Silva, O.N.; Martins, A.P.; Pereira, E.S.; Sampaio, L.S. Tolerance to waterlogging in young *Euterpe oleracea* plants. *Photosynthetica* **2014**, *52*, 186–192. [CrossRef]
82. Horiguchi, G.; Nemoto, K.; Yokoyama, T.; Hirotsu, N. Photosynthetic acclimation of terrestrial and submerged leaves in the amphibious plant *Hygrophila difformis*. *AoB Plants* **2019**, *11*, plz009. [CrossRef] [PubMed]
83. De Pedro, L.F.; Mignolli, F.; Scartazza, A.; Colavita, J.P.M.; Bouzo, C.A.; Vidoz, M.L. Maintenance of photosynthetic capacity in flooded tomato plants with reduced ethylene sensitivity. *Physiol. Plant.* **2020**, *170*, 202–217. [CrossRef] [PubMed]
84. Hole, D.J.; Cobb, B.G.; Hole, P.S.; Drew, M.C. Enhancement of anaerobic respiration in root tips of *Zea mays* following low-oxygen (hypoxic) acclimation. *Plant Physiol.* **1992**, *99*, 213–218. [CrossRef]
85. Fan, X.; Zhang, Z.; Gao, H.; Yang, C.; Liu, M.; Li, Y.; Li, P. Photoinhibition-like damage to the photosynthetic apparatus in plant leaves induced by submergence treatment in the dark. *PLoS ONE* **2014**, *9*, e89067. [CrossRef] [PubMed]
86. Waszczak, C.; Carmody, M.; Kangasjarvi, J. Reactive oxygen species in plant signaling. *Annu. Rev. Plant Biol.* **2018**, *69*, 209–236. [CrossRef]
87. Polle, A. Dissecting the superoxide dismutase-ascorbate-glutathione-pathway in chloroplasts by metabolic modeling. Computer simulations as a step towards flux analysis. *Plant Physiol.* **2001**, *126*, 445–462. [CrossRef]

88. Miller, G.; Suzuki, N.; Ciftci-Yilmaz, S.; Mittler, R. Reactive oxygen species homeostasis and signaling during drought and salinity stresses. *Plant Cell Environ.* **2010**, *33*, 453–467. [CrossRef]
89. Wu, S.W.; Hu, C.X.; Tan, Q.L.; Xu, S.J.; Sun, X.C. Nitric oxide mediates molybdenum-induced antioxidant defense in wheat under drought stress. *Front Plant Sci.* **2017**, *8*, 1085. [CrossRef]
90. Ahmad, S.; Kamran, M.; Ding, R.X.; Meng, X.P.; Wang, H.Q.; Ahmad, I.; Fahad, S.; Han, Q.F. Exogenous melatonin confers drought stress by promoting plant growth, photosynthetic capacity and antioxidant defense system of maize seedlings. *Peer J.* **2019**, *7*, e7793. [CrossRef]
91. Laxa, M.; Liebthal, M.; Telman, W.; Chibani, K.; Dietz, K.J. The role of the plant antioxidant system in drought tolerance. *Antioxidants* **2019**, *8*, 94. [CrossRef]
92. Wang, W.B.; Kim, Y.H.; Lee, H.S.; Kim, K.Y.; Deng, X.P.; Kwak, S.S. Analysis of antioxidant enzyme activity during germination of alfalfa under salt and drought stresses. *Plant Physiol. Biochem.* **2009**, *47*, 570–577. [CrossRef] [PubMed]
93. Rajinder SDhindsa, W.M. Drought tolerance in two mosses: Correlated with enzymatic defence against lipid peroxidation. *J. Exp. Bot.* **1981**, *32*, 79–91.
94. Huang, Y.M.; Zou, Y.N.; Wu, Q.S. Alleviation of drought stress by mycorrhizas is related to increased root H_2O_2 efflux in trifoliate orange. *Sci. Rep.-UK* **2017**, *7*, 42335. [CrossRef] [PubMed]
95. Zhang, H.H.; Xu, N.; Teng, Z.Y.; Wang, J.R.; Ma, S.L.; Wu, X.Y.; Li, X.; Sun, G.Y. 2-Cys Prx plays a critical role in scavenging H_2O_2 and protecting photosynthetic function in leaves of tobacco seedlings under drought stress. *J. Plant Interact.* **2019**, *14*, 119–128. [CrossRef]
96. Jiang, Y.; Huang, B. Drought and heat stress injury to two cool-season turfgrasses in relation to antioxidant metabolism and lipid peroxidation. *Crop Sci.* **2001**, *41*, 436. [CrossRef]
97. Vaseva, I.; Akiscan, Y.; Simova-Stoilova, L.; Kostadinova, A.; Nenkova, R.; Anders, I.; Feller, U.; Demirevska, K. Antioxidant response to drought in red and white clover. *Acta Physiol. Plant.* **2012**, *34*, 1689–1699. [CrossRef]
98. Di, T.M.; Zhao, L.; Chen, H.M.; Qian, W.J.; Wang, P.Q.; Zhang, X.F.; Xia, T. Transcriptomic and metabolic insights into the distinctive effects of exogenous melatonin and gibberellin on terpenoid synthesis and plant hormone signal transduction pathway in camellia sinensis. *J. Agric. Food Chem.* **2019**, *67*, 4689–4699. [CrossRef]
99. Song, S.; Chang, J.; Ma, C.J.; Tan, Y.W. Single-molecule fluorescence methods to study plant hormone signal transduction pathways. *Front. Plant Sci.* **2017**, *8*, 1888. [CrossRef]
100. Takeuchi, J.; Fukui, K.; Seto, Y.; Takaoka, Y.; Okamoto, M. Ligand-receptor interactions in plant hormone signaling. *Plant J.* **2021**, *105*, 290–306. [CrossRef]
101. Goodger, J.Q.D.; Sharp, R.E.; Marsh, E.L.; Schachtman, D.P. Relationships between xylem sap constituents and leaf conductance of well-watered and water-stressed maize across three xylem sap sampling techniques. *J. Exp Bot.* **2005**, *56*, 2389–2400. [CrossRef] [PubMed]
102. Fereres, E.; Soriano, M.A. Deficit irrigation for reducing agricultural water use. *J. Exp. Bot.* **2007**, *58*, 147–159. [CrossRef] [PubMed]
103. Schachtman, D.P.; Goodger, J.Q.D. Chemical root to shoot signaling under drought. *Trends Plant Sci.* **2008**, *13*, 281–287. [CrossRef]
104. Yin, D.M.; Chen, S.M.; Chen, F.D.; Guan, Z.Y.; Fang, W.M. Morphological and physiological responses of two chrysanthemum cultivars differing in their tolerance to waterlogging. *Environ. Exp. Bot.* **2009**, *67*, 87–93. [CrossRef]
105. Bulgakov, V.P.; Wu, H.C.; Jinn, T.L. Coordination of ABA and chaperone signaling in plant stress responses. *Trends Plant Sci.* **2019**, *24*, 636–651. [CrossRef] [PubMed]
106. Qian, D.; Zhang, Z.; He, J.X.; Zhang, P.; Ou, X.B.; Li, T.; Niu, L.P.; Nan, Q.; Niu, Y.; He, W.L. *Arabidopsis* ADF5 promotes stomatal closure by regulating actin cytoskeleton remodeling in response to ABA and drought stress. *J. Exp. Bot.* **2019**, *70*, 435–446. [CrossRef] [PubMed]
107. Gong, P.J.; Zhang, J.H.; Li, H.X.; Yang, C.X.; Zhang, C.J.; Zhang, X.H.; Khurram, Z.; Zhang, Y.Y.; Wang, T.T.; Fei, Z.J. Transcriptional profiles of drought-responsive genes in modulating transcription signal transduction, and biochemical pathways in tomato. *J. Exp. Bot.* **2010**, *61*, 3563–3575. [CrossRef]
108. Urbanavičiūtė, I.; Bonfiglioli, L.; Pagnotta, M.A. One hundred candidate genes and their roles in drought and salt tolerance in wheat. *Int. J. Mol. Sci.* **2021**, *22*, 6378. [CrossRef]
109. Sheng, H.Y.; Xiang, L.N.; Di, L.; Chang, D.; Chen, X.Y.; Wei, T.; Lu, B.R.; Liu, Y.S. Functional characterization of an aldehyde dehydrogenase homologue in rice. *J. Integr. Agr.* **2012**, *11*, 1434–1444.
110. Shang, X.G.; Yu, Y.J.; Zhu, L.J.; Liu, H.Q.; Chai, Q.C.; Guo, W.Z. A cotton NAC transcription factor *GhirNAC2* plays positive roles in drought tolerance via regulating ABA biosynthesis. *Plant Sci.* **2020**, *296*, 110498. [CrossRef]
111. Jensen, M.K.; Lindemose, S.; Masi, F.; Reimer, J.J.; Nielsen, M.; Perera, V.; Workman, C.T.; Turck, F.; Grant, M.R.; Mundy, J. *ATAF1* transcription factor directly regulates abscisic acid biosynthetic gene *NCED3* in *Arabidopsis thaliana*. *FEBS Open Bio* **2013**, *3*, 321–327. [CrossRef] [PubMed]
112. Sato, H.; Takasaki, H.; Takahashi, F.; Suzuki, T.; Iuchi, S.; Mitsuda, N.; Ohme-Takagi, M.; Ikeda, M.; Seo, M.; Yamaguchi-Shinozaki, K. Arabidopsis thaliana *NGATHA1* transcription factor induces ABA biosynthesis by activating *NCED3* gene during dehydration stress. *Proc. Natl. Acad. Sci. USA* **2018**, *115*, E11178–E11187. [CrossRef] [PubMed]
113. Zhao, N.A.; He, M.J.; Li, L.; Cui, S.L.; Hou, M.Y.; Wang, L.; Mu, G.J.; Liu, L.F.; Yang, X.L. Identification and expression analysis of *WRKY* gene family under drought stress in peanut (*Arachis hypogaea* L.). *PLoS ONE* **2020**, *15*, e0231396. [CrossRef] [PubMed]

114. Wang, N.N.; Xu, S.W.; Sun, Y.L.; Liu, D.; Zhou, L.; Li, Y.; Li, X.B. The cotton *WRKY* transcription factor (*GhWRKY33*) reduces transgenic *Arabidopsis* resistance to drought stress. *Sci. Rep.* **2019**, *42*, 1471–1485. [CrossRef]
115. Zhang, L.; Song, Z.Q.; Li, F.F.; Li, X.X.; Ji, H.K.; Yang, S.S. The specific MYB binding sites bound by *TaMYB* in the *GAPCp2/3* promoters are involved in the drought stress response in wheat. *BMC Plant Biol.* **2020**, *20*, 366. [CrossRef]
116. Tan, W.; Zhang, D.; Zhou, H.; Zheng, T.; Yin, Y.; Lin, H. Transcription factor *HAT1* is a substrate of *SnRK2.3* kinase and negatively regulates ABA synthesis and signaling in *Arabidopsis* responding to drought. *PLoS Genet.* **2018**, *14*, e1007336. [CrossRef]
117. Ding, Y.; Avramova, Z.; Fromm, M. The *Arabidopsis trithorax*-like factor *ATX1* functions in dehydration stress responses via ABA-dependent and ABA-independent pathways. *Plant J.* **2011**, *66*, 735–744. [CrossRef]
118. Zhao, P.X.; Miao, Z.Q.; Zhang, J.; Chen, S.Y.; Liu, Q.Q.; Xiang, C.B. *Arabidopsis* MADS-box factor *AGL16* negatively regulates drought resistance via stomatal density and stomatal movement. *J. Exp. Bot.* **2020**, *71*, 6092–6106. [CrossRef]
119. Yan, H.R.; Jia, H.H.; Chen, X.B.; Hao, L.L.; An, H.L.; Guo, X.Q. The Cotton *WRKY* transcription factor *GhWRKY17* functions in drought and salt stress in transgenic *Nicotiana benthamiana* through ABA signaling and the modulation of reactive oxygen species production. *Plant Cell Physiol.* **2014**, *55*, 2060–2076. [CrossRef]
120. Chen, T.Z.; Li, W.J.; Hu, X.H.; Guo, J.R.; Liu, A.M.; Zhang, B.L. A Cotton MYB transcription factor, *GbMYB5*, is positively involved in plant adaptive response to drought stress. *Plant Cell Physiol.* **2015**, *56*, 917–929. [CrossRef]
121. Yang, W.; Liu, X.D.; Chi, X.J.; Wu, C.A.; Li, Y.Z.; Song, L.L.; Liu, X.M.; Wang, Y.F.; Wang, F.W.; Zhang, C.A. Dwarf apple *MbDREB1* enhances plant tolerance to low temperature, drought, and salt stress via both ABA-dependent and ABA-independent pathways. *Planta* **2011**, *233*, 219–229. [CrossRef] [PubMed]
122. Bandeppa, S.; Paul, S.; Thakur, J.K.; Chandrashekar, N.; Umesh, D.K.; Aggarwal, C.; Asha, A.D. Antioxidant, physiological and biochemical responses of drought susceptible and drought tolerant mustard (*Brassica juncea* L) genotypes to rhizobacterial inoculation under water deficit stress. *Plant Physiol. Bioch.* **2019**, *143*, 19–28. [CrossRef]
123. Thirumalaikumar, V.P.; Devkar, V.; Mehterov, N.; Ali, S.; Ozgur, R.; Turkan, I.; Mueller-Roeber, B.; Balazadeh, S. NAC transcription factor *JUNGBRUNNEN1* enhances drought tolerance in tomato. *Plant Biotechnol. J.* **2018**, *16*, 354–366. [CrossRef] [PubMed]
124. Chen, Y.X.; Huang, L.K.; Yan, H.D.; Zhang, X.; Xu, B.; Ma, X. Cloning and characterization of an ABA-independent DREB transcription factor gene, *HcDREB2*, in *Hemarthria compressa*. *Hereditas* **2016**, *153*, 3. [CrossRef] [PubMed]
125. Jimenez, J.D.; Cardoso, J.A.; Kotula, L.; Veneklaas, E.J.; Pedersen, O.; Colmer, T.D. Root length is proxy for high-throughput screening of waterlogging tolerance in *Urochloa* spp. grasses. *Funct. Plant Biol.* **2021**, *48*, 411–421. [CrossRef] [PubMed]
126. Ye, H.; Song, L.; Chen, H.T.; Valliyodan, B.; Cheng, P.; Ali, L.; Vuong, T.; Wu, C.J.; Orlowski, J.; Buckley, B. A major natural genetic variation associated with root system architecture and plasticity improves waterlogging tolerance and yield in soybean. *Plant Cell Environ.* **2018**, *41*, 2169–2182. [CrossRef]
127. Hu, B.; Cao, J.J.; Ge, K.; Li, L. The site of water stress governs the pattern of ABA synthesis and transport in peanut. *Sci. Rep.* **2016**, *6*, 32143. [CrossRef]
128. Najeeb, U.; Tan, D.K.Y.; Bange, M.P.; Atwell, B.J. Protecting cotton crops under elevated CO_2 from waterlogging by managing ethylene. *Funct. Plant Biol.* **2018**, *45*, 340–349. [CrossRef]
129. Zhang, Q.; Liu, X.F.; Zhang, Z.F.; Liu, N.F.; Li, D.Z.; Hu, L.X. Melatonin improved waterlogging tolerance in Alfalfa (*Medicago sativa*) by reprogramming polyamine and ethylene metabolism. *Front. Plant Sci.* **2019**, *10*, 44. [CrossRef]
130. Ismond, K.P.; Dolferus, R.; de Pauw, M.; Dennis, E.S.; Good, A.G. Enhanced low oxygen survival in *Arabidopsis* through increased metabolic flux in the fermentative pathway. *Plant Physiol.* **2003**, *132*, 1292–1302. [CrossRef]
131. Christianson, J.A.; Llewellyn, D.J.; Dennis, E.S.; Wilson, I.W. Comparisons of early transcriptome responses to low-oxygen environments in three dicotyledonous plant species. *Plant Signal Behav.* **2010**, *5*, 1006–1009. [CrossRef]
132. Yue, M.F.; Shen, H.; Li, W.H.; Chen, J.F.; Ye, W.H.; Tian, X.S.; Yin, A.G.; Cheng, S.M. Waterlogging tolerance of *Bidens pilosa* translates to increased competitiveness compared to native *Bidens biternata*. *Plant Soil* **2019**, *437*, 301–311. [CrossRef]
133. Zaman, M.S.U.; Malik, A.I.; Erskine, W.; Kaur, P. Changes in gene expression during germination reveal pea genotypes with either "quiescence" or "escape" mechanisms of waterlogging tolerance. *Plant Cell Environ.* **2019**, *42*, 245–258. [CrossRef] [PubMed]
134. Rzewuski, G.; Sauter, M. Ethylene biosynthesis and signaling in rice. *Plant Sci.* **2008**, *175*, 32–42. [CrossRef]
135. Zhou, X.; Zhang, Z.L.; Park, J.; Tyler, L.; Yusuke, J.; Qiu, K.; Nam, E.A.; Lumba, S.; Desveaux, D.; McCourt, P. The ERF11 transcription factor promotes internode elongation by activating gibberellin biosynthesis and signaling. *Plant Physiol.* **2016**, *171*, 2760–2770. [CrossRef]
136. Zhai, F.F.; Li, H.D.; Zhang, S.W.; Li, Z.J.; Liu, J.X.; Qian, Y.Q.; Ju, G.S.; Zhang, Y.X.; Liu, L.; Han, L. Male and female plants of *Salix viminalis* perform similarly to flooding in morphology, anatomy, and physiology. *Forests* **2020**, *11*, 321. [CrossRef]
137. Zhang, Y.; Liu, G.; Dong, H.; Li, C. Waterlogging stress in cotton: Damage, adaptability, alleviation strategies, and mechanisms. *Crop J.* **2021**, *9*, 257–270. [CrossRef]
138. Avila-Lovera, E.; Zerpa, A.J.; Santiago, L.S. Stem photosynthesis and hydraulics are coordinated in desert plant species. *New Phytol.* **2017**, *216*, 1119–1129. [CrossRef]
139. Choi, D. Molecular events underlying coordinated hormone action in submergence escape response of deepwater rice. *J. Plant Biol.* **2011**, *54*, 365–372. [CrossRef]
140. Yu, F.; Liang, K.; Fang, T.; Zhao, H.L.; Han, X.S.; Cai, M.J.; Qiu, F.Z. A group VII ethylene response factor gene, *ZmEREB180*, coordinates waterlogging tolerance in maize seedlings. *Plant Biotechnol. J.* **2019**, *17*, 2286–2298. [CrossRef]

141. Wei, X.N.; Xu, H.J.; Rong, W.; Ye, X.G.; Zhang, Z.Y. Constitutive expression of a stabilized transcription factor group VII ethylene response factor enhances waterlogging tolerance in wheat without penalizing grain yield. *Plant Cell Environ.* **2019**, *42*, 1471–1485. [CrossRef]
142. Geisler-Lee, J.; Caldwell, C.; Gallie, D.R. Expression of the ethylene biosynthetic machinery in maize roots is regulated in response to hypoxia. *J. Exp. Bot.* **2010**, *61*, 857–871. [CrossRef]
143. Peng, Y.Q.; Zhu, J.; Li, W.J.; Gao, W.; Shen, R.Y.; Meng, L.J. Effects of grafting on root growth, anaerobic respiration enzyme activity and aerenchyma of bitter melon under waterlogging stress. *Sci. Hortic.* **2020**, *261*, 108977. [CrossRef]
144. Zhang, Y.J.; Chen, Y.Z.; Lu, H.Q.; Kong, X.Q.; Dai, J.L.; Li, Z.H.; Dong, H.Z. Growth, lint yield and changes in physiological attributes of cotton under temporal waterlogging. *Field Crop Res.* **2016**, *194*, 83–93. [CrossRef]
145. Licausi, F.; Dongen, J.T.; Giuntoli, B.; Novi, G.; Santaniello, A.; Geigenberger, P.; Perata, P. HRE1 and HRE2, two hypoxia-inducible ethylene response factors, affect anaerobic responses in *Arabidopsis thaliana*. *Plant J.* **2010**, *62*, 302–315. [CrossRef]
146. Eysholdt-Derzso, E.; Sauter, M. Hypoxia and the group VII ethylene response transcription factor HRE2 promote adventitious root elongation in *Arabidopsis*. *Plant Biol.* **2019**, *21*, 103–108. [CrossRef]
147. Peng, H.P.; Chan, C.S.; Shih, M.C.; Yang, S.F. Signaling events in the hypoxic induction of alcohol dehydrogenase gene in *Arabidopsis*. *Plant Physiol.* **2001**, *126*, 742–749. [CrossRef]
148. Min, T.; Yin, X.R.; Shi, Y.N.; Luo, Z.R.; Yao, Y.C.; Grierson, D.; Ferguson, I.B.; Chen, K.S. Ethylene-responsive transcription factors interact with promoters of ADH and PDC involved in persimmon (*Diospyros kaki*) fruit de-astringency. *J. Exp. Bot.* **2012**, *63*, 6393–6405. [CrossRef]
149. Zhang, Z.G.; Luan, D.T.; Chen, G.; Xiao, F.; Yin, D.M.; Ni, D.; Li, X.R.; Qi, L.H. Isolation and characterization of a waterlogging-responsive gene involved in ethylene biosynthesis in *Chrysanthemum*. *Acta Physiol. Plant* **2018**, *40*, 100. [CrossRef]
150. Hinz, M.; Wilson, I.W.; Yang, J.; Buerstenbinder, K.; Llewellyn, D.; Dennis, E.S.; Sauter, M.; Dolferus, R. Arabidopsis *RAP2.2*, An ethylene response transcription factor that is important for hypoxia survival. *Plant Physiol.* **2010**, *153*, 757–772. [CrossRef]
151. Yang, C.Y.; Hsu, F.C.; Li, J.P.; Wang, N.N.; Shih, M.C. The *AP2/ERF* transcription factor *AtERF73/HRE1* modulates ethylene responses during hypoxia in *Arabidopsis*. *Plant Physiol.* **2011**, *156*, 202–212. [CrossRef]
152. Juntawong, P.; Butsayawarapat, P.; Songserm, P.; Pimjan, R.; Vuttipongchaikij, S. Overexpression of *Jatropha curcas* ERFVII2 transcription factor confers low oxygen tolerance in transgenic *Arabidopsis* by modulating expression of metabolic enzymes and multiple stress-responsive genes. *Plants* **2020**, *9*, 1068. [CrossRef] [PubMed]
153. Kaspary, T.E.; Roma-Burgos, N.; Merotto, A. Snorkeling strategy: Tolerance to flooding in rice and potential application for weed management. *Genes* **2020**, *11*, 975. [CrossRef]
154. Fukao, T.; Xiong, L.Z. Genetic mechanisms conferring adaptation to submergence and drought in rice: Simple or complex? *Curr. Opin. Plant Biol.* **2013**, *16*, 196–204. [CrossRef]
155. Khan, M.I.R.; Trivellini, A.; Chhillar, H.; Chopra, P.; Ferrante, A.; Khan, N.A.; Ismail, A.M. The significance and functions of ethylene in flooding stress tolerance in plants. *Environ. Exp. Bot.* **2020**, *179*, 104188. [CrossRef]
156. Locke, A.M.; Barding, G.A.; Sathnur, S.; Larive, C.K.; Bailey-Serres, J. Rice *SUB1A* constrains remodelling of the transcriptome and metabolome during submergence to facilitate post-submergence recovery. *Plant Cell Environ.* **2018**, *41*, 721–736. [CrossRef] [PubMed]
157. Fukao, T.; Bailey-Serres, J. Submergence tolerance conferred by Sub1A is mediated by *SLR1* and *SLRL1* restriction of gibberellin responses in rice. *Proc. Natl. Acad. Sci. USA* **2008**, *105*, 16814–16819. [CrossRef]
158. Schmitz, A.J.; Folsom, J.J.; Jikamaru, Y.; Ronald, P.; Walia, H. *SUB1A*-mediated submergence tolerance response in rice involves differential regulation of the brassinosteroid pathway. *New Phytol.* **2013**, *198*, 1060–1070. [CrossRef]
159. Hattori, Y.; Nagai, K.; Furukawa, S.; Song, X.J.; Kawano, R.; Sakakibara, H.; Wu, J.Z.; Matsumoto, T.; Yoshimura, A.; Kitano, H. The ethylene response factors *SNORKEL1* and *SNORKEL2* allow rice to adapt to deep water. *Nature* **2009**, *460*, 1026–1116. [CrossRef]
160. Fukushima, A.; Kuroha, T.; Nagai, K.; Hattori, Y.; Kobayashi, M.; Nishizawa, T.; Kojima, M.; Utsumi, Y.; Oikawa, A.; Seki, M. Metabolite and phytohormone profiling illustrates metabolic reprogramming as an escape strategy of deepwater rice during partially submerged stress. *Metabolites* **2020**, *10*, 68. [CrossRef]

forests

MDPI

Article

Transcriptome Analysis of Apricot Kernel Pistils Reveals the Mechanisms Underlying ROS-Mediated Freezing Resistance

Xiaojuan Liu, Yingying Yang, Huihui Xu, Dan Yu, Quanxin Bi and Libing Wang *

State Key Laboratory of Tree Genetics and Breeding, Research Institute of Forestry, Chinese Academy of Forestry, Beijing 100091, China

* Correspondence: wlibing@caf.ac.cn; Tel.: +010-62888593

Abstract: Spring frost is a major limiting factor in the production and cultivation of apricot kernels, an ecological and economic dry-fruit tree in China. The frequent occurrence of spring frost often coincides with the blooming period of apricot kernels, resulting in significant damage to floral organs and reductions in yield. We investigated the molecular signature of pistils from two apricot kernel cultivars with different frost-resistance levels using transcriptome data. A total of 3223 differently expressed genes (DEGs) were found between two apricot kernel cultivars under freezing stress, including the bHLH and AP2/ERF-ERF transcription factors. Based on KEGG analysis, DEGs were mostly enriched in the biosynthesis of the secondary metabolites, in the metabolic pathways, and in plant-hormone signal transduction. The co-expression network, which included 81 hub genes, revealed that transcription factors, protein kinases, ubiquitin ligases, hormone components, and Ca^{2+}-related proteins coregulated the ROS-mediated freezing response. Moreover, gene interaction relationships, such as ERF109-HMGCR1, ERF109-GRXC9, and bHLH13-JAZ8, were predicted. These findings revealed the regulatory factors for differences in frost resistance between the two tested apricot kernel cultivars and contributed to a deeper understanding of the comprehensive regulatory program during freezing stress. Some of the hub genes identified in this work provide new choices and directions for breeding apricot kernels with a high frost resistance.

Keywords: freezing stress; apricot kernel; transcriptome; transcription factors; ROS; regulatory network

Citation: Liu, X.; Yang, Y.; Xu, H.; Yu, D.; Bi, Q.; Wang, L. Transcriptome Analysis of Apricot Kernel Pistils Reveals the Mechanisms Underlying ROS-Mediated Freezing Resistance. *Forests* **2022**, *13*, 1655. https://doi.org/10.3390/f13101655

Academic Editor: Bryce Richardson

Received: 18 August 2022
Accepted: 6 October 2022
Published: 9 October 2022

1. Introduction

Frost is a common meteorological disaster which refers to a sudden drop in the air and surface temperature to below 0 °C. It has been recognized as a major threat to plant growth, development, and agricultural and forestry productivity [1,2]. Since climate warming has increased temperatures in early spring, perennial plants have become increasingly vulnerable to lower temperatures due to phenological shifts, such as advanced flowering time [3]. Spring frost has been shown to cause irreparable losses to vegetables, fruit trees, and crops [4].

An apricot kernel, an apricot (*Prunus armeniaca* L.) plant with almonds as its main use, is also a fresh fruit with a unique taste and mainly includes a big flat apricot (*Armeniaca vulgaris* × *sibirica*) and Siberian apricot (*Armeniaca sibirica* L.). It is an important raw material for the food and pharmaceutical industries and is mainly distributed in northern China. Spring frost frequently occurs during their flowering time between late March and mid-April [5]. Among the apricot flower organs, the freezing resistance of the pistils is the weakest, followed by the stamen and petals [6]. Spring frost causes severe damage to apricot kernels' reproductive organs, resulting in significant yield loss [7]. Frost injury is the main limiting factor in apricot kernel production.

Cold stress is an environmental stress that can be divided into chilling stress (0–15 °C) and freezing stress (<0 °C) (e.g., spring frost) [8]. Plants that suffer from freezing stress have developed sophisticated cold-acclimation mechanisms that improve their freezing tolerance upon exposure to nonlethal lower temperatures [9,10]. A series of cellular responses and molecular strategies are initiated when a plant perceives freezing stress, such as the production of ROS and osmolytes, and changes in the cytosolic Ca^{2+} concentration, hormone content, and gene expression [10–12]. *CBF* genes, which are rapidly induced by low temperatures, play central roles in cold acclimation. Many transcription factors (TFs) (e.g., ICEs, CAMTAs, and MYB15) act as upstream regulators that regulate the expression of *CBFs* [8,10]. Among them, ICE1, a bHLH TF, is the best-characterized transcriptional activator of *CBF* genes [13]. Post-translational modifications, such as ubiquitination and phosphorylation, are important for the function of ICE1 in cold tolerance. For example, the protein kinases OST1 and MPK3/6 can phosphorylate ICE1, affecting its transcriptional activity to regulate *CBF* expression and cold tolerance [14].

In plants, ROS act as a double-edged sword. Excessive ROS accumulation due to stress induces oxidative stress, which can damage plant cells; at this time, ROS-scavenging systems consist of an endogenous defensive mechanism that comprises different enzymatic (e.g., superoxide dismutase, SOD, and catalase, CAT) and nonenzymatic (e.g., glutathione, GSH) antioxidants that are activated to maintain ROS levels [15–17]. Moreover, ROS, as messenger molecules, participate in acclimation responses to freezing stress. ROS can interact with different hormones (e.g., ET, JA, and BRs) to control gene expression and induce physiological changes in response to cold stress [18]. In addition, emerging evidence indicates that some key signaling components participate in ROS-mediated stress response processes, such as messenger molecules (e.g., Ca^{2+} and NO), protein kinases (e.g., CIPKs and MAPKs), and TFs (e.g., MYC and MYB) [18,19].

In apricot kernels, research on freezing resistance primarily focuses on the physiological level and differences between different varieties [20,21]. It has been reported that the activities of antioxidant enzymes, such as SOD, in apricots show a change pattern in which they first increase and then decrease under freezing stress, with their activities higher in variety with a strong freezing resistance [7,22]. Our previous study found that many regulators (such as TFs and protein kinases) and some of the genes involved in the oxidation reduction process were regulated in apricot under natural spring frost conditions via transcriptome analysis [7]. However, the underlying relationship and functional mechanism of differentially expressed genes (DEGs) are still largely unknown. We investigated the comparative transcriptome of pistils in this work based on two apricot kernel varieties ('Weixuan 1' and 'Longwangmao') (*Armeniaca vulgaris* × *sibirica*) with different frost-resistance levels under simulated spring frost conditions. Our study aims to analyze the DEGs and different biological processes between 'Weixuan 1' and 'Longwangmao' and elucidate ROS-mediated molecular mechanisms in response to freezing stress. This may provide new insights into the response mechanisms underlying freezing stress in apricot kernels.

2. Materials and Methods

2.1. Plant Materials and Treatment

For the analysis of apricot kernel pistils' freezing resistance, two main cultivated varieties were selected, namely 'Weixuan 1' and 'Longwangmao'. 'Weixuan 1' is a frost-resistant variety selected from 'Longwangmao' through bud mutation [23].

During hibernation, flower branches from cold-tolerant 'Weixuan 1' (CtW) and cold-sensitive 'Longwangmao' (CsL) were collected from the Apricot Germplasm Resource nursery in Shanxi, China. Flower branches were brought to full bloom via hydroponic cultures in an incubator (20 °C) and were treated with temperatures of −2 °C, −3 °C, and −4 °C for 1 h. The cooling method involved reducing the temperature from 20 °C to 2 °C at a rate of 10 °C/0.5 h, and then reducing the temperature from 2 °C to the treatment temperature at a rate of 3 °C/h in a low-temperature incubator. The flower branches were then placed in the incubator (20 °C) to recover for 3 h; that is, when the browning of the pistils showed no significant change. The pistils undergoing and not undergoing freezing treatments were collected for transcriptome sequencing and quantitative real-time PCR (qRT-PCR) analysis in liquid nitrogen and were stored at −80 °C.

2.2. *RNA-seq*

The pistils of CsL and CtW undergoing and not undergoing freezing treatments were used for RNA exaction. According to the differences in the temperature treatments between CsL and CtW, the samples were named CsL1 (20 °C), CsL2 (−2 °C), CsL3 (−3 °C), CsL4 (−4 °C), CtW1 (20 °C), CtW2 (−2 °C), CtW3 (−3 °C), and CtW4 (−4 °C), and every sample contained three biological replicates (every five flower branches represented a biological repeat, containing 50–70 pistils). Total RNA was extracted with the RNAprep Pure Plant Plus Kit (polysaccharide- and polyphenolic-rich; Tiangen, Beijing, China). RNA quality estimation, including purity, integrity, and concentration, was checked by 1% agarose gel, a Nano Photometer spectrophotometer (Implen, Westlake Village, CA, USA), a Qubit 2.0 Fluorometry (Life Technologies, Carlsbad, CA, USA), and an Agilent 2100 Bioanalyzer (Agilent Technologies, CA, USA). High-quality RNAs were used to construct cDNA libraries sequenced by Illumina paired-end sequencing technology on an Illumina HiSeq platform (Illumina, San Diego, CA, USA) at Metwell Biotechnology Co., Ltd. (Wuhan, China). The raw data were cleaned by removing adapter sequences, the reads with more than 10% N bases, and low-quality reads with a percentage of base quality value ≤ 5 exceeding 50%. The clean reads were mapped onto the apricot (*Prunus armeniaca* L.) reference genome using the HISAT2 software using the default parameters [24].

2.3. *Identification of DEGs and Enrichment Analyses*

The number of reads mapped to each gene were counted using HTSeq v0.6.1, and the fragments per kilobase of transcript per million fragments mapped (FPKM) of each gene were calculated based on the gene length and the read count mapped to the gene. The DEG analysis between two groups was performed using the DESeq2 R package. The false discovery rate (FDR) was obtained from *p*-values adjusted using the Benjamini–Hochberg method. The DEGs were screened with |log2Fold Change| >= 1 and FDR < 0.01. KEGG enrichment analyses of the DEGs were conducted using KOBAS software.

2.4. *Weighted Gene Co-Expression Network and Hub Genes Analysis*

The antioxidant enzymes (POD, SOD, and CAT) were determined using water-soluble tetrazolium salt-(WST-1), guaiacol-, and hydrogen-peroxide-based methods, respectively [25]. The mixed pistils undergoing and not undergoing freezing treatments were used to analyze the activities of the antioxidant enzymes, and each experiment contained four biological replicates. SPSS 25 was used to analyze the data of the enzyme activities by one-way ANOVA and Duncan's multiple comparison analysis.

The activity of the antioxidant enzymes and DEGs was used for weighted gene co-expression network analysis (WGCNA). WGCNA was conducted in R using the default parameters. The FPKM values of the DEGs were normalized, and Pearson's correlation coefficient was calculated for each pair of genes to construct an adjacency matrix. Gene modules were identified based on TOM and converted by the adjacency matrix, using the WGCNA package. The correlation between modules and traits was estimated. The hub genes within a selected module were screened by kME (intra-module connectivity) > 0.9 and GS (gene significance) > 0.2. Cytoscape (version 2.7.2) was used to visualize the relationships between hub genes.

2.5. *Protein–Protein Interaction (PPI) Network Prediction*

For the PPI analysis, hub genes associated with the activity of antioxidant enzymes were used to retrieve genes interacting with STRING 11 in *Prunus armeniaca* var. *bungo*. A required confidence score (combined score) greater than 0.4 was used as the threshold for the interaction. The disconnected genes were hidden in the network. The PPI network was constructed using STRING and was further analyzed using Cytoscape (version 2.7.2).

2.6. Verification of qRT-PCR Analysis

Nine genes were selected from the hub genes related to the antioxidant enzyme for qRT-PCR analysis. Total RNA extraction was performed with a TIANGEN kit (Beijing, China), and cDNA was synthesized using a reverse transcription kit (Takara Dalian, Japan). A qRT-PCR test containing three biological replicates was conducted using the KAPA SYBR FAST qPCR Master Mix (Kapa Biosystems, Boston, MA, USA) on a LightCycler 480 II Real-time PCR Instrument (Roche) according to the manufacturer's protocols. In that test, *18S* was used as a reference gene. The primers are listed in Table S1.

3. Results

3.1. Transcriptome Analysis of Apricot Kernel Pistils under Freezing Stress

RNA-seq data were generated for eight different freezing-treated and untreated pistils to explore the molecular mechanism of the freezing resistance between CsL and CtW pistils. In total, 120.78 Gb of clean reads were obtained from 24 libraries, ranging from 4.16 to 6.02 Gb per library, with an average GC content of 45.81% (Table 1). The high Q30 (>91.73%) represents bases with error rates < 0.1% for libraries showing high-quality RNA-seq. The rate of clean reads mapped to the apricot genome ranged from 87.55% to 94.74%, in which uniquely mapped reads exceeded 84.85%.

Table 1. Summary of mapping transcriptome reads to reference sequence.

Variety	Treatment	Sample	Clean Reads	GC (%)	Q20 (%)	Q30 (%)	Read Mapped (%)	Unique Mapped (%)
'Longwangmao' (CsL)	CsL1	CsL11	60,245,518	45.92	97.21	92.15	94.14	91.47
		CsL12	48,251,936	45.93	97.01	91.73	93.41	90.81
		CsL13	48,950,240	46.25	97.06	91.86	93.94	91.31
	CsL2	CsL21	47,482,406	45.86	97.66	93.19	94.74	92.12
		CsL22	45,957,672	45.72	97.67	93.26	93.90	90.94
		CsL23	54,500,124	45.50	97.50	92.88	93.69	90.63
	CsL3	CsL31	56,350,610	45.30	97.64	93.21	93.29	90.70
		CsL32	56,506,172	45.78	97.65	93.25	92.27	89.72
		CsL33	55,646,854	45.85	97.67	93.31	92.45	89.86
	CsL4	CsL41	54,666,004	45.78	97.66	93.34	87.55	84.85
		CsL42	44,453,966	45.72	97.60	93.14	93.55	91.11
		CsL43	49,831,556	45.90	97.58	93.12	94.34	91.61
'Weixuan 1' (CtW)	CtW1	CtW11	41,850,108	45.93	97.30	92.29	92.11	89.62
		CtW12	41,637,772	46.20	97.32	92.34	94.16	91.59
		CtW13	50,142,906	46.10	97.29	92.29	93.79	91.23
	CtW2	CtW21	45,960,022	45.49	97.62	93.05	94.17	91.66
		CtW22	55,251,738	45.38	97.75	93.40	94.46	91.98
		CtW23	51,158,772	45.88	97.61	93.14	94.37	91.71
	CtW3	CtW31	43,518,372	45.87	97.72	93.45	93.96	91.45
		CtW32	60,004,540	45.87	97.44	92.79	93.50	91.03
		CtW33	60,219,916	45.69	97.55	93.03	94.03	91.44
	CtW4	CtW41	45,356,754	45.84	97.65	93.25	90.25	87.83
		CtW42	46,029,912	45.81	97.83	93.67	94.45	91.78
		CtW43	43,832,166	45.84	97.70	93.37	93.16	90.58

3.2. Identification and Functional Analysis of DEGs of Apricot Kernel Pistils under Freezing Stress

In total, 5206, 2032, and 3223 DEGs were identified under freezing treatment in CsL, CtW, and CsL vs. CtW. Among all of the DEGs, 509 shared DEGs that were found in the three comparison groups (CsL, CtW, and CsL vs. CtW). In the three CsL freezing-treatment groups, 438 genes were up-regulated, and 68 genes were down-regulated. In the three CtW freezing-treatment groups, 367 genes were up-regulated, and 142 genes were down-regulated; in CsL vs. CtW, 355 genes were up-regulated, and 157 genes were down-regulated (Figure 1a).

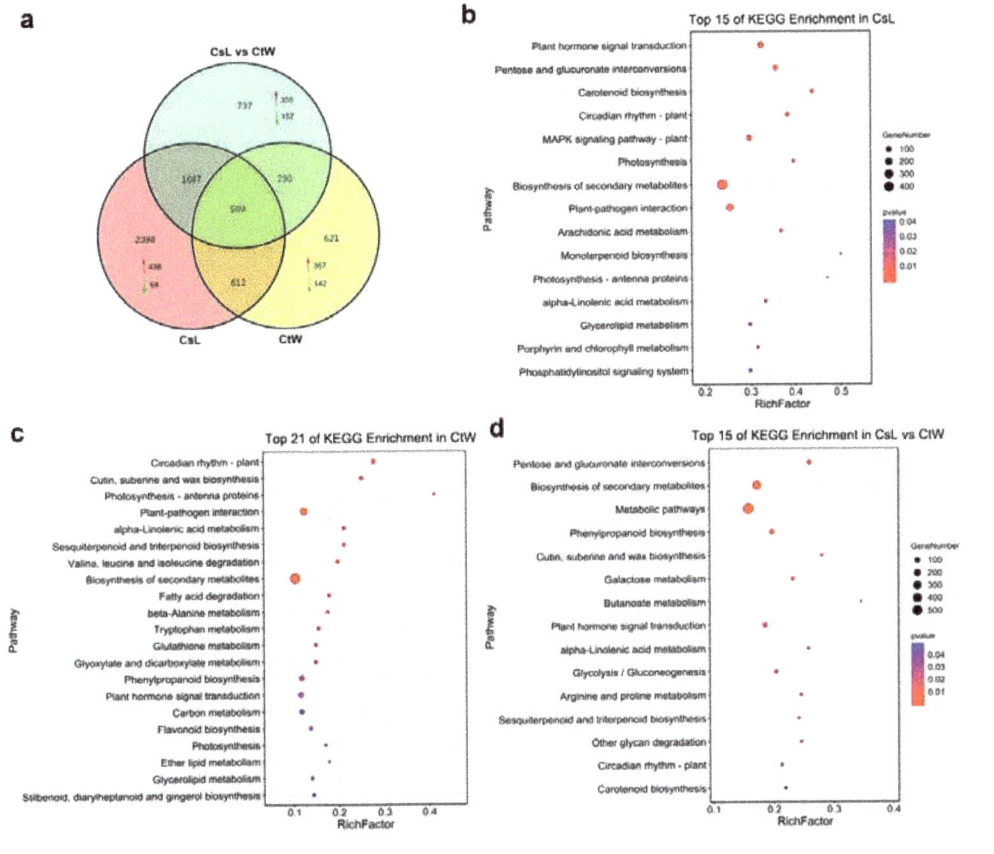

Figure 1. Venn diagram and significantly enriched KEGG pathways of differently expressed genes (DEGs) under freezing treatment. (**a**) Venn diagram showing the shared and specific number of up-regulated and down-regulated DEGs identified in CtW, CsL, and CtW vs. CsL. CtW: cold-tolerant 'Weixuan 1'; CsL: cold-sensitive 'Longwangmao'. (**b**,**c**) Significantly enriched KEGG pathways of DEGs in CtW (**b**), CsL (**c**), and CtW vs. CsL (**d**).

DEGs were characterized using KEGG databases to understand their biological roles in CsL and CtW. There were 15, 21, and 15 significantly enriched pathways ($p < 0.05$) in CsL, CtW, and CsL vs. CtW, respectively. In CsL, most of the DEGs were enriched in plant-hormone signal transduction (7.17%), the MAPK signaling pathway (4.89%), the biosynthesis of secondary metabolites (25.40%), and plant–pathogen interaction (9.83%) (Figure 1b). In CtW, most DEGs were enriched in the biosynthesis of secondary metabolites (26.96%), plant–pathogen interaction (11.65%), plant-hormone signal transduction (6.28%), and phenylpropanoid biosynthesis (5.76%) (Figure 1c). In CsL vs. CtW, more DEGs were enriched in the biosynthesis pathways of the secondary metabolites (27.90%), in the metabolic pathways (45.42%), in plant-hormone signal transduction (6.24%), and in phenylpropanoid biosynthesis (6.00%) (Figure 1d).

3.3. Differentially Expressed Transcription Factors of Apricot Kernel Pistils under Freezing Stress

The transcriptional regulation of cold stress has been widely studied in model plants. To identify the TFs involved in apricot kernels' response to freezing stress, we analyzed differentially expressed TFs in CsL and CtW under freezing stress. As shown in Figure S1, 423 DEGs were assigned to 50 TF families in CsL; 151 DEGs were assigned to 33 TF families in CtW; and 242 DEGs were assigned to 48 TF families in CsL vs. CtW. Genes belonging to the NAC, AP2/ERF-ERF, MYB, WRKY, and bHLH TF families in the CsL, CtW, and CsL vs. CtW groups accounted for more than 40%. Furthermore, there were 44 shared TFs, including 10 AP2/ERF-ERF TFs, 6 bHLH TFs, 5 MYB TFs, 3 NAC TFs, and 2 WRKY TFs in the three groups (Figure 2). Out of these TFs, the expression level of most genes under freezing stress was up-regulated compared to the control, except for the down-regulated *PARG01786* (*RAP2.4*), *PARG12349* (*MYB6*), *PARG06699* (*MYB21*), and *PARG30216* (*NAC25*) TFs in CtW and the *PARG29164* (*WRKY70*) TF in CsL (Figure 2b). These findings demonstrate that these TFs, especially NAC, AP2/ERF-ERF, MYB, WRKY, and bHLH, may govern the transcriptional changes through both transcriptional activation and repression in response to freezing stress.

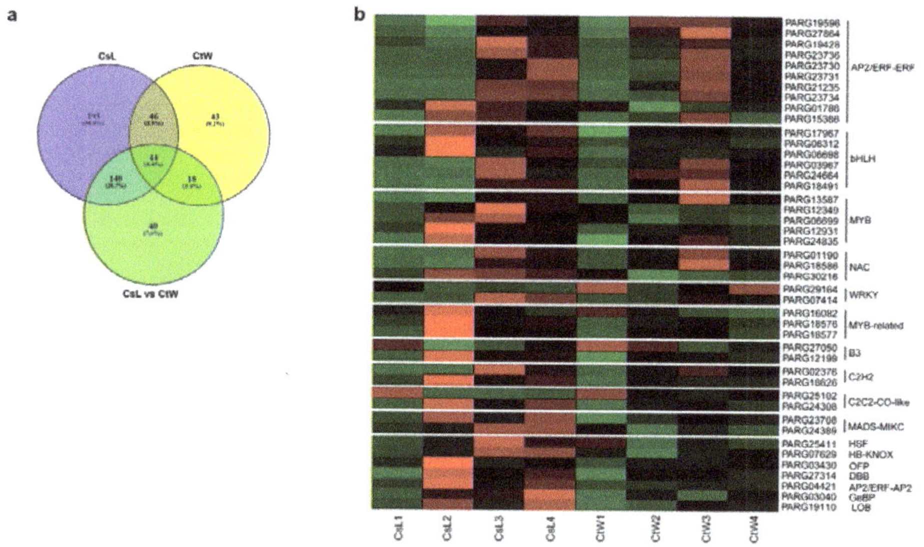

Figure 2. Analysis of differently expressed transcription factors (TFs) under freezing treatment. (**a**) Venn diagram showing the shared and specific number (and ratio) of differently expressed TFs identified in CtW, CsL, and CtW vs. CsL. (**b**) Heatmap of overlapping TFs between CtW, CsL, and CtW vs. CsL. Each column represents the gene expression at different temperatures (20 °C, −2 °C, −3 °C, and −4 °C) in CtW and CsL.

3.4. The Co-Expression Network Analysis of DEGs Related to the Antioxidant Enzyme Activity

Freezing stress produces excessive ROS scavenged by antioxidant mechanisms, such as enzymatic and nonenzymatic systems, to regulate cold resistance in plants [26,27]. We employed WGCNA to detect the co-expressed genes associated with antioxidant enzyme activities, including POD, SOD, and CAT, that may be involved in regulating freezing resistance in CsL and CtW (Table S2). The network was constructed with 3223 DEGs, and nine co-expressed gene modules labeled with different colors were determined (Figure 3a). Then, the correlations between the module eigengenes (MEs) and antioxidant enzyme parameters were analyzed. Only the magenta module with 173 genes was significantly associated with SOD activity ($r = 0.73$, $p = 0.04$) (Figure 3b). In addition, the module membership and GS of the magenta module were highly correlated (Table 2), further demonstrating that genes in the magenta module were significantly associated with SOD activity.

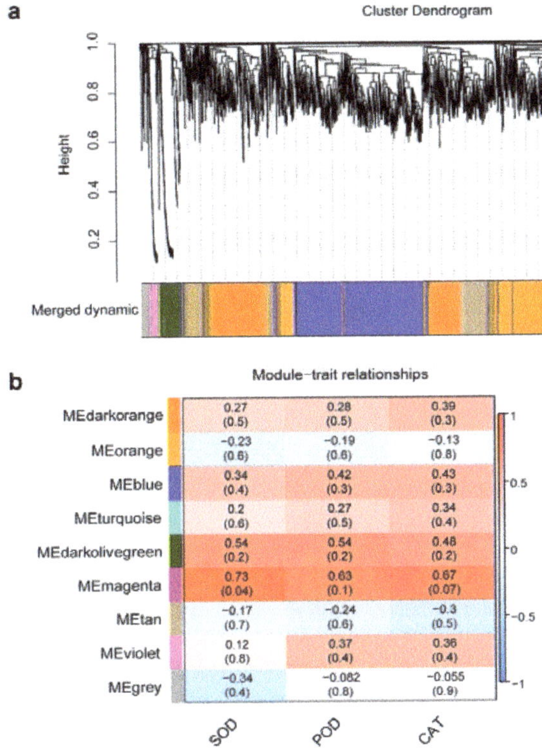

Figure 3. WGCNA of DEGs and antioxidant enzyme activity. (**a**) Cluster dendrogram presents nine co-expression modules labeled with different colors. (**b**) Correlation heatmap between DEGs and antioxidant enzyme activity. Rows correspond to modules. The color (from blue to red) indicates the correlation value (from −1 to 1).

Table 2. Correlations between module membership and gene significance of modules.

Module	No. of Genes	Module Membership vs. Gene Significance	
		Correlation (r^2)	p-Value
Orange	997	−0.0094	0.77
Dark orange	526	−0.2	3.8×10^{-6}
Blue	734	−0.65	2.4×10^{-89}
Tan	270	−0.51	2.8×10^{-19}
Violet	51	−0.36	0.0095
Dark olive green	121	0.027	0.77
Turquoise	230	0.04	0.55
Magenta	173	0.58	6.2×10^{-17}
Grey	121	0.036	0.7

Subsequently, the DEGs in the magenta module that exhibited the strongest correlation with the SOD parameter were further analyzed. These genes were significantly up-regulated in CsL3 (treated with −3 °C) and in CtW2 (treated with −2 °C) (Figure S2), implying that the genes in CtW were modulated earlier than those in CsL in order to adapt to and defend against freezing stress when suffering from lower temperatures. Hub genes within the magenta module, referring to the most highly connected nodes within the module, were used to construct the gene network. A total of 81 hub genes containing 77 up-regulated and 4 down-regulated genes in CsL vs. CtW were identified based on kME > 0.9 and GS > 0.2 (Figure 4; Table S3). The co-expression network had ten hormone-related genes and five Ca^{2+}-related genes: ERF025, ERF109, JAZ8, JAZ10B, CAMBP25, and PBP1-like genes were highly connected within the module. Moreover, twelve TFs, including one CBF/DREB subfamily member, three WRKY members, three bHLH members, and two NAC members, were related to the SOD parameter, and NAC090, bHLH35, ZAT11, and GTE12 had higher connections within the module. In addition, twelve post-translational modified proteins with five protein kinases, one phospholipase PP2C25, five E3 ubiquitin-protein ligases, and the F-box protein SKIP27, were also connected with the SOD parameter; the kME of the PUB21-like, SKIP27, EFR, and PARG02353 (LRR receptor-like serine/threonine-protein kinase At3g47570) were high. Some structural functional genes, such as the redox genes HMCGR1 and GRXC9, and the jasmonic acid (JA) synthesis genes, namely AOS1 and OPR3, were correlated with the SOD parameter (Figure 4; Table S3).

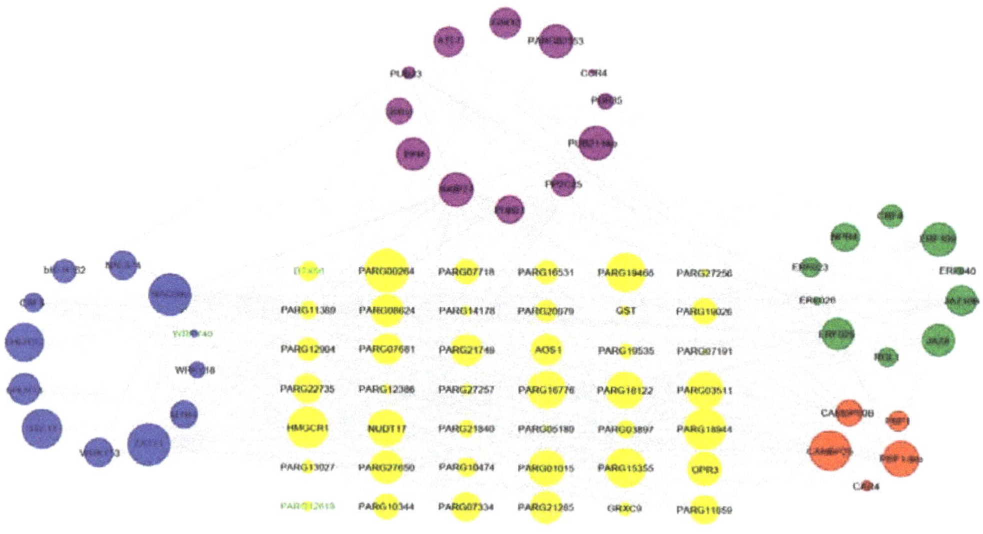

Figure 4. The co-expression network of hub genes in the magenta module. Red nodes refer to Ca^{2+}-related genes; green nodes refer to genes involved in hormone signaling; purple nodes refer to genes involved in post-translational modification; blue nodes refer to TFs. Black words indicate up-regulated genes; green words indicate down-regulated genes. The size of the nodes is based on intra-modular connectivity (kME). The width of the edges is based on weight.

In addition, the co-expression relationship of TF–TF, TF–post-translational modified protein, and TF–structural functional genes were identified in the gene network (Figure 4). The TFs ERF025, WRK18, and ERF109, as well as the E3 ubiquitin-protein ligases ATL7 and PUB23, showed a higher correlation with CBF3. The repressor proteins JAZ8 and JAZ10B in JA signaling were co-expressed with NAC090, WRKY53, ERF109, PUB21-like, and SKIP27. A co-expression relationship between HMCGR1 and the TFs (ERF109, WRKY40, NAC090, bHLH35, and ZAT11) was found in the gene network. These relationships can act as a reference for the research on the freezing response.

3.5. Interaction Network Analysis of Hub Genes Related to Antioxidant Enzyme Activity

A PPI regulatory network of 81 hub genes in the magenta module was constructed to explain the potential regulatory mechanism in apricot kernels responding to freezing stress. As shown in Figure 5, JAZ8 has a direct relationship with bHLH13, WRKY18/40, ERF109, and GATA25, and especially with bHLH13, in addition to the genes in the JA signal pathways (JAZ1/10/12, NINJA, and COI1) and the JA biosynthesis genes (OPR3, AOC3/4, and AOS). bHLH13 plausibly interacts with the JA-related genes (JAZ1/8/10, NINJA, COL1, OPR3, and AOS1), GATA25, and bHLH92, whereas bHLH35 may only interact with JAZ1 and bHLH92. There may also be interactions between ERF109 and WRKY18/40, HMGCR1, GRXC9, and PP2C25. Moreover, WRKY18/40 and bHLH92, GRXC9 and JAZ12, and RGI3 and RGL1 were found to have direct relationships. PP2C25, NPR4, CCR4, ZAT11, and PBP1 were all related to PP2Cc. The JA-related genes (JAZ8, AOS1, and OPR3) and bHLH13 were considered to be key genes in the network, suggesting that they may play an important role in the ROS-mediated freezing response.

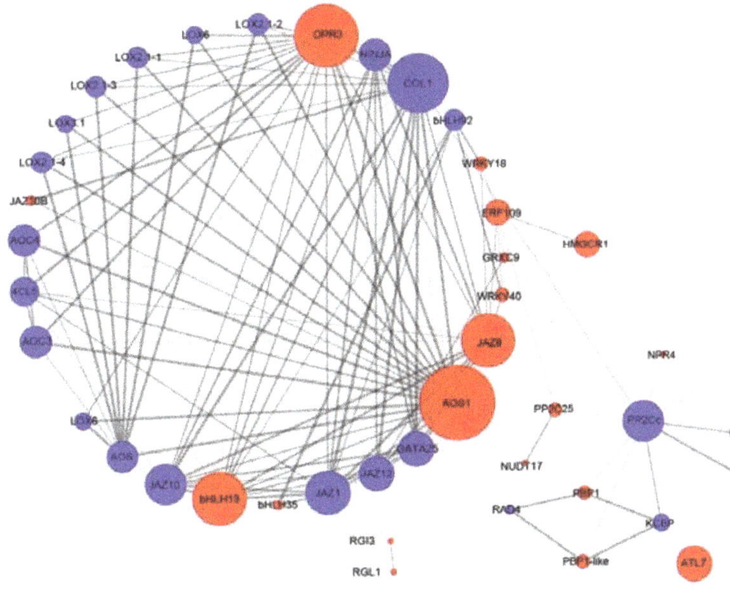

Figure 5. PPI regulatory network of hub genes in the magenta module. Red nodes refer to hub genes identified in the magenta module; blue nodes refer to genes in the genome that interacted with hub genes. The size of the nodes is based on degree. The width of the edges is based on the confidence score.

3.6. qRT-PCR Validation of Key DEGs Involved in the Response of Apricot Kernels to Freezing Stress

We selected nine key genes in the magenta module for qRT-PCR analysis to verify the expression pattern's accuracy for these hub genes in the transcriptome data. With the exception of *NAC090* in CtW3 and CtW4, the expression levels of *ERF109*, *ZAT11*, *PBP1*, *NAC090*, *PP2C25*, and *PUB21* were higher under lower temperatures, higher in CsL3 or CsL4 in 'Longwangmao', and higher in CtW2 in 'Weixuan 1' (Figure 6). *JAZ8* was induced by freezing stress in CsL, whereas it had no significant change or decrease in CtW. In addition, the expression levels of *bHLH35* and *OPR3* obviously increased in CsL4 in 'Longwangmao' and decreased in CtW3 in 'Weixuan 1'. The qRT-PCR results were consistent with the RNA-Seq results, indicating the reliability of the RNA-Seq data.

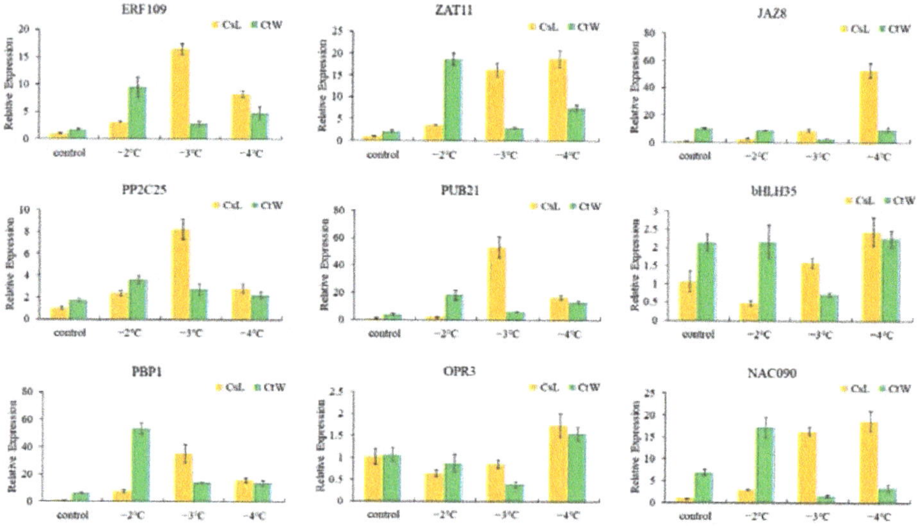

Figure 6. The relative expression levels of nine hub genes by qRT-PCR. *18S* was used as the internal control. Error bars indicate SD.

4. Discussion

4.1. Transcriptional Regulation in Freezing Response of Apricot Kernels

TFs are important regulators for controlling gene expression to modulate the stress response. Many TFs regulate the expression of cold-stress-responsive genes (CORs) and adapt the tolerance of plants to cold stress, such as the AP2/ERF, NAC, WRKY, MYB, and bZIP TFs [28]. In the pistils of CsL and CtW, these families were also the main differentially expressed TFs under freezing stress (Figure S1). Recent studies show that CBFs elevate antioxidant enzymes to regulate cold tolerance [29,30]. Consistent with these observations, one CBF/DREB TF, CBF3, was regulated by freezing stress and was found to be associated with antioxidant enzyme activity (SOD) in apricot kernels (Figure 4). No ICEs, which are key inducers of *CBFs* expression, were differentially expressed in CtW and CsL under freezing stress, whereas other bHLH TFs (such as bHLH13 and bHLH35) were induced by freezing stress and were correlated with antioxidant enzyme activity (Figures 2 and 4). Moreover, bHLH13 and bHLH35 were predicted to interact with JAZ8 or JAZ1, suggesting that they link with JA signaling to regulate the SOD-mediated freezing response.

In addition to the ICE1-CBF pathway, many other TFs regulate plant resistance in a CBF-dependent or CBF-independent way [10]. For example, MYB15 inhibits the expression of *CBFs* and negatively regulates the freezing tolerance of *Arabidopsis thaliana* [31]. In soybean, the overexpression of *GmNAC20* increases the activity of antioxidant enzymes and enhances cold tolerance via the CBF-COR pathway [32,33]. On the contrary, some CBF-independent TFs, such as MYB73, WRKY33, and ZAT10, function parallel to CBFs and modulate COR expression [10,28,30]. In this work, some TFs, including ERF109, bHLH35, and WRKY18, were found to be related to antioxidant enzyme activity and to be co-expressed with CBF3 (Figure 4). Furthermore, ERF109 may directly regulate the expression of redox proteins HMGCR1 and GRXC9 in response to freezing stress (Figure 5). Our results suggest that these TFs regulate the antioxidant-enzyme-mediated freezing response by modulating *CBFs*.

4.2. Post-Translational Regulation in the Freezing Response of Apricot Kernels

Besides transcriptional regulation, post-translational modifications, such as phosphorylation and ubiquitination, also play important roles in the responses of plants to cold [10]. Many protein kinases and phosphatases have been confirmed to regulate the cold tolerance of plants, such as OST1/SnRK2.6, MPK3, MPK6, and BIN2 [34]. In Arabidopsis, a type 2C phosphatase, ABI1, dephosphorylates protein kinase OST1 to repress kinase activity and to negatively regulate freezing tolerance [35]. In the pistils of apricot kernels, five protein kinases (RGI3, EFR, CCR4, GSO2, and At3g47570-like) and one phosphatase (PP2C25) were identified as hub genes in the magenta module: they were highly correlated with antioxidant enzyme activity under freezing stress (Figures 3 and 4). Previous studies have shown that OsCPK24 functions as a positive regulator of cold tolerance by phosphorylating and inhibiting OsGrx10 to improve glutathione (an antioxidant) levels [36]. The PPI network predicted the direct relationship between PP2C25-ERF109, PP2Cc-ZAT11, and RGI3-RGL1 (Figure 5), indicating that these protein kinases and phosphatase may control the tolerance of apricot kernels to freezing stress by regulating the phosphorylation level of the targets related to the antioxidant process.

Ubiquitination and protein-degradation mediated by the E3 ubiquitin ligases are also important for cold signaling. E3 ubiquitin ligases such as HOS1, PUB25, ATL78, and ATL80 have been extensively studied to be involved in cold stress in plants [34,37,38]. Two U-box type E3 ubiquitin ligases, PUB25 and PUB26, improve their E3 ligases activity by the phosphorylation of OST1 and function as negative regulators in response to cold stress by targeting MYB15 for degradation [39]. Four PUBs E3 ubiquitin ligases (PUB21, PUB21-like, PUB23, and PUB35) were differentially expressed in apricot kernel pistils during freezing stress and were highly related to antioxidant enzyme activity (Figure 4). OsSRFP1, a RING finger E3 ligase, negatively regulates the activity of antioxidant enzymes and cold stress tolerance in *Oryza sativa* [40]. However, there are few studies on the mechanism in ubiquitin ligase that regulates antioxidant enzyme activity. In addition to PUB35, other E3 ubiquitin ligases have co-expressive relationships with ERF109 (Figure 4), suggesting that these E3 ubiquitin ligases regulate ROS homeostasis and the freezing resistance of apricot kernels through ERF109.

4.3. Ca^{2+} and Hormone Signaling in Freezing Response of Apricot Kernels

Ca^{2+} is an important secondary messenger in plant response to cold stress. Previous research has shown that calmodulin (CAM) activity is essential for the expression of *CORs* and CAMTAs, which harbor conserved CAM-binding sites and activate *CBF* expression [10,41]. In this study, two CAM-binding proteins, CAMBP60B and CAMBP25, were induced by freezing stress and were found to be associated with antioxidant enzyme activity in apricot kernels (Figure 4). However, our results did not observe a co-expression relationship between CAMBPs and CBFs. There is the possibility that CAMBP60B and CAMBP25 are involved in the freezing response in apricot kernel pistils via a CBF-independent pathway.

Plant hormones that play key roles in cold stress tolerance by regulating ROS balance have been found [18]. Cold-activated BZR1, a positive transcriptional factor in BRs signaling, directly promotes *RBOH1* expression and H_2O_2 production [42]. In Arabidopsis and peas, SLs positively regulate chilling tolerance via increasing glutathione and ascorbate accumulation [43]. Two JAZ proteins (JAZ8 and JAZ10B), two ethylene-responsive TFs (ERF109 and CRF4), one DELLA protein RGL1, and the SA receptor NPR4 were found to be highly associated with antioxidant enzyme activity in apricot kernel pistils under freezing stress (Figure 4). Consistent with these findings, JA, SA, ET, and GA were also involved in the response to cold stress; PtrERF109 directly promotes the expression of *Ptr-Prx1* to improve peroxidase activity [12,18,44]. Moreover, the direct relationships between JAZ8 and bHLH13, ERF109, WRKY18, and WRKY40 were predicted, suggesting that JA signaling cooperates with other TFs and hormones to affect the ROS-mediated freezing response. These potential functional genes involved in freezing stress can provide a choice for genetic-engineering-assisted breeding through gene-editing technology and a direction reference for molecular-marker-assisted breeding.

5. Conclusions

In the present study, we investigated gene co-expression networks in the SOD-mediated response to freezing stress in the pistils of two apricot kernel cultivars with a different level of frost resistance. One gene network was identified to correlate with the antioxidant enzyme SOD activity under freezing stress. The direct relationship of regulatory and functional hub genes within this network were predicted. Our study confirmed some novel hub genes and potential mechanisms underlying the variation in the freezing resistance of apricot kernels.

Supplementary Materials: The following supporting information can be downloaded at: https://www.mdpi.com/article/10.3390/f13101655/s1, Figure S1: The distribution of differently expressed transcription factor families in CsL (a), CtW (b), and CsL vs. CtW (c). CtW: cold-tolerant 'Weixuan 1'; CsL: cold-sensitive 'Longwangmao'. Figure S2: The expression analysis of DEGs in the magenta module. (A) Heatmap of DEGs in the magenta module. (B) The overall expression level of eigengenes identified in the magenta module for each sample. Table S1: List of primers used for qRT-PCR. Table S2: The antioxidant enzyme activities in apricot kernel pistils under freezing stress. Table S3: The annotation of hub genes in the magenta module.

Author Contributions: X.L. and L.W. designed the research. X.L. performed the experiments and analyzed the data. X.L. and Y.Y. wrote the manuscript. H.X., D.Y. and Q.B. provided helpful comments on the work and manuscript. All authors have read and agreed to the published version of the manuscript.

Funding: This work was financially supported by the National Key Research and Development Program of China (2018YFD1000606-3-3) and the Central Public-Interest Scientific Institution Basal Research Fund (CAFYBB2019SY005).

Data Availability Statement: The data supporting the results are concluded in the article and supplementary information files. The RNA-seq data were deposited in NCBI SRA under the accession number PRJNA832066.

Acknowledgments: We thank the forestry and seedling workstation in Yuyang District, Yulin City, for the 'Weixuan 1' and 'Longwangmao' material.

Conflicts of Interest: The authors declare no conflict of interest.

References

1. Vitasse, Y.; Schneider, L.; Rixen, C.; Christen, D.; Rebetez, M. Increase in the risk of exposure of forest and fruit trees to spring frosts at higher elevations in Switzerland over the last four decades. *Agric. For. Meteorol.* **2018**, *248*, 60–69. [CrossRef]
2. Yang, J.; Huo, Z.; Wang, P.; Wu, D.; Ma, Y.; Yao, S.; Dong, H. Process-based indicators for timely identification of apricot frost disaster on the warm temperate Zone, China. *Theor. Appl. Climatol.* **2021**, *1463*, 1143–1155. [CrossRef]
3. Chen, L.J.; Xiang, H.Z.; Miao, Y.; Zhang, L.; Guo, Z.F.; Zhao, X.H.; Lin, J.W.; Li, T.L. An overview of cold resistance in plants. *J. Agron. Crop. Sci.* **2014**, *200*, 237–245. [CrossRef]
4. Augspurger, C.K. Reconstructing patterns of temperature, phenology, and frost damage over 124 years: Spring damage risk is increasing. *Ecology* **2013**, *94*, 41–50. [CrossRef] [PubMed]
5. Liu, Y.C.; Chen, W.Z.; Liu, W.S.; Liu, N.; Liu, S. Palynological study on the origin and systematic evolution of kernel-using apricots. *Acta Hortic. Sin.* **2010**, *37*, 1377–1387. [CrossRef]
6. Ning, C.; Meng, Q.; Li, S.; Li, Y.; Yang, J. Comparative study of cold resistance in Kernel-apricot germplasm floral organs. *J. Agric. Univ. Hebei* **2010**, *33*, 37–41. [CrossRef]
7. Yu, D.; Liu, X.; Cui, Y.; Bi, Q.; Zhao, Y.; Li, D.; Yu, H.; Wang, L. Comparative transcriptome combined with morpho-physiological analyses revealed candidate genes potentially for differential cold tolerance in two contrasting apricot *Prunus armeniaca* L. cultivars. *Trees* **2020**, *34*, 1205–1217. [CrossRef]
8. Shi, Y.; Ding, Y.; Yang, S. Molecular regulation of CBF signaling in cold acclimation. *Trends Plant Sci.* **2018**, *23*, 623–637. [CrossRef] [PubMed]
9. Thomashow, M.F. Plant cold acclimation: Freezing tolerance genes and regulatory mechanisms. *Annu. Rev. Plant Biol.* **1999**, *50*, 571–599. [CrossRef] [PubMed]

10. Ding, Y.; Shi, Y.; Yang, S. Advances and challenges in uncovering cold tolerance regulatory mechanisms in plants. *New Phytol.* **2019**, *222*, 1690–1704. [CrossRef] [PubMed]
11. Liu, C.; Cao, M.; Zhou, D.; Peng, W. Drought, salt and temperature stress-induced metabolic changes in plant. *Biotechnol. Bull.* **2013**, *4*, 1–7. [CrossRef]
12. Shi, Y.; Ding, Y.; Yang, S. Cold signal transduction and its interplay with phytohormones during cold acclimation. *Plant Cell Physiol.* **2015**, *561*, 7–15. [CrossRef] [PubMed]
13. Chinnusamy, V.; Ohta, M.; Kanrar, S.; Lee, B.H.; Hong, X.; Agarwal, M.; Zhu, J.K. ICE1: A regulator of cold-induced transcriptome and freezing tolerance in *Arabidopsis*. *Genes Dev.* **2003**, *17*, 1043–1054. [CrossRef]
14. Ding, Y.; Shi, Y.; Yang, S. Molecular regulation of plant responses to environmental temperatures. *Mol Plant.* **2020**, *13*, 544–564. [CrossRef] [PubMed]
15. Choudhury, F.K.; Rivero, R.M.; Blumwald, E.; Mittler, R. Reactive oxygen species, abiotic stress and stress combination. *Plant J.* **2017**, *90*, 856–867. [CrossRef] [PubMed]
16. Hasanuzzaman, M.; Bhuyan, M.H.M.B.; Zulfiqar, F.; Raza, A.; Mohsin, S.M.; Mahmud, J.A.; Fujita, M.; Fotopoulos, V. Reactive oxygen species and antioxidant defense in plants under abiotic stress: Revisiting the crucial role of a universal defense regulator. *Antioxidants* **2020**, *9*, 681. [CrossRef]
17. Wani, K.I.; Naeem, M.; Castroverde, C.D.M.; Kalaji, H.M.; Albaqami, M.; Aftab, T. Molecular mechanisms of nitric oxide NO signaling and reactive oxygen species ROS homeostasis during abiotic stresses in plants. *Int. J. Mol. Sci.* **2021**, *2217*, 9656. [CrossRef]
18. Devireddy, A.R.; Zandalinas, S.I.; Fichman, Y.; Mittler, R. Integration of reactive oxygen species and hormone signaling during abiotic stress. *Plant J.* **2021**, *105*, 459–476. [CrossRef]
19. Zandalinas, S.I.; Fichman, Y.; Devireddy, A.R.; Sengupta, S.; Azad, R.K.; Mittler, R. Systemic signaling during abiotic stress combination in plants. *Proc. Natl. Acad. Sci. USA* **2020**, *117*, 13810–13820. [CrossRef]
20. Tian, J.; Wei, A.; Liu, Y.; Wang, S.H.; Liu, Y. Research progress on cold resistance of kernel apricot. *J. Nor. For. Univ.* **2018**, *331*, 174–178. [CrossRef]
21. Kaya, O.; Kose, C.; Esitken, A.; Gecim, T.; Donderalp, V.; Taskin, S.; Turan, M. Frost tolerance in apricot *Prunus armeniaca* L. receptacle and pistil organs: How is the relationship among amino acids, minerals, and cell death points? *Int. J. Biometeorol.* **2021**, *65*, 2157–2170. [CrossRef] [PubMed]
22. Yang, Y.; Yang, F.; Song, Y.; Li, J.; Lü, Z. Principal component analysis on cold resistance of different apricot varieties in Shanxi. *J. Shanxi Agric. Sci.* **2020**, *48*, 1913–1915. [CrossRef]
23. Li, Y.; Gao, L.; Ren, S.; Fu, Y.; Yang, F.; Yang, J. A new cold-resistant kernel apricot variety "Waixuan1". *Acta Hortic. Sin.* **2010**, *371*, 155–156. [CrossRef]
24. Jiang, F.; Zhang, J.; Wang, S.; Yang, L.; Luo, Y.; Gao, S.; Zhang, M.; Wu, S.; Hu, S.; Wang, Y. The apricot (*Prunus armeniaca* L.) genome elucidates Rosaceae evolution and beta-carotenoid synthesis. *Hortic. Res.* **2019**, *6*, 12. [CrossRef]
25. Yu, D.; Liu, X.; Bi, Q.; Zhao, Y.; Ju, J.; Li, D.; Yu, H.; Wang, L. Response of pistils in two contrasting apricot *Prunus armeniaca* L. cultivars to low temperature stress and recovery. *J. Nor. For. Univ.* **2021**, *491*, 1–5. [CrossRef]
26. Li, K.; Du, Y.; Miao, Y. Future challenges in understanding ROS in plant responses to abiotic stress. *Sci. China Life Sci.* **2016**, *5912*, 1343–1344. [CrossRef]
27. Dreyer, A.; Dietz, K.J. Reactive oxygen species and the redox-regulatory network in cold stress acclimation. *Antioxidants* **2018**, *711*, 169. [CrossRef]
28. Mehrotra, S.; Verma, S.; Kumar, S.; Kumari, S.; Mishra, B.N. Transcriptional regulation and signalling of cold stress response in plants: An overview of current understanding. *Environ. Exp. Bot.* **2020**, *180*, 104243. [CrossRef]
29. Maurya, N.K.; Goswami, A.K.; Singh, S.K.; Prakash, J.; Goswami, S.; Chinnusamy, V.; Talukdar, A.; Pradhan, S.; Kumari, A. Studies on expression of *CBF1* and *CBF2* genes and anti-oxidant enzyme activities in papaya genotypes exposed to low temperature stress. *Sci. Horticult.* **2020**, *261*, 108914. [CrossRef]
30. Ritonga, F.N.; Ngatia, J.N.; Wang, Y.R.; Khoso, M.A.; Farooq, U.; Chen, S. AP2/ERF, an important cold stress-related transcription factor family in plants: A review. *Physiol. Mol. Biol. Plants* **2021**, *279*, 1953–1968. [CrossRef]
31. Agarwal, M.; Hao, Y.; Kapoor, A.; Dong, C.H.; Fujii, H.; Zheng, X.; Zhu, J.K. A R2R3 type MYB transcription factor is involved in the cold regulation of *CBF* genes and in acquired freezing tolerance. *J. Biol. Chem.* **2006**, *281*, 37636–37645. [CrossRef] [PubMed]
32. Hao, Y.J.; Wei, W.; Song, Q.X.; Chen, H.W.; Zhang, Y.Q.; Wang, F.; Zou, H.F.; Gang, L.; Tian, A.G.; Zhang, W.K. Soybean NAC transcription factors promote abiotic stress tolerance and lateral root formation in transgenic plants. *Plant J.* **2011**, *682*, 302–313. [CrossRef] [PubMed]
33. Yarra, R.; Wei, W. The NAC-type transcription factor GmNAC20 improves cold, salinity tolerance, and lateral root formation in transgenic rice plants. *Funct. Integr. Genom.* **2021**, *213*, 473–487. [CrossRef] [PubMed]
34. Chen, X.; Ding, Y.; Yang, Y.; Song, C.; Wang, B.; Yang, S.; Guo, Y.; Gong, Z. Protein kinases in plant responses to drought, salt, and cold stress. *J. Integr. Plant Biol.* **2021**, *63*, 53–78. [CrossRef] [PubMed]
35. Ding, Y.; Li, H.; Zhang, X.; Xie, Q.; Gong, Z.; Yang, S. OST1 kinase modulates freezing tolerance by enhancing ICE1 stability in *Arabidopsis*. *Dev. Cell* **2015**, *32*, 278–289. [CrossRef]
36. Liu, Y.; Xu, C.; Zhu, Y.; Zhang, L.; Chen, T.; Zhou, F.; Chen, H.; Lin, Y. The calcium-dependent kinase OsCPK24 functions in cold stress responses in rice. *J. Integr. Plant Biol.* **2018**, *60*, 95–110. [CrossRef]

37. Kim, S.J.; Kim, W.T. Suppression of *Arabidopsis* RING E3 ubiquitin ligase AtATL78 increases tolerance to cold stress and decreases tolerance to drought stress. *FEBS Lett.* **2013**, *587*, 2584–2590. [CrossRef]
38. Suh, J.Y.; Kim, W.T. *Arabidopsis* RING E3 ubiquitin ligase AtATL80 is negatively involved in phosphate mobilization and cold stress response in sufficient phosphate growth conditions. *Biochem. Biophys. Res. Commun.* **2015**, *4634*, 793–799. [CrossRef]
39. Wang, X.; Ding, Y.L.; Li, Z.Y.; Shi, Y.T.; Wang, J.L.; Hua, J.; Gong, Z.Z.; Zhou, J.M.; Yang, S.H. PUB25 and PUB26 promote plant freezing tolerance by degrading the cold signaling negative regulator MYB15. *Dev. Cell* **2019**, *51*, 222–235. [CrossRef]
40. Fang, H.; Meng, Q.; Xu, J.; Tang, H.; Tang, S.; Zhang, H.; Huang, J. Knock-down of stress inducible OsSRFP1 encoding an E3 ubiquitin ligase with transcriptional activation activity confers abiotic stress tolerance through enhancing antioxidant protection in rice. *Plant Mol. Biol.* **2015**, *874–875*, 441–458. [CrossRef]
41. Doherty, C.J.; Van Buskirk, H.A.; Myers, S.J.; Thomashow, M.F. Roles for *Arabidopsis* CAMTA transcription factors in cold-regulated gene expression and freezing tolerance. *Plant Cell* **2009**, *21*, 972–984. [CrossRef] [PubMed]
42. Fang, P.; Yan, M.; Chi, C.; Wang, M.; Zhou, Y.; Zhou, J.; Shi, K.; Xia, X.; Foyer, C.H.; Yu, J. Brassinosteroids act as a positive regulator of photoprotection in response to chilling stress. *Plant Physiol.* **2019**, *180*, 2061–2076. [CrossRef] [PubMed]
43. Cooper, J.W.; Hu, Y.; Beyyoudh, L.; Yildiz Dasgan, H.; Kunert, K.; Beveridge, C.A.; Foyer, C.H. Strigolactones positively regulate chilling tolerance in pea and in *Arabidopsis*. *Plant Cell Environ.* **2018**, *41*, 1298–1310. [CrossRef] [PubMed]
44. Wang, M.; Dai, W.; Du, J.; Ming, R.; Dahro, B.; Liu, J.H. ERF109 of trifoliate orange *Poncirus trifoliata* L. Raf. contributes to cold tolerance by directly regulating expression of *Prx1* involved in antioxidative process. *Plant Biotechnol. J.* **2019**, *17*, 1316–1332. [CrossRef] [PubMed]

forests

MDPI

Article

Identification of Aquaporin Gene Family in Response to Natural Cold Stress in *Ligustrum* × *vicaryi* Rehd.

Jiahui Dong [1,†], Shance Niu [1,†], Ji Qian [1,†], Juan Zhou [2], Mengnan Zhao [1], Yu Meng [3] and Bao Di [1,*]

1 College of Horticulture, Hebei Agriculture University, Baoding 071000, China; djh285170105@126.com (J.D.); niushance@163.com (S.N.); qianji167@163.com (J.Q.); zhaomengnan526@163.com (M.Z.)
2 College of Mechanical and Electric Engineering, Hebei Agriculture University, Baoding 071000, China; zhoujuan1129@163.com
3 College of Landscape and Tourism, Hebei Agriculture University, Baoding 071000, China; my131sohu@126.com
* Correspondence: dibao666@126.com; Tel.: +86-312-7528313
† These authors contributed equally to this work.

Abstract: Plants are susceptible to a variety of abiotic stresses during the growing period, among which low temperature is one of the more frequent stress factors. Maintaining water balance under cold stress is a difficult and critical challenge for plants. Studies have shown that aquaporins located on the cytomembrane play an important role in controlling water homeostasis under cold stress, and are involved in the tolerance mechanism of plant cells to cold stress. In addition, the aquaporin gene family is closely related to the cold resistance of plants. As a major greening tree species in urban landscaping, *Ligustrum*× *vicaryi* Rehd. is more likely to be harmed by low temperature after a harsh winter and a spring with fluctuating temperatures. Screening the target aquaporin genes of *Ligustrum* × *vicaryi* responding to cold resistance under natural cold stress will provide a scientific theoretical basis for cold resistance breeding of *Ligustrum* × *vicaryi*. In this study, the genome-wide identification of the aquaporin gene family was performed at four different overwintering periods in September, November, January and April, and finally, 58 candidate *Ligustrum* × *vicaryi* aquaporin (LvAQP) genes were identified. The phylogenetic analysis revealed four subfamilies of the LvAQP gene family: 32 PIPs, 11 TIPs, 11 NIPs and 4 SIPs. The number of genes in PIPs subfamily was more than that in other plants. Through the analysis of aquaporin genes related to cold stress in other plants and LvAQP gene expression patterns identified 20 LvAQP genes in response to cold stress, and most of them belonged to the PIPs subfamily. The significantly upregulated LvAQP gene was *Cluster-9981.114831*, and the significantly downregulated LvAQP genes were *Cluster-9981.112839*, *Cluster-9981.107281*, and *Cluster-9981.112777*. These genes might play a key role in responding to cold tolerance in the natural low-temperature growth stage of *Ligustrum* × *vicaryi*.

Keywords: *Ligustrum* × *vicaryi* Rehd.; aquaporin; natural cold stress; cold resistance

Citation: Dong, J.; Niu, S.; Qian, J.; Zhou, J.; Zhao, M.; Meng, Y.; Di, B. Identification of Aquaporin Gene Family in Response to Natural Cold Stress in *Ligustrum* × *vicaryi* Rehd. *Forests* **2022**, *13*, 182. https://doi.org/10.3390/f13020182

Academic Editor: Yuepeng Song and Carol A. Loopstra

Received: 9 December 2021
Accepted: 24 January 2022
Published: 26 January 2022

1. Introduction

Aquaporin is a protein located on the cytomembrane that controls the entry and exit of water in cells. Water uptake, transport across membranes and tissues are essential for plants growth and development, and the transmembrane transport of water molecules is mainly regulated by aquaporins. In biological membranes, plant aquaporins have a highly conserved Asn-Pro-Ala (NPA) motif structure, which plays a crucial role in the formation of water-selective channels [1]. It has been reported that AEFXXT motif located in the first helix (TM1) in plant aquaporins is highly conserved in almost all major intrinsic proteins (MIPs), but the exact function of the AEFXXT motif is still unclear [2]. The previous studies based on genomic data revealed that aquaporins constitute a huge gene family in plants. These aquaporins are divided into five main subfamilies according to their amino acid sequence [3]: plasma membrane intrinsic proteins (PIPs), tonoplast intrinsic proteins

(TIPs), nodulin26-like intrinsic proteins (NIPs), small basic intrinsic proteins (SIPs) and uncharacterized X intrinsic proteins (XIPs). In recent research, there have been studies related to the identification and expression analysis of the whole aquaporin gene family in more than 20 plants, such as *Arabidopsis thaliana* [4], *Oryza sativa* [5], *Zea mays* [6], *Hordeum vulgare* [7], *Glycine max* [8], *Gossypium hirsutum* [9], *Citrullus lanatus* [10], *Brassica rapa* [11], and so on. However, a considerable number of studies have confirmed that it is a difficult but critical challenge for plants to maintain water balance under various adversities. Therefore, aquaporins have a great effect on maintaining water homeostasis in plants under different environmental stress [12–18].

Plants are susceptible to a variety of abiotic stresses during the growing period, especially cold stress [19,20]. Plants respond to cold stress by increasing root water absorption [21]. As an important regulator of water absorption and transport, aquaporins play a key role in regulating water balance in plants at low temperature. For instance, Azad et al. showed that temperature changes could induce AQP phosphorylation and dephosphorylation, thus affecting water transport [22]. Many studies have shown that aquaporins play a crucial role in resisting cold stress [23]. For instance, under cold stress, *OsPIP2;4* and *OsPIP2;5* were abundantly expressed in the root system in order to enhance cold resistance in rice [24]. The *OsPIP2;5* and *OsPIP2;7* of *Oryza sativa* were engaged with rapid water transport and with maintaining water balance during the cold stress stage, which played a major role in regulating water channel opening under cold stress [25]. The overexpression of *AtPIP1;4* or *AtPIP2;5* in transgenic plants of *Arabidopsis thaliana* could improve water conductivity and promote germination [26]. The overexpression of *TaTIP2;2* in transgenic plants of wheat could make plants grow normally under cold conditions as well [27]. Overexpressing or repressing expression of related aquaporin genes to enhance cold resistance of *Oryza sativa* [28], *Hordeum vulgare* [29], *Musa acuminata* [30–32], *Populus trichocarpa* [33], *Sorghum bicolor* [34], *Triticum aestivum* [35,36] and *Brassica rapa* [11] have been investigated under cold stress. Among these plants, 11, 11, 8, 6, 9, 2 and 8 AQP genes showed significant correlations with cold stress. Numerous studies have shown that the aquaporin gene family is closely related to the cold resistance of plants.

With golden yellow leaves, *Ligustrum* × *vicaryi* is widely used in China, the United States, and Canada along with *Berberis thunbergii* var. *atropurpurea* and *Buxus megistophylla* Levl., but it is susceptible to low temperature injury during the seedling stage [37]. In this study, we aimed to identify the *Ligustrum* × *vicaryi* aquaporin (LvAQP) gene family, and its expression pattern was analyzed, and the expression changes of the LvAQP gene family in different periods were investigated; the screened target aquaporin genes responded to cold resistance under natural low temperature stress. The results of this research will lay the foundation for further biological function verification of cold resistance-related aquaporin candidate genes in *Ligustrum* × *vicaryi*, especially in the PIPs subfamily, and they will provide a theoretical basis for improving seedling quality and breeding of *Ligustrum* × *vicaryi*.

2. Materials and Methods

2.1. Plant Materials and Treatment

One-year-old container seedlings of *Ligustrum* × *vicaryi* (txid1133299) were obtained from the Beijing Florascape Co., Ltd. (Beijing, China) (40°11′ N, 116°48′ E). We obtained the permission to collect the plant samples from the Beijing Florascape Co., Ltd. (Beijing, China), and the plant materials were formally identified by senior engineer Ju Chen of the company and were later identified by professor Gang Zhang of Hebei Agricultural University. The *Ligustrum* × *vicaryi* were cultivated in Specimen Park (38°50′ N, 115°26′ E) of Hebei Agricultural University, Baoding City, Hebei Province, in September 2019. In the experimental setup, the container seedlings were divided into three replicates for the measurements at each sampling time, with 25 plants in each replicate. The spacing within the row and the spacing between rows were 25 × 50 cm, which were consistent cultivation conditions and conventional maintenance management.

The seedlings overwintered naturally in the open field. The seedlings were sampled on the 24th of each month in September and November 2019 and in January and April 2020.

2.2. *Determination of Cold Resistance of Ligustrum × vicaryi*

The seedlings were taken at four sampling times. The seedlings were placed in an artificial climate chamber for low temperature treatment using the method of Di [38] (Table 1). During the cooling period, the seedlings were kept at −3 °C for 5 h to keep the soil and air temperature consistent and were then continued to cool at the same temperature. Each set target temperature was maintained for 4 h. After that treatment, the seedlings were placed at 4 °C for 5 days and room temperature for 1 day, and the cold resistance of roots was measured by relative electrolyte leakage (REL).

Table 1. The temperatures for the measurement of cold hardiness after controlled freezing tests.

Date (m)	Temperature/°C					
September	4	−3	−8	−18	−23	−35
November	4	−6	−12	−23	−35	−45
January	4	−8	−18	−30	−48	−60
April	4	−6	−12	−23	−35	−45

2.3. *RNA−Seq*

2.3.1. RNA Extraction and Detection

The seedlings were taken at four sampling times. The roots of the plants were washed by tap water to remove the soil, and the fine roots were rinsed with tap water, distilled water and ultrapure water in turn (the water was placed in Specimen Park in advance and the temperature was kept consistent with the environment), and then they were frozen in liquid nitrogen and stored in an ultra-low temperature freezer at −80 °C.

Material was sequenced at Tianjin Novogene Biotechnology Co., Ltd., (Tianjin, China). Total RNA was extracted by Omniplant RNA Kit (DNase I).

The RNA integrity was detected by agarose gel electrophoresis with 2% concentration, 150 V, 150 mA. The concentration of each RNA sample and its optical density in the wavelength range of 260 and 280 nm were measured by Nano Drop one spectrophotometer, and the OD_{260}/OD_{280} value was calculated to detect the purity of RNA; then the RNA was stored in an ultra-low temperature freezer at −80°C.

2.3.2. cDNA Library Construction

First, magnetic beads with Oligo(dT) were used to enrich eukaryotic mRNA. Second, mRNA was broken into short fragments by adding fragmentation buffer. One-stranded cDNA was synthesized with six-base random hexamers using mRNA as a template. Third, double-stranded cDNA was formed by adding buffer, dNTPs, DNA polymerase I, and RNase H, which was purified by AMPure XP beads. The purified double-stranded cDNA was end-repaired and dA-tailed to ligate to sequencing connectors, and then fragment size selection was performed with AMPure XP beads. Finally, polymerase chain reaction (PCR) amplification was conducted, and the PCR products were purified with AMPure XP beads to obtain the final cDNA library. After the completion of the cDNA library construction, the initial quantification was operationalized by using Qubit 2.0, and then the library was diluted. Subsequently, the insert size of the library was tested. When the insert size met the expectation, the effective concentration of the library was accurately quantified by the Q-PCR method (effective library concentration > 2nM) to ensure the quality of the cDNA library.

2.3.3. RNA Data Analysis

The raw image data generated by the sequencer were transformed into raw data or raw reads by base calling. The results were stored in fastq format, which were part of the original file, including the sequence of reads and the sequencing quality of reads. Raw reads were processed to obtain clean reads by removing reads containing an adaptor, reads containing more than 10% of N and reads containing a small amount of low-quality sequences (the number of bases with quality value Q < 5 accounts for more than 50% of the entire reads).

The transcriptome data were assembled by Trinity v2.4.0 software with the following commands and parameters: Trinity—seq Type fq—max_memory 300G—left file_1. fq—right file_2.fq—CPU 50—full_clean up—KMER_SIZE 30—min_kmer_cov 5. Among the genes containing multiple transcripts, the sequence with the longest transcript was used as the basis for calculating expression, RSEM worked as the method for transcript abundance calculation, and trimmed mean of M-values (TMM) was used as the method for normalization between samples.

KEGG PATHWAY enrichment analysis on the results of variance analysis was performed by kobas software.

2.4. Identification of LvAQP Gene Family

Based on the transcriptome of *Ligustrum* × *vicaryi* and the study on the model plant *Arabidopsis thaliana*, the gene sequences of 35 *Arabidopsis thaliana* aquaporin genes (Table A1) were downloaded from NCBI (https://www.mhttps//www.ncbi.nlm.nih.gov/, accessed on 2 June 2021). The transcriptome database of *Ligustrum* × *vicaryi* was searched by blast homology retrieval method, and the LvAQP gene family was identified. The LvAQP genes were screened by MAFFT comparison software and a manual correction process using CD-HIT Suite (http://weizhong-lab.ucsd.edu/cdhit-web-server/cgi-bin/index.cgi?cmd=cd-hit-est, accessed on 10 June 2021) to remove redundancy. The screened LvAQP gene family protein sequences were analyzed for physicochemical properties such as number of amino acids, molecular weight, theoretical pI, aliphatic index and grand average of hydropathicity, using the online software ExPASy (https://web.expasy.org/protparam/, accessed on 25 June 2021).

2.5. Construction of Phylogenetic Tree

Multiple sequence alignment of candidate genes was performed by MAFFT software's E-INS-I strategy with necessary manual corrections. MEGA and PhyML3.0 software were used to construct the phylogenetic tree, and iTOL online software (https://itol.embl.de/upload.cgi, accessed on 24 July 2021) was used for phylogenetic tree beautification. The constructed phylogenetic tree was analyzed by Alrt detection method and WAG model.

2.6. Conserved Motifs Analysis

LvAQP gene family was extracted using the MEME tool (https://meme-suite.org/meme/tools/meme, accessed on 26 July 2021), with the following parameters: motif sequences, sites, width, E-value for each motif.

2.7. Gene Expression Profile Analysis

The LvAQP gene family expression profile was constructed using MeV software. The candidate genes were initially selected by differential gene expression analysis between natural low temperature treatment and non-low temperature treatment. Combining the constructed phylogenetic tree of LvAQP genes and the known cold resistance aquaporin genes of other plants aimed to find the homologs of known cold resistance genes of other plants within *Ligustrum* × *vicaryi*.

2.8. LvAQP Gene Family Quantitative Real-Time PCR Validation

2.8.1. RNA Extraction

Refer to Section 2.3.1. for RNA extraction.

2.8.2. Reverse Transcription of RNA into cDNA

The UEIris II RT-PCR System for First-Strand cDNA Synthesis (with dsDNase) reverse transcription kit was used as follows: The RNA was denatured thermally at 65 °C for 5 min, immediately iced for more than 3 min, and then the reaction system was prepared as 20 µL of the reverse transcription system: 2 µL total RNA, 4 µL UEIris II RT MasterMix (5×), 13 µL RNase-free Water, 1 µL dsDNase.

Reaction conditions: reverse transcription 37 °C for 2 min; 55 °C for 10 min; 85 °C for 10 s. After the reaction, it was stored at −20 °C.

2.8.3. Design of Primers for Quantitative Real-Time qRT-PCR

Nine LvAQP genes related to cold stress were selected for qRT-PCR. Primers were designed by Primer3 Plus and synthesized by Sangon Biotech Co., Ltd. (Shanghai, China), and *Ligustrum* × *vicaryi LvEF-1α* was selected as an internal reference gene; the primer information is shown in Table S1.

2.8.4. qRT-PCR Reaction System and Reaction Conditions

According to the instructions of AugeGreen™ qPCR Master Mix reagent, Roche fluorescence quantitative PCR instrument lightcycl96 was used to detect the expression of target genes. Preparation of reaction solution for 20 µL reaction system: 10 µL 2× AugeGreen™ Master Mix, 7 µL ddH2O, 1 µL Forward Primer, 1 µL Reverse Primer, 1 µL cDNA template. qRT-PCR reaction procedure was as follows: 95 °C for 2 min; 40 cycles of 15 s at 95 °C and 60 s at 58 °C; 95 °C for 10 s; 65 °C for 60 s; 97 °C for 1 s.

The last step was to analyze the solubility curve of the amplification products to determine the specificity of the primers. There were 3 technical replicates and 3 biological replicates for each sample during qRT-PCR reactions. The results of qRT-PCR were calculated by the $2^{-\Delta\Delta CT}$ method to obtain gene expression.

3. Results

3.1. Cold Resistance of *Ligustrum* × *vicaryi* during Natural Cold Stress Period

During the overwintering period, the cold resistance ability of *Ligustrum* × *vicaryi* gradually increased with the change of temperature in the early overwintering period, and gradually decreased in the late overwintering period (Figure 1). The strongest cold resistance appeared in January, reaching −22.1 °C and 11.4 °C higher than that in September (−10.7 °C). The cold resistance in November was −15.2 °C, and the cold resistance in April was −13.0 °C, which was close to that in September.

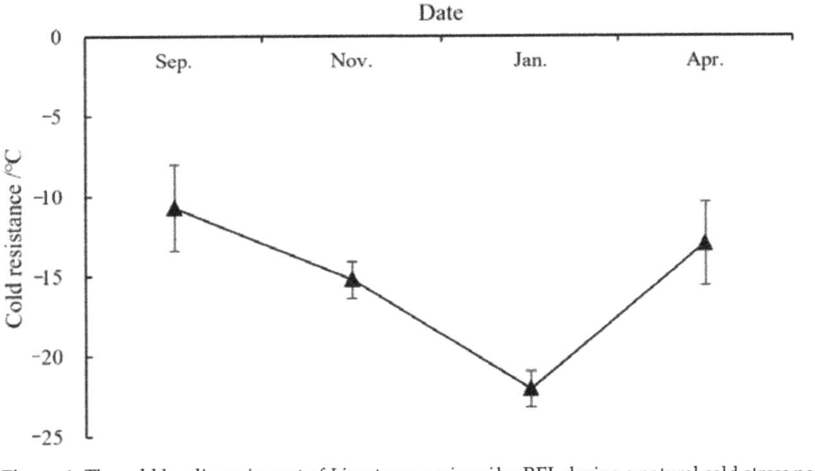

Figure 1. The cold hardiness in root of *Ligustrum* × *vicaryi* by REL during a natural cold stress period.

3.2. LvAQP Gene Family Identification

Based on the *Arabidopsis thaliana* aquaporin gene family (Table A1), 58 candidate LvAQP genes were identified (Table 2). According to sequence alignment, the correlation of characteristic proteins and phylogenetic relationship, the 58 LvAQP genes were classified into four subfamilies: PIPs, TIPs, NIPs and SIPs, which contained 32, 11, 11 and 4 genes, respectively. The results of the analysis of the physicochemical properties of the LvAQP gene family proteins (Table 2) showed that there were differences in the number of amino acids, molecular weight, theoretical pI, aliphatic index and grand average of hydropathicity of LvAQP gene family proteins. The number of amino acids in the protein sequence of the LvAQP gene family was 137–341. The molecular weight of the LvAQP gene family protein sequence was 13986.15–36029.94 Da. The theoretical pI of the LvAQP gene family protein sequences ranged from 5.34–9.91, and 40 of the 58 LvAQP proteins had a theoretical pI of greater than 7.5, indicating that most of the LvAQP gene family proteins are basic proteins. The aliphatic index of the LvAQP gene family protein sequences ranged from 89.25–116.37. The grand average of hydropathicity of the LvAQP gene family proteins was positive, revealing that they were hydrophobic proteins.

Table 2. Physicochemical properties of LvAQP gene family in *Ligustrum × vicaryi*.

Subfamily	Gene Name	Number of Amino Acids (aa)	Molecular Weight (Da)	Theoretical pI	Aliphatic Index	Grand Average of Hydropathicity
PIP	*Cluster-9981.112265*	262	27,966.27	7.22	91.26	0.255
	Cluster-9981.115068	298	32,266.44	9.33	93.05	0.237
	Cluster-9981.109600	289	31,046.09	8.60	95.57	0.348
	Cluster-9981.29850	308	32,910.19	9.43	91.66	0.332
	Cluster-19966.0	308	32,956.20	9.32	91.62	0.307
	Cluster-19036.0	311	32,859.87	7.20	95.08	0.341
	Cluster-9981.112839	268	28,664.31	9.34	89.25	0.204
	Cluster-9981.117133	278	30,018.76	9.14	93.74	0.188
	Cluster-9981.29849	299	31,461.40	8.32	100.94	0.510
	Cluster-9981.21661	300	32,170.31	7.30	101.20	0.512
	Cluster-9981.200292	299	31,419.32	8.32	99.97	0.503
	Cluster-9981.111171	306	32,894.05	6.48	101.73	0.414
	Cluster-9981.109034	321	34,345.95	8.51	103.40	0.451
	Cluster-9981.89369	290	30,949.85	8.25	96.55	0.461
	Cluster-9981.170229	290	30,804.69	7.84	98.28	0.494
	Cluster-9981.689	287	30,215.31	9.46	101.74	0.516
	Cluster-9981.690	289	30,357.47	9.46	101.73	0.525
	Cluster-9981.110451	301	32,265.67	9.05	106.05	0.487
	Cluster-9981.198491	321	34,046.79	9.66	101.93	0.428
	Cluster-9981.691	319	33,904.63	9.66	101.94	0.419
	Cluster-9981.154931	288	30,489.49	8.94	96.98	0.476
	Cluster-9981.154932	290	30,758.77	8.94	95.97	0.503
	Cluster-9981.114832	307	33,106.57	9.24	100.75	0.255
	Cluster-9981.114831	286	30,393.47	8.81	104.76	0.492
	Cluster-9981.47893	286	30,590.74	9.36	99.65	0.513
	Cluster-9981.47892	288	30,860.03	9.36	98.61	0.540
	Cluster-9981.111170	279	30,122.78	5.76	99.28	0.379
	Cluster-9981.21660	182	19,444.61	9.33	101.37	0.513
	Cluster-9981.118516	276	30,359.41	8.44	101.81	0.446
	Cluster-9981.107281	190	20,225.65	9.75	99.16	0.581
	Cluster-48310.0	200	21,799.73	9.91	99.50	0.405
	Cluster-9981.86061	148	15,695.98	9.43	92.43	0.336
TIP	*Cluster-9981.112777*	262	27,264.72	5.66	110.27	0.822
	Cluster-9981.35612	268	28,185.59	7.22	100.11	0.691
	Cluster-9981.111753	267	27,893.40	6.41	103.82	0.742
	Cluster-9981.115801	254	26,223.37	5.57	104.96	0.772
	Cluster-24993.0	252	25,967.19	6.49	107.26	0.779
	Cluster-9981.112790	263	27,555.92	5.36	107.98	0.835
	Cluster-9981.112789	261	26,692.22	6.05	115.17	0.970
	Cluster-20432.0	250	25,340.33	5.34	105.40	0.881
	Cluster-9981.122691	234	23,815.83	5.79	116.37	0.918
	Cluster-9981.172823	262	27,677.53	6.39	111.64	0.719
	Cluster-9981.111196	137	13,986.15	6.68	96.93	0.596

Table 2. *Cont.*

Subfamily	Gene Name	Number of Amino Acids (aa)	Molecular Weight (Da)	Theoretical pI	Aliphatic Index	Grand Average of Hydropathicity
NIP	*Cluster-9981.78169*	290	30,914.92	9.43	104.55	0.423
	Cluster-9981.133629	316	33,888.57	9.34	104.02	0.450
	Cluster-9981.133630	320	34,289.04	9.34	105.16	0.463
	Cluster-51933.0	291	30,748.42	6.71	100.52	0.463
	Cluster-9981.178700	240	25,532.72	7.84	107.21	0.555
	Cluster-36924.0	250	26,540.04	8.83	107.24	0.667
	Cluster-9981.104986	341	36,029.94	9.51	94.13	0.332
	Cluster-9981.123071	308	31,953.21	9.53	98.90	0.378
	Cluster-9981.97911	284	30,038.99	8.34	99.61	0.544
	Cluster-28512.0	285	30,427.38	6.57	113.68	0.472
	Cluster-9981.54345	159	16,867.60	9.89	107.92	0.487
SIP	*Cluster-9981.120365*	245	25,817.37	8.46	105.59	0.790
	Cluster-9981.105938	142	15,367.01	9.14	95.56	0.411
	Cluster-9981.88037	286	31,510.94	9.54	97.48	0.234
	Cluster-9981.88036	155	16,945.07	9.47	103.81	0.405

3.3. Phylogenetic Analysis of LvAQP Gene Family

By constructing phylogenetic tree, the distribution and development of the 58 candidate aquaporin genes of the four subfamilies of the LvAQP gene family can be known (Figure 2). The internal genes of the PIPs subfamily were more similar than that of the TIPs subfamily, NIPs subfamily, and SIPs subfamily. The PIPs subfamily had the largest genes, which were due to tandem repeats of some genes with similar structures on the chromosome. Among the four subfamilies, the PIPs subfamily contained seven pairs of tandem repeats genes, while only one tandem repeat gene in the NIPs and SIPs subfamilies (Table 3).

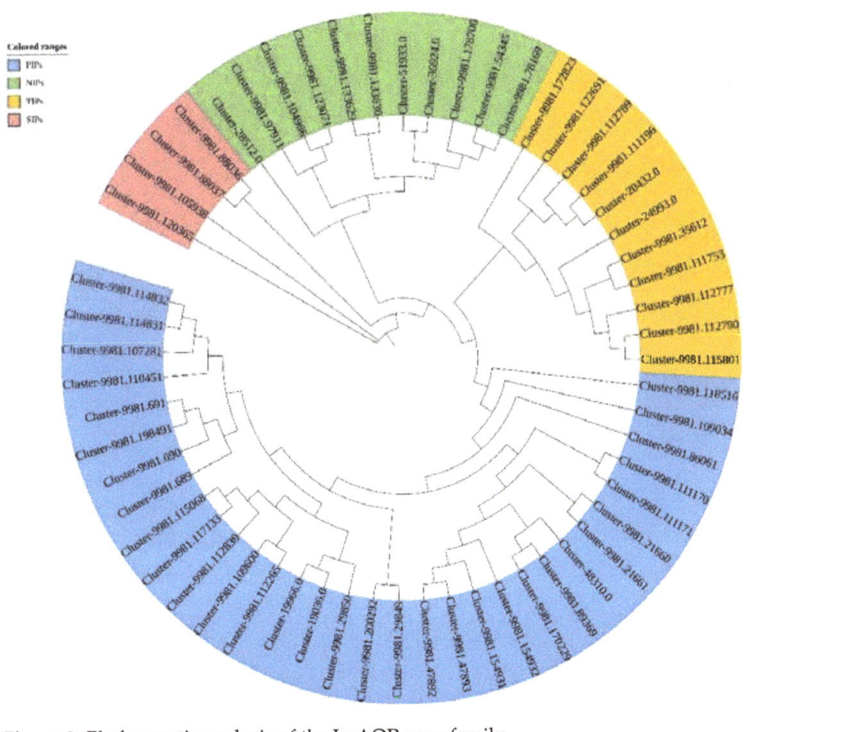

Figure 2. Phylogenetic analysis of the LvAQP gene family.

Table 3. Tandem repeats genes in the LvAQP gene family.

Subfamily	Number of Subfamily Tandem Repeats	Tandem Repeats Gene Name		
PIPs	7	*Cluster-9981.114831*		*Cluster-9981.114832*
		Cluster-9981.689	*Cluster-9981.690*	*Cluster-9981.691*
		Cluster-9981.47892		*Cluster-9981.47893*
		Cluster-9981.154931		*Cluster-9981.154932*
		Cluster-9981.21660		*Cluster-9981.21661*
		Cluster-9981.111170		*Cluster-9981.111171*
		Cluster-9981.29849		*Cluster-9981.29850*
NIPs	1	*Cluster-9981.133629*		*Cluster-9981.133630*
SIPs	1	*Cluster-9981.88036*		*Cluster-9981.88037*

Comparing the aquaporin gene family of Monocotyledonous *Oryza sativa*, *Zea mays* and *Musa acuminata*, and dicotyledonous *Arabidopsis thaliana* and *Brassica rapa* with that of *Ligustrum* × *vicary*, the results showed that the number of the PIPs gene subfamily was the largest, while the number of the SIPs gene subfamily was the smallest (Table 4). Furthermore, the distribution of the four gene subfamilies of the LvAQP gene family was generally the same as that of the subfamily members of other plants. *Ligustrum* × *vicaryi* is a dicotyledonous plant. Unlike other plants, in *Ligustrum* × *vicaryi*, the number of aquaporin genes in the PIPs subfamily was nearly two times higher than that of TIPs subfamily and NIPs subfamily, while it was similar to that of the TIPs and NIPs subfamilies in other plants. PIPs located on the plasma membrane were highly selective to the transport matrix, and they are critical for maintaining the water balance of cells in plants [39]. Thus, it isspeculated that the *Ligustrum* × *vicaryi* PIPs subfamily (LvPIPs) may play a major role in maintaining its own water balance of cells.

Table 4. Distribution of subfamily members of the AQP gene family in various plants.

Plants	PIPs	TIPs	NIPs	SIPs	Total
Arabidopsis thaliana	13	10	9	3	35
Oryza sativa	11	10	10	2	33
Zea mays	13	11	4	3	31
Brassica rapa	22	16	15	6	59
Musa acuminata	18	17	9	3	47
Ligustrum × *vicaryi*	32	11	11	4	58

In this study, a phylogenetic tree was constructed based on 35 *Arabidopsis thaliana* aquaporin genes, 35 *Oryza sativa* aquaporin genes, 58 candidate LvAQP genes and aquaporin genes related to cold stress in other plants (Figure 3). Most of the aquaporin genes related to cold resistance were distributed in the PIPs subfamily (Figure 3), while the number of genes of the PIPs subfamily in *Arabidopsis thaliana* and *Oryza sativa* were relatively small (Table 4). There was only a pair of tandem repeats genes (At2G37170 and At2G37180) in the PIPs subfamily in *Arabidopsis thaliana* (Table A1), while there were seven pairs of tandem repeats genes in the LvPIPs subfamily. Therefore, the reason for the large number of LvPIPs may be that genes are relatively tightly distributed on chromosomes, and tandem duplication led to gene amplification.

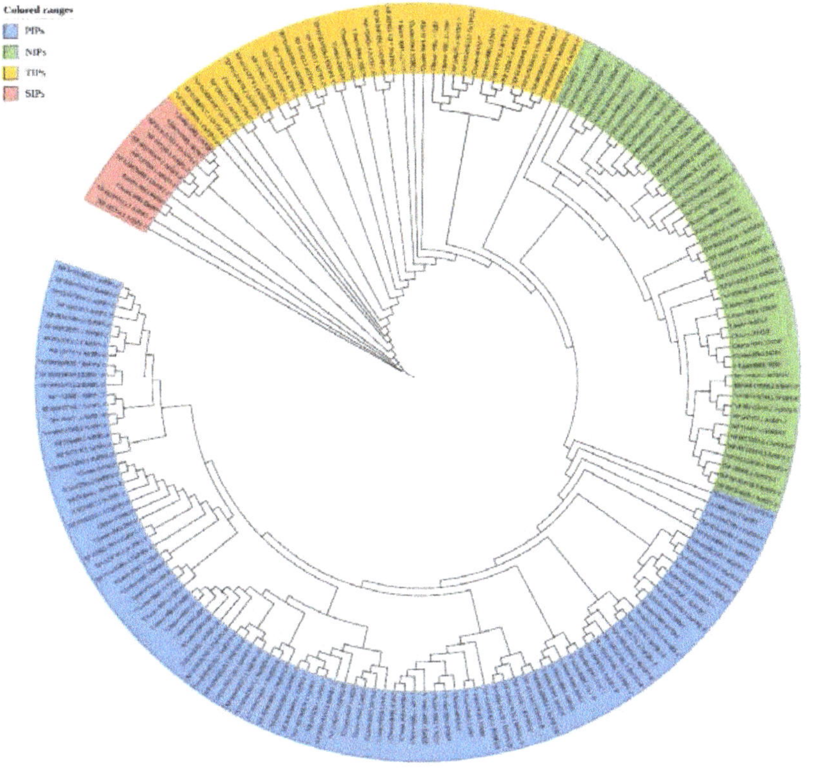

Figure 3. Phylogenetic analysis of LvAQP genes and cold stress-related aquaporin genes in other plants.

3.4. LvAQP Gene Family Conserved Motifs Analysis

The identified LvAQP genes all contained conserved domains. Table 5 showes that the LvAQP gene family containes 19 main conserved motifs. The distribution of 58 LvAQP conservative motifs isshown in Figure 4, and the four subfamilies share common conservative motifs, such as motif1. Each subfamily has similar conserved sites, and the members of each subfamily contain similar conserved motifs, even the same, such as *Cluster-9981.115068* and *Cluster-9981.109600* of the PIPs subfamily. Each subfamily contains its own unique motifs. For example, all members of the PIPs subfamily contain motif 3, motif 4, motif 5, and motif 11. All members of the TIPs subfamily contain motif 7 and motif 19. All members of the NIPs subfamily contain motif 12, motif 14, and motif 15. The connected motif 17 and motif 1 are in the SIPs subfamily. Each LvAQP subfamily was highly conserved during the process of evolution, which was beneficial to the phylogeny of the LvAQP gene family.

Table 5. The 19 conserved motif information of LvAQP genes.

Motif Type	Motif Sequences	Sites	Width	E-Value
Motif 1	KRSARDSHVPVLAPLPIGFAVFMVHLATIPITGTGINPARSFGAAVIYNK	54	50	7.7e-1602
Motif 2	VYCTAGISGGHINPAVTFGLFLARKVSLIRAIMYIVAQCLGAICGVGLVK	50	50	2.5e-1531
Motif 3	KDYKDPPPAPLFDAGELKKWSFYRALIAEFIATLLFLYITVLTVIGYKSQ	28	50	9.2e-1070
Motif 4	YQKYGGGANELADGYSKGTGLGAEIIGTFVLVYTVFSATDP	32	41	6.3e-911
Motif 5	KAWDDHWIFWVGPFIGAAIAAFYHQYILR	26	29	4.3e-568
Motif 6	DKCGGVGILGIAWAFGGMIFV	35	21	2.1e-374
Motif 7	KAALAEFISTLIFVFAGEGSGMAYNKLTGBAPLTPAGLVAAAVAHAFALF	9	50	2.8e-167
Motif 8	KGIWVYWVGPLIGAGLAAWVY	25	21	5.9e-162
Motif 9	SDWZALVVEIIITFGLVFTVY	22	21	3.0e-168
Motif 10	AMENKEEDVRLGANKYSERQPJGTAAQSD	8	29	3.4e-125
Motif 11	AIKALGSFRSS	19	11	2.1e-090
Motif 12	KHGNSSGCSLLTLSFIQKIIAEILGTYFLIFAGCAAVVVNA	8	41	5.2e-079
Motif 13	MAKDVEEEPEG	19	11	6.4e-049
Motif 14	NIIRFTDKPLREITKS	6	16	2.2e-045
Motif 15	LLFTGKHDHFSGTLP	7	15	5.0e-037
Motif 16	AFQKSY	24	6	2.3e-035
Motif 17	TPVIPAPYPDILRGPSLNVDLKSGALAEGLLTFAITF	6	37	2.3e-027
Motif 18	LRQQGHIFNPSJPKPSHKAPNAFLLNRSRPPKSRFLFDSVQ	4	47	1.2e-025
Motif 19	FFINHSHEPLPSSEY	7	15	1.7e-019

Motif sequences represent the motif consensus in this experiment. Sites stand for the number of occurrences of this motif in 58 LvAQP genes. Width represents the width of the motif. E-value represents the statistical significance of the motif. The smaller the E-value, the more reliable the result.

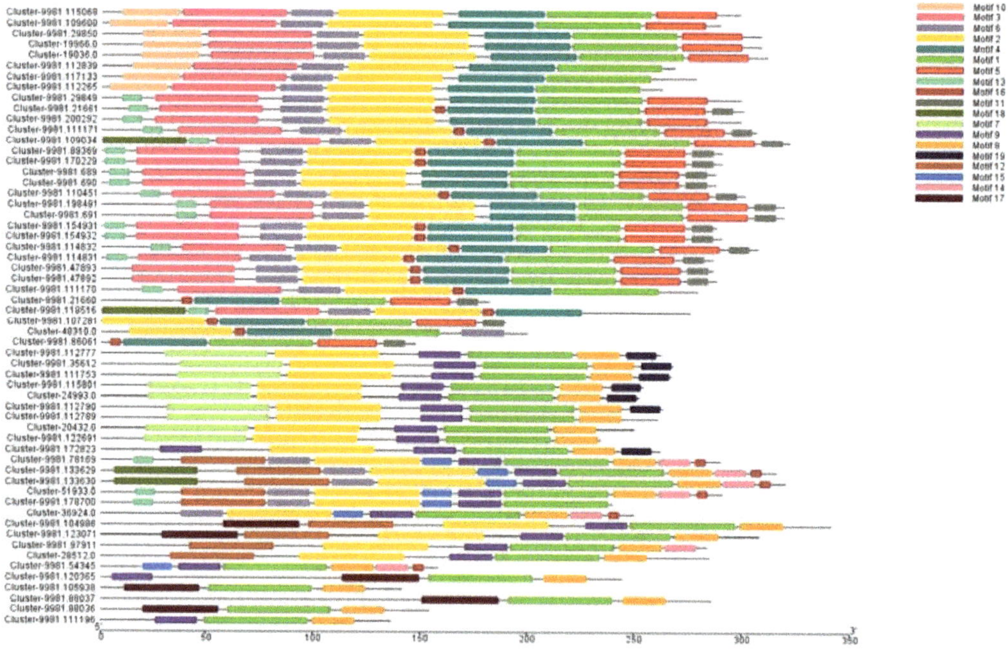

Figure 4. Conserved motif distribution map of LvAQP genes.

All the known aquaporin genes related to cold stress contain several common gene sequence fragments (Figure 5), namely IAEFXXT, GIAW, GGMI, LVYCTAG, SGGHINPAVT, GTFVLVYTVF and ATD, which may play a key role in resisting cold. The above fragments in LvAQP came from motif 2, motif 3, motif 4, and motif 6. Most of these motifs were distributed in the PIPs subfamily of LvAQP. Therefore, the PIPs subfamily might be important for *Ligustrum* × *vicaryi* under cold stress.

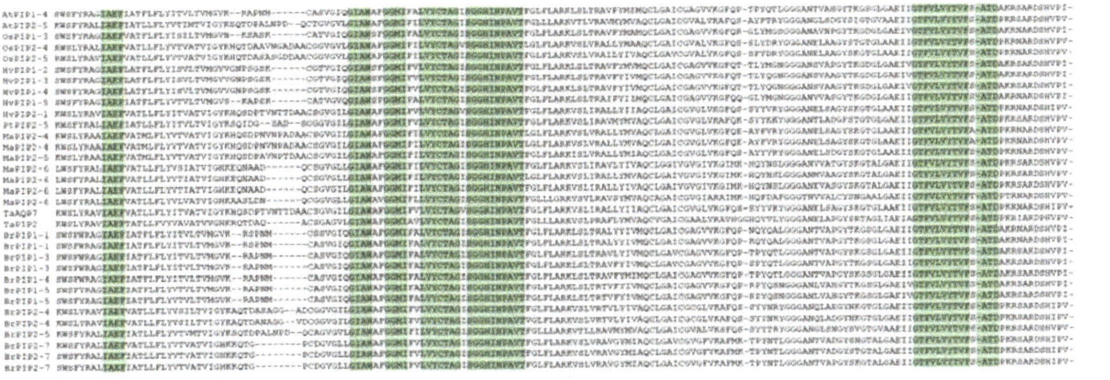

Figure 5. Special gene sequence fragments of aquaporins related to cold stress in plants.

3.5. Analysis of LvAQP Gene Expression Pattern

The transcript abundance of LvAQP was analyzed in four sampling times, and combining the phylogenetic relationship between cold stress aquaporin genes of various plants and of LvAQP genes (Figure 3) was helpful in identifying the specific expression patterns of individual genes of the LvAQP gene family.

The expression of 58 LvAQP genes changed in September, November, January and April (Figure 6). Transcriptional analysis showed that the PIPs subfamily and TIPs subfamily contained a relatively high expression in four sampling times. Compared to September, 8% of LvAQP gene expression increased in November and January and decreased in April; 21% of LvAQP gene expression decreased in November and January and increased in April; 24% of LvAQP gene expression increased in November, decreased in January, and increased in April; 17% of LvAQP gene expression decreased in November, increased in January, and decreased in April; 21% of LvAQP gene expression increased in November and decreased in January and April; 3.4% of LvAQP gene expression decreased in November and increased in January and April; and 5% of LvAQP gene expression decreased consecutively in November, January, and April. According to relevant research, the overexpression of *MusaPIP1;2* in Musa acuminata enhanced plant freezing resistance [40]; in *Arabidopsis thaliana*, the overexpression of *AtPIP1;4* and *AtPIP2;5*, along with repressed expression of other PIPs family members enhanced plant cold resistance [25]; in *Oryza sativa*, there was increased expression of *OsPIP2;5* and *OsPIP2;7* and decreased expression of *OsPIP1;3*, which helped to improve cold resistance [28,41]. Studies had shown that plants enhance cold resistance by overexpressing or inhibiting the expression of aquaporin genes under cold stress. Therefore, in this study, the researchers selected LvAQP genes, whose gene expression increased in November and January and decreased in April, and whose gene expression decreased in November and January and increased in April in four sampling times, as the target genes.

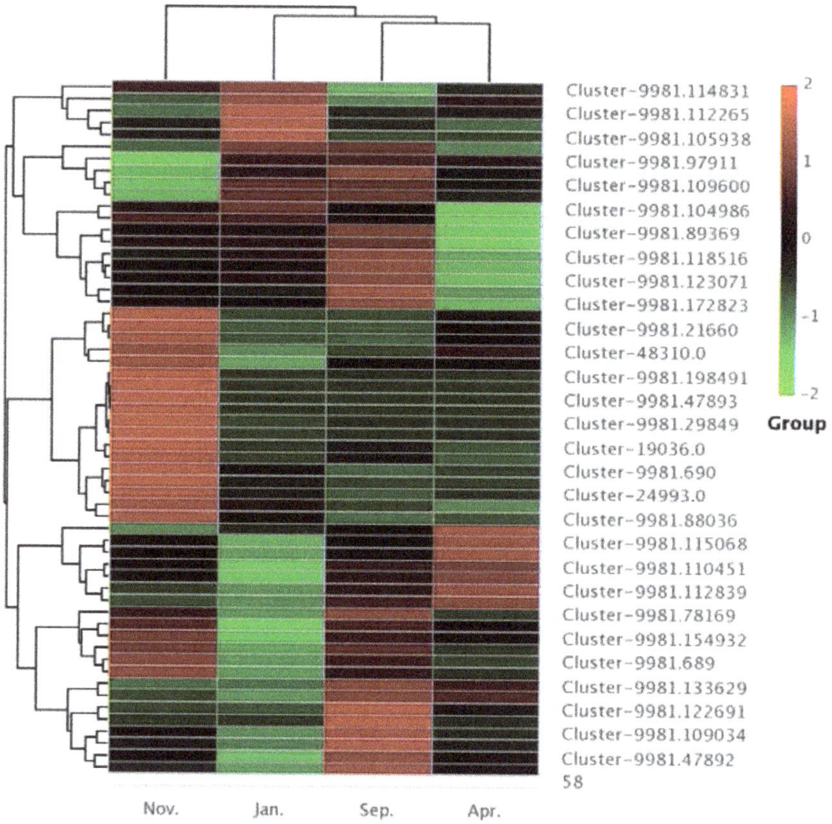

Figure 6. Relative transcript abundance profiles of LvAQP genes during a natural cold stress period.

By analyzing the relative transcript abundance profile of LvAQP genes (Figure 6) and the phylogenetic relationship between cold stress aquaporin genes of various plants and LvAQP genes (Figure 3), 20 LvAQP genes that responded to cold stress were determined: *Cluster-9981.109600*, *Cluster-9981.112839*, *Cluster-9981.112265*, *Cluster-9981.111171*, *Cluster-9981.109034*, *Cluster-9981.89369*, *Cluster-9981.110451*, *Cluster-9981.114832*, *Cluster-9981.114831*, *Cluster-9981.107281*, *Cluster-9981.86061*, *Cluster-9981.112777*, *Cluster-9981.111753*, *Cluster-9981.115801*, *Cluster-9981.112789*, *Cluster-9981.122691*, *Cluster-9981.104986*, *Cluster-9981.123071*, *Cluster-9981.120365* and *Cluster-9981.8803*. Among the determined 20 LvAQP genes, 11 genes were part of the PIPs subfamily, fivegenes were part of the TIPs subfamily, and two genes were part of the NIPs subfamily and SIPs subfamily, separately.

Among the 20 LvAQP genes identified in response to cold stress, the expression of *Cluster-9981.114831* was significantly up-regulated during the two periods of lowest natural temperature in November and January and the most cold-resistant period in January, while the expression of three genes was significantly downregulated, namely, *Cluster-9981.112839*, *Cluster-9981.107281* and *Cluster-9981.112777*. All the significantly upregulated genes contained motif 6, and all the significantly downregulated genes contained motif 1 and motif 2, which were basically consistent with the common special motifs reported in aquaporin genes related to cold stress. It was speculated that the key role of some AQP genes in *Ligustrum* × *vicaryi* for cold resistance might respond to the presence of these specific modular motifs.

3.6. KEGG Enrichment Analysis of Differentially Expressed Genes

KEGG pathway enrichment analysis of differentially expressed genes was conducted under natural cold stress in *Ligustrum* × *vicaryi* (Table 6). A total of 12,872 differentially expressed genes were distributed in 338 pathways, and 10 of them showed significant differences ($p < 0.05$). The differentially expressed genes were significantly enriched in ribosome (ko03010), starch and sucrose metabolism (ko00500), plant hormone signal transduction (ko04075).

Table 6. KEGG pathway enrichment analysis in DEGs of *Ligustrum* × *vicaryi*.

Pathway ID	Pathway	*p* Value	Gene Number
ko03010	Ribosome	1.51E-07	448
ko00940	Phenylpropanoid biosynthesis	1.18E-06	131
ko00460	Cyanoamino acid metabolism	5.77E-06	85
ko00500	Starch and sucrose metabolism	8.38E-05	208
ko04075	Plant hormone signal transduction	2.11E-03	170
ko05322	Systemic lupus erythematosus	7.93E-03	48
ko05034	Alcoholism	8.15E-03	87
ko04915	Estrogen signaling pathway	1.16E-02	68
ko04612	Antigen processing and presentation	2.38E-02	72
ko00520	Amino sugar and nucleotide sugar metabolism	3.49E-02	122

3.7. Expression Verification of Cold-Responsive LvAQP Target Ggenes

The nine screened LvAQP genes that target differentially expressed gene were verified by real-time PCR. The qRT-PCR results were first calculated by the $2^{-\triangle\triangle CT}$ method, followed by log calculation based 2. The change of $\log_2$ multiples for the real-time fluorescence quantification of the nine target genes areshown in Table 7.

Table 7. Fluorescence quantification of nine cold-responsive LvAQP target genes.

Gene Name	Types of Aquaporin Genes	$\log_2$ Fold Change	
		RNA-Seq	Real-Time
Cluster-9981.109600	PIPs	−1.8929	−0.76
Cluster-9981.112839	PIPs	−1.5391	−1.18
Cluster-9981.111171	PIPs	−2.0145	−0.63
Cluster-9981.109034	PIPs	−2.2498	−2.82
Cluster-9981.114831	PIPs	2.5535	2.86
Cluster-9981.107281	PIPs	−3.7174	−0.79
Cluster-9981.112777	TIPs	−3.4918	−4.25
Cluster-9981.115801	TIPs	−1.9979	−0.38
Cluster-9981.122691	TIPs	−4.1174	−3.16

Although the real-time PCR results of individual genes deviated from the RNA-Seq results in terms of differential fold, the up-regulated and down-regulated expression trends between them were consistent (Figure 7). In addition, the correlation analysis between the results of the qRT-PCR analysis and RNA-seq sequencing results showed that the correlation coefficient R^2 reached 0.70 (Figure 8), indicating that the transcriptome sequencing results of *Ligustrum* × *vicaryi* cold resistance were reliable.

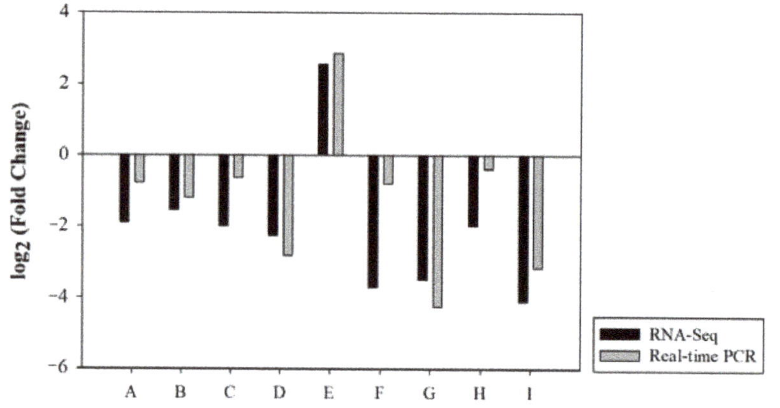

Figure 7. Comparison between RNA-seq and real-time PCR results. A: *Cluster-9981.109600;* B: *Cluster-9981.112839;* C: *Cluster-9981.111171;* D: *Cluster-9981.109034;* E: *Cluster-9981.114831;* F: *Cluster-9981.107281;* G: *Cluster-9981.112777;* H: *Cluster-9981.115801;* I: *Cluster-9981.122691.*

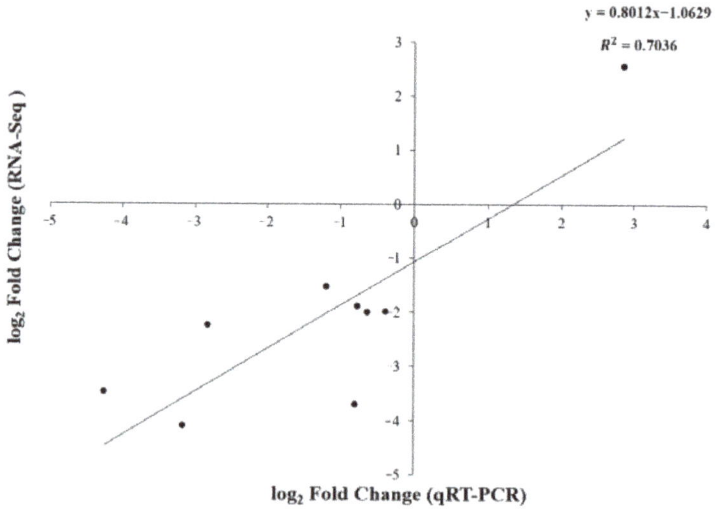

Figure 8. Correlation analysis between RNA-seq and qRT-PCR.

4. Discussion

The number of genes encoding aquaporin of *Ligustrum × vicaryi* was more than that in *Arabidopsis thaliana,* especially in the PIPs subfamily due to gene amplification. In this study, 58 candidate LvAQP genes were found. Phylogenetic analysis showed that these 58 LvAQP genes can be divided into four subfamilies: PIPs, TIPs, NIPs and SIPs. A fifth subfamily has also been reported: XIPs, which is a class of atypical non-specific intrinsic aquaporins. It was absent in *Arabidopsis thaliana, Oryza sativa, Zea mays* and *Ligustrum× vicaryi.* Plasma membrane intrinsic proteins are highly selective for the transporting matrix, and they play an important role in maintaining cell water balance under various adversities [15]. Studies have shown that plants resist abiotic stress by regulating the expression and activity of PIPs in the plasma membrane [39,42–44]. Plants mainly regulate their response to stress through the expression or inhibition of PIPs genes of the aquaporin family. Under natural low temperature adversity, maintaining water balance in the body is a considerable challenge to *Ligustrum × vicaryi.* At this time, the transmembrane transport of water in

Ligustrum × *vicaryi* mainly depends on the PIPs subfamily of the aquaporin family. This study found that the number of PIPs subfamily was the largest in the LvAQP gene family, which was consistent with the results of previous studies [4–10]. The differences from previous studies were that *Arabidopsis thaliana* and *Oryza sativa* have 13 and 11 PIPs genes, respectively, while this study found 32 PIPs subfamily genes in *Ligustrum* × *vicaryi*. The number of genes in the LvPIPs subfamily was much higher than that of other plants. When a certain gene family has obvious gene clusters on the chromosome, it is often accompanied by the gene expansion mechanism of tandem replication [45]. The large number of the PIPs subfamilies of LvAQP gene family was caused by the expansion and tandem duplication of some genes with similar structure in the gene cluster. In this study, the number of LvPIPs genes was higher than that of other plants. There were 11 of the 20 aquaporin genes screened that were related to low temperature stress belonged to the PIPs subfamily. The result of 11 genes belonged to the PIPs subfamily was in accordance with previous studies on aquaporins in response to cold stress, suggesting that the PIPs subfamily of aquaporin might play a major role in responses to cold stress in *Ligustrum* × *vicaryi* [11,28–30,32,33]. Unlike in previous studies, two genes of the SIPs subfamily in the LvAQP gene family also responded to cold stress.

In the face of cold stress, plants generally respond to stress by regulating water homeostasis in the body, in which aquaporin proteins are one of the key pathways of water transport [46–49]. The expression patterns of aquaporins in various plant tissues are different, which indicates that aquaporins may have different functions in plants [50]. After freezing treatment, the low-temperature-tolerant *Zea mays* variety z7 maintained root hydraulic conductivity and water transport by expressing a large amount of aquaporins to reduce freezing damage [51]. The aquaporins PIP1 and PIP2 of *Arabidopsis thaliana* cooperated synergistically in the roots under cold stress to affect root hydraulic conductivity and to regulate plant cold resistance [52]. Overexpression of *PtPIP2;5*, *PtPIP2;1* and *PtPIP2;3* in *Populus trichocarpa* affected its response to cold stress and osmotic stress [53]. Under cold stress, the overexpression of banana *MaPIP2;7* lowered the MDA content and electrolyte leakage in the plant, while the content of chlorophyll, proline, soluble sugar and ABA was higher, thereby enhancing the tolerance to various stresses such as the cold [54]. The overexpression of *MaSIP2;1*, *OsPIP2;7*, and *TaAQP7* (*PIP2*) regulated the osmotic balance in plants, reduced membrane damage and oxidation, and adjusted the levels of hormones such as ABA and GA to improve the cold tolerance of plants [30,36,43].

In this study, the phylogenetic comparison between LvAQP genes and reported aquaporin genes related to cold stress in other plants as well as the changes of aquaporin genes transcription abundance in four sampling times was conducted to identify the specific expression patterns of individual genes of the gene family under natural cold stress. The 20 aquaporin genes that responded to cold stress were screened from the 58 LvAQP genes; eleven belonged to the LvPIPs subfamily, five belonged to the LvTIPs subfamily, and two belonged to the LvNIPs subfamily and LvSIPs subfamily, separately, which indicated that genes of the PIPs subfamily played a major role in response to natural cold stress. Among these 20 aquaporin genes of *Ligustrum* × *vicaryi* that responded to cold stress, all the significantly upregulated genes contained motif 6, while all the significantly downregulated genes contained motif 1 and motif 2. It was speculated that motif 1, motif 2 and motif 6 might play an important role in response to cold stress when *Ligustrum* × *vicaryi* is under a natural low temperature. In the analysis of the reported cold stress-related AQP gene sequences of other plants, we found that the gene sequence SGGHINPAVT was present in motif 2 and GIAW and GGMI were present in motif 6; thus, it was further speculated that the gene sequences of SGGHINPAVT, GIAW and GGMI might play a major role in the response to cold stress in *Ligustrum* × *vicaryi*. However, the AEFXXT motif, which was conserved in almost all MIPs in previous studies, was not conserved in all significantly upregulated and significantly downregulated genes in *Ligustrum* × *vicaryi* in response to cold stress. Therefore, we speculated that the AEFXXT motif might not be the key motif in genes responding to cold stress. From the determination of cold resistance of *Ligustrum* × *vicaryi*, it can be seen that *Ligustrum* × *vicaryi* was most resistant to cold in January during the natural overwintering process, and the cold resistance of the plant changed with the change of time. In this study, 75% of the LvAQP genes that were significantly related to cold stress decreased in November and January, and their expression increased in April, which is consistent with the results of transcriptome analysis of *Arabidopsis thaliana*, *Oryza sativa*, and the roots and leaves of *Zea mays* [25,43,51]. In winter, low temperatures can easily lead to freeze thaw embolism of plants, which blocks water transport and leads to withering. At this time, aquaporin may be involved in embolization repair. Low soil temperature limits the absorption of water by roots, leading to a water imbalance. Low soil temperature can reduce or increase the activity of aquaporin in roots, but appropriate low temperature acclimation can promote the abundance of AQP in roots. In the process of natural cold stress, with the enhancement of cold resistance, *Ligustrum* × *vicaryi* regulated the decrease or increase in the expression of aquaporin genes and the corresponding protein activity, and adjusted root hydraulic conductivity, thus maintaining the water balance in the plant, resisting the effects of natural low-temperature stress, and ensuring normal life activities.

Aquaporins are important membrane functional proteins in many physiological reactions, which play a key role mainly through transcriptional regulation, post-translational modification and subcellular localization [55,56]. Plasma membrane intrinsic proteins and tonoplast intrinsic proteins are located on the inner chloroplast membrane and thylakoid membrane [3]. KEGG enrichment analysis of *Ligustrum* × *vicaryi* genes showed that they responded to cold stress mainly through the sucrose metabolism pathway and plant hormone signal transduction pathway. It was speculated that some genes of the PIPs and TIPs subfamilies on the plasma membrane and in the chloroplast were upregulated or downregulated, which would enhance the cold resistance of *Ligustrum* × *vicaryi* by regulating the synthesis and transformation of soluble sugar or starch. After feeling a natural low temperature, differentially expressed genes related to hormone signaling were enriched, and pathways such as ABA signaling were turned on under low temperature stress, thus inducing the expression of downstream regulatory genes. Then, the expression of AQP genes changed in order to regulate the synthesis of corresponding proteins and other macromolecules, to stabilize the membrane structure, and to reduce the water transport rate to avoid low temperature damage of *Ligustrum* × *vicaryi*.

5. Conclusions

In this research, the gene expression of LvAQP under natural cold stress was studied. We identified 58 candidate LvAQP genes. Based on phylogenetic analysis, the 58 candidate LvAQP genes were divided into four subfamilies: 32 belonged to the PIPs subfamily, 11 belonged to the TIPs subfamily, 11 belonged to the NIPs subfamily and 4 belonged to the SIPs subfamily. The number of genes in the PIPs subfamily was nearly twice as large as that in other plants. The LvAQP gene family contained nine pairs of tandem repeats genes, which had high conservatism in the process of evolution by searching for conserved motifs. We obtained 20 differentially expressed LvAQP genes under natural cold stress. Among the 20 differentially expressed genes, 11 belonged to the LvPIPs subfamily. Among the 20 differentially expressed genes, the significantly up-regulated gene was *Cluster-9981.114831*; the significantly down-regulated genes were *Cluster-9981.112839*, *Cluster-9981.107281* and *Cluster-9981.112777*. These four LvAQP genes might play important roles in response to low temperature stress. The results laid the foundation for further exploration of cold resistant aquaporin genes and biological function verification of *Ligustrum* × *vicaryi*.

Supplementary Materials: The following are available online at https://www.mdpi.com/article/10.3390/f13020182/s1, Table S1: The genes and primers used for qRT-PCR analysis.

Author Contributions: J.D., S.N., J.Q. and B.D. designed the experiments, J.D., S.N., J.Q., J.Z., M.Z. and Y.M. performed the experiments and collected the data. J.D., S.N., J.Q., J.Z. and M.Z. analyzed the data. J.D. wrote the manuscript. S.N., J.Q. and B.D. revised the manuscript. All authors have read and agreed to the published version of the manuscript.

Funding: This research was supported by the Science and Technology Development Center Project of National Forestry and Grassland Administration (KJZXSA202036) and Financial Aid Project for the Introduction of Overseas Students in Hebei Province (CN201711).

Acknowledgments: We would like to thank Huan Sun for her help in the details of the experiment, and Lanbo Wei and Wenjia Shi for their extensive participation in the experiment.

Conflicts of Interest: All the authors declare no conflict of interest.

Abbreviations

MIPs: major intrinsic proteins; LvAQP: *Ligustrum* × *vicaryi* aquaporins; PIPs: plasma membrane intrinsic proteins; TIPs: tonoplast intrinsic proteins; NIPs: nodulin26-like intrinsic proteins; SIPs: small basic intrinsic proteins; XIPs: uncharacterized X intrinsic proteins; TMM: trimmed mean of M-values; qRT-PCR: quantitative real-time PCR; LvPIPs: *Ligustrum* × *vicaryi* plasma membrane intrinsic proteins; LvTIPs: *Ligustrum* × *vicaryi* tonoplast intrinsic proteins; LvNIPs: *Ligustrum* × *vicaryi* nodulin26-like intrinsic proteins; LvSIPs: *Ligustrum* × *vicaryi* small basic intrinsic proteins; REL: relative electrolyte leakage.

Appendix A

Table A1. List of 35 Aquaporins of *Arabidopsis thaliana*.

Subfamily	Name	Synonyms	NCBI Reference Sequence
PIPs	*PIP1;1*	PIP1A	AT3G61430
	PIP1;2	PIP1B; TMPA	AT2G45960
	PIP1;3	PIP1C; TMPB	AT1G01620
	PIP1;4	TMPC	AT4G00430
	PIP1;5	PIP1D	AT4G23400
	PIP2;1	PIP2A	AT3G53420
	PIP2;2	PIP2B; TMB2B	AT2G37170
	PIP2;3	RD28; TMP2C	AT2G37180
	PIP2;4	PIP2F	AT5G60660
	PIP2;5	PIP2D	AT3G54820
	PIP2;6	PIP2E	AT2G39010
	PIP2;7	PIP3; SIMIP	AT4G35100
	PIP2;8	PIP3B	AT2G16850
TIPs	*TIP1;1*	GAMMA-TIP	AT2G36830
	TIP1;2	TIP2	AT3G26520
	TIP1;3	GAMMA-TIP1	AT4G01470
	TIP2;1	DELTA-TIP	AT3G16240
	TIP2;2	DELTA-TIP2	AT4G17340
	TIP2;3	DELTA-TIP3	AT5G47450
	TIP3;1	α-TIP	AT1G73190
	TIP3;2	BETA-TIP	AT1G17810
	TIP4;1		AT2G25810
	TIP5;1		AT1G17820
NIPs	*NIP1;1*	NLM1	AT4G19030
	NIP1;2	NLM2	AT4G18910
	NIP2;1		AT2G34390
	NIP3;1		AT1G31885
	NIP4;1		AT5G37810
	NIP4;2		AT5G37820
	NIP5;1		AT4G10380
	NIP6;1		AT1G80760
	NIP7;1		AT3G06100

Table A1. *Cont.*

Subfamily	Name	Synonyms	NCBI Reference Sequence
SIPs	*SIP1;1*	SIP1A	AT3G04090
	SIP1;2		AT5G18290
	SIP2;1		AT3G56950

References

1. Kapilan, R.; Vaziri, M.; Zwiazek, J.J. Regulation of aquaporins in plants under stress. *Biol. Res.* **2018**, *51*, 2227–2236. [CrossRef] [PubMed]
2. Hove, R.M.; Bhave, M. Plant aquaporins with non-aqua functions: Deciphering the signature sequences. *Plant Mol. Biol.* **2011**, *75*, 413–430. [CrossRef] [PubMed]
3. Zunaira, A.; Howton, T.; Sun, Y.; Mukhtar, M.S. The Roles of Aquaporins in Plant Stress Responses. *J. Dev. Biol.* **2016**, *4*, 9. [CrossRef]
4. Quigley, F.; Rosenberg, J.M.; Shachar-Hill, Y.; Bohnert, H.J. From genome to function: The Arabidopsis aquaporins. *Genome Biol.* **2002**, *3*, 97–108. [CrossRef]
5. Sakurai, J.; Ishikawa, F.; Yamaguchi, T.; Uemura, M.; Maeshima, M. Identification of 33 rice aquaporin genes and analysis of their expression and function. *Plant Cell Physiol.* **2005**, *46*, 1568–1577. [CrossRef] [PubMed]
6. Chaumont, F.; Barrieu, F.; Wojcik, E.; Chrispeels, M.J.; Jung, R. Aquaporins Constitute a Large and Highly Divergent Protein Family in Maize. *Plant Physiol.* **2001**, *125*, 1206–1215. [CrossRef]
7. Hove, R.M.; Ziemann, M.; Bhave, M. Identification and Expression Analysis of the Barley (*Hordeum vulgare* L.) Aquaporin Gene Family. *PLoS ONE* **2015**, *10*, e0128025. [CrossRef]
8. Da, Y.Z.; Ali, Z.; Chang, B.W.; Xu, L.; Yi, J.X.; Xu, Z.L.; Liu, X.Q.; He, X.L.; Huang, Y.H.; Ma, H.X.; et al. Genome-Wide Sequence Characterization and Expression Analysis of Major Intrinsic Proteins in Soybean (*Glycine max* L.). *PLoS ONE* **2013**, *8*, e56312. [CrossRef]
9. Park, W.; Scheffler, B.E.; Bauer, P.J.; Campbell, B.T. Identification of the family of aquaporin genes and their expression in upland cotton (*Gossypium hirsutum* L.). *BMC Plant Biol.* **2010**, *10*, 142. [CrossRef]
10. Zhou, Y.; Tao, J.; Ahammed, G.J.; Li, J.; Yang, Y. Genome-wide identification and expression analysis of aquaporin gene family related to abiotic stress in watermelon. *Genome* **2019**, *62*, 643–656. [CrossRef]
11. Kayum, M.A.; Park, J.I.; Nath, U.K.; Biswas, M.K.; Kim, H.T.; Nou, I.S. Genome-wide expression profiling of aquaporin genes confer responses to abiotic and biotic stresses in *Brassica rapa*. *BMC Plant Biol.* **2017**, *17*, 23. [CrossRef] [PubMed]
12. Song, J.; Ye, G.; Qian, Z.; Ye, Q. Virus-induced plasma membrane aquaporin *PsPIP2;1* silencing inhibits plant water transport of *Pisum sativum*. *Bot. Stud.* **2016**, *57*, 1–10. [CrossRef] [PubMed]
13. Gitto, A.; Fricke, W. Zinc treatment of hydroponically grown barley plants causes a reduction in root and cell hydraulic conductivity and isoform-dependent decrease in aquaporin gene expression. *Physiol. Plant.* **2018**, *164*, 176–190. [CrossRef] [PubMed]
14. Macho-Rivero, M.A.; Begoa, H.R.M.; Ramona, B.; Schäffner, A.R.; Tanaka, N.; Fujiwara, T.; Gonzalez-Fontez, A.; Camacho-Cristóbal, J.J.; Camacho-Cristóbal, J.J. Boron Toxicity Reduces Water Transport from Root to Shoot in Arabidopsis Plants. Evidence for a Reduced Transpiration Rate and Expression of Major PIP Aquaporin Genes. *Plant Cell Physiol.* **2018**, *59*, 836–844. [CrossRef] [PubMed]
15. Pawłowicz, I.; Rapacz, M.; Perlikowski, D.; Gondek, K.; Kosmala, A. Abiotic stresses influence the transcript abundance of PIP and TIP aquaporins in Festuca species. *J. Appl. Genet.* **2017**, *58*, 421–435. [CrossRef]
16. Ayadi, M.; Brini, F.; Masmoudi, K. Overexpression of a Wheat Aquaporin Gene, *TdPIP2;1*, Enhances Salt and Drought Tolerance in Transgenic Durum Wheat cv. *Maali*. *Int. J. Mol. Sci.* **2019**, *20*, 2389. [CrossRef]
17. Braz, L.; Fernandes, P.D.; Barbosa, D.D.; Dutra, W.F.; Silva, C.R.; Lima, L.M.; Cavalcanti, J.J.V.; Farias, F.J.C.; Santos, R.C. Expression of aquaporin subtypes (*GhPIP1;1*, *GhTIP2;1* and *GhSIP1;3*) in cotton (*Gossypium hirsutum*) submitted to salt stress. *Agric. Week* **2020**, *11*, 72–84. [CrossRef]
18. Feng, Z.J.; Liu, N.; Zhang, G.W.; Niu, F.G.; Xu, S.C.; Gong, Y.M. Investigation of the AQP Family in Soybean and the Promoter Activity of *TIP2;6* in Heat Stress and Hormone Responses. *Int. J. Mol. Sci.* **2019**, *20*, 262. [CrossRef]
19. Belanger, G.; Rochette, P.; Castonguay, Y.; Bootsma, A.; Mongrain, D.; Ryan, D.A. Climate change and winter survival of perennial forage crops in eastern Canada. *Agron. J.* **2002**, *94*, 1120–1130. [CrossRef]
20. Ogren, E. Effects of climatic warming on cold hardiness of some northern woody plants assessed from simulation experiments. *Physiol. Plant.* **2001**, *112*, 71–77. [CrossRef]
21. Lee, S.H.; Singh, A.P.; Chung, G.C. Rapid accumulation of hydrogen peroxide in cucumber roots due to exposure to low temperature appears to mediate decreases in water transport. *J. Exp. Bot.* **2004**, *55*, 1733–1741. [CrossRef] [PubMed]
22. Azad, A.K.; Sawa, Y.; Ishikawa, T.; Shibata, H. Phosphorylation of plasma membrane aquaporin regulates temperature dependent opening of tulip petals. *Plant Cell Physiol.* **2004**, *45*, 608–617. [CrossRef] [PubMed]
23. Lee, S.H.; Chung, G.C.; Steudle, E. Gating of aquaporins by low temperature in roots of chilling sensitive cucumber and chilling tolerant figleaf gourd. *J. Exp. Bot.* **2005**, *56*, 985–995. [CrossRef] [PubMed]

24. Kuwagata, T.; Ishikawa-Sakurai, J.; Hayashi, H.; Nagasuga, K.; Fukushi, K.; Ahamed, A.; Takasugi, K.; Katsuhara, M.; Murai-Hatano, M. Influence of Low Air Humidity and Low Root Temperature on Water Uptake, Growth and Aquaporin Expression in Rice Plants. *Plant Cell Physiol.* **2012**, *53*, 1418–1431. [CrossRef]
25. Jang, J.Y. Transgenic Arabidopsis and tobacco plants overexpressing an aquaporin respond differently to various abiotic stresses. *Plant Mol. Biol.* **2007**, *64*, 621–632. [CrossRef] [PubMed]
26. Ahamed, A.; Murai-Hatano, M.; Ishikawa-Sakurai, J.; Hayashi, H.; Kawamura, Y.; Uemura, M. Cold stress-induced acclimation in rice is mediated by root-specific aquaporins. *Plant Cell Physiol.* **2012**, *53*, 1445–1456. [CrossRef]
27. Xu, C.; Wang, M.; Zhou, L.; Quan, T.; Xia, G. Heterologous Expression of the Wheat Aquaporin Gene *TaTIP2;2* Compromises the Abiotic Stress Tolerance of *Arabidopsis thaliana*. *PLoS ONE* **2013**, *8*, e79618. [CrossRef]
28. Matsumoto, T.; Lian, H.L.; Su, W.A.; Tanaka, D.; Liu, C.W.; Iwasaki, I.; Kitagawa, Y. Role of the aquaporin PIP1 subfamily in the chilling tolerance of rice. *Plant Cell Physiol.* **2009**, *50*, 216–229. [CrossRef]
29. Duan, R.J.; Xiong, H.Y.; Katsuhara, M. Expression of Root *HvPIPs* of Barley Seedling under Chilling Stress. *Plant Physiol. J.* **2014**, *50*, 1203–1208. [CrossRef]
30. Xu, Y.; Li, J.; Song, S.; Liu, J.; Hou, X.; Li, Y.; Jin, Z.; Wang, A.; Huang, D.; Xu, Z.; et al. A novel aquaporin gene MaSIP2-1 confers tolerance to drought and cold stresses in transgenic banana. *Mol. Breed.* **2020**, *40*, 439–447. [CrossRef]
31. Hu, W.; Hou, X.; Chao, H.; Yan, Y.; Tie, W.; Ding, Z.; Jin, Z.; Wei, Y.; Liu, J.; Miao, H.; et al. Genome-Wide Identification and Expression Analyses of Aquaporin Gene Family during Development and Abiotic Stress in Banana. *Int. J. Mol. Sci.* **2015**, *16*, 19728–19751. [CrossRef] [PubMed]
32. He, W.D.; Gao, J.; Dou, T.X.; Shao, X.H.; Bi, F.C.; Sheng, O.; Deng, G.M.; Li, C.Y.; Hu, C.H.; Liu, J.H.; et al. Early Cold-Induced Peroxidases and Aquaporins Are Associated with High Cold Tolerance in Dajiao (*Musa* spp. 'Dajiao'). *Front. Plant Sci.* **2018**, *9*, 282. [CrossRef] [PubMed]
33. Ranganathan, K.; El Kayal, W.; Cooke, J.E.; Zwiazek, J.J. Responses of hybrid aspen over-expressing a *PIP2;5* aquaporin to low root temperature. *J. Plant Physiol.* **2016**, *192*, 98–104. [CrossRef] [PubMed]
34. Reddy, P.S.; Sharma, K.K.; Vadez, V. Genome-wide identification and characterization of the aquaporin gene family in *Sorghum bicolor* (L.). *Plant Gene.* **2015**, *1*, 18–28. [CrossRef]
35. Forrest, K.L.; Bhave, M. The PIP and TIP aquaporins in wheat form a large and diverse family with unique gene structures and functionally important features. *Funct. Integr. Genom.* **2008**, *8*, 115–133. [CrossRef]
36. Huang, C.; Zhou, S.; Hu, W.; Deng, X.; Wei, S.; Yang, G.; He, G. The Wheat Aquaporin Gene *TaAQP7* Confers Tolerance to Cold Stress in Transgenic Tobacco. *Z. Naturforschung. C J. Biosci.* **2014**, *69*, 142–148. [CrossRef]
37. Hao, M.Z.; Han, M.H.; Peng, F.R.; Liang, Y.W. Comparison of the Cold-resistance Capabilities of Four Ligustrum Cultivars. *Med. Plant* **2012**, *3*, 111–115.
38. Di, B.; Luoranen, J.; Lehto, T.; Himanen, K.; Silvennoinen, M.; Silvennoinen, R.; Repo, T. Biophysical changes in the roots of Scots pine seedlings during cold acclimation and after frost damage. *For. Ecol. Manag.* **2018**, *431*, 63–72. [CrossRef]
39. Kumar, K.; Mosa, K.A.; Meselhy, A.G.; Dhankher, O.P. Molecular insights into the plasma membrane intrinsic proteins roles for abiotic stress and metalloids tolerance and transport in plants. *Indian J. Plant Physiol.* **2018**, *23*, 721–730. [CrossRef]
40. Sreedharan, S.; Shekhawat, U.; Ganapathi, T.R. Transgenic banana plants overexpressing a native plasma membrane aquaporin *MusaPIP1;2* display high tolerance levels to different abiotic stresses. *Plant Biotechnol. J.* **2013**, *11*, 942–952. [CrossRef]
41. Li, G.W.; Zhang, M.H.; Cai, W.M.; Sun, W.N.; Su, W.A. Characterization of *OsPIP2;7*, a water channel protein in rice. *Plant Cell Physiol.* **2008**, *49*, 1851–1858. [CrossRef]
42. Wang, X.; Zhang, J.L.; Feng, X.X.; Li, H.J.; Zhang, G.F. The molecular mechanisms of plant plasma membrane intrinsic proteins trafficking and stress response. *Yi Chuan Hered.* **2017**, *39*, 293–301. [CrossRef]
43. Yu, X.; Peng, Y.H.; Zhang, M.H.; Shao, Y.J.; Su, W.A.; Tang, Z.C. Water relations and an expression analysis of plasma membrane intrinsic proteins in sensitive and tolerant rice during chilling and recovery. *Cell Res.* **2006**, *16*, 599–608. [CrossRef] [PubMed]
44. Jiang, C.; Song, X.; He, H.; Chu, L.; Zhou, H.; Zhao, Y.; Xu, Y.; Zeng, W.; Lin, X.; Lu, M.Z. Genome-wide identification of plasma membrane aquaporin gene family in Populus and functional identification of *PIP1;1* involved in osmotic stress. *Environ. Exp. Bot.* **2020**, *179*, 104200. [CrossRef]
45. Simes, M.S.; Carvalho, G.G.; Ferreira, S.S.; Hernandes-Lopes, J.; Setta, N.; Cesarino, I. Genome-wide characterization of the laccase gene family in *Setaria viridis* reveals members potentially involved in lignification. *Planta* **2020**, *251*, 46. [CrossRef] [PubMed]
46. Pawłowicz, I.; Masajada, K. Aquaporins as a link between water relations and photosynthetic pathway in abiotic stress tolerance in plants. *Gene* **2019**, *687*, 166–172. [CrossRef] [PubMed]
47. Cheng, H.Q.; Ding, Y.E.; Shu, B.; Zou, Y.N.; Wu, Q.S.; Kuča, K. Plant Aquaporin Responses to Mycorrhizal Symbiosis under Abiotic Stress. *Int. J. Agric. Biol.* **2020**, *5*, 786–794.
48. Tang, N.; Shahzad, Z.; Lonjon, F.; Loudet, O.; Vailleau, F.; Maurel, C. Natural variation at XND1 impacts root hydraulics and trade-off for stress responses in Arabidopsis. *Nat. Commun.* **2018**, *9*, 1321–1358. [CrossRef]
49. Jang, J.Y.; Kim, D.G.; Kim, Y.O.; Kim, J.S.; Kang, H. An Expression Analysis of a Gene Family Encoding Plasma Membrane Aquaporins in Response to Abiotic Stresses in *Arabidopsis Thaliana*. *Plant Mol. Biol.* **2004**, *54*, 713–725. [CrossRef]
50. Yang, S.D.; Guo, D.L.; Pei, M.S.; Liu, H.; Wei, T.; Yu, Y. Identification of Grapevine AQP Family and prediction of transcriptional regulatory network under drought stress. *J. Fruit Sci.* **2021**, *38*, 1638–1652. [CrossRef]

51. Aroca, R.; Amodeo, G.; Fernández-Illescas, S.; Herman, E.M.; Chaumont, F.; Chrispeels, M.J. The Role of Aquaporins and Membrane Damage in Chilling and Hydrogen Peroxide Induced Changes in the Hydraulic Conductance of Maize Roots. *Plant Physiol.* **2005**, *137*, 341–353. [CrossRef] [PubMed]
52. Rahman, A.; Kawamura, Y.; Maeshima, M.; Rahman, A.; Uemura, M. Plasma Membrane Aquaporin Members PIPs Act in Concert to Regulate Cold Acclimation and Freezing Tolerance Responses in *Arabidopsis thaliana*. *Plant Cell Physiol.* **2020**, *61*, 787–802. [CrossRef] [PubMed]
53. Ranganathan, K. Over-Expression of *PIP2;5* Aquaporin in a *Populus tremula* × *P. albaclone* and its Effects on Plant Responses to Low Root Temperature and Osmotic Stress. Ph.D. Thesis, University of Alberta, Edmonton, AB, Canada, 2015. [CrossRef]
54. Yi, X.; Hu, W.; Liu, J.; Song, S.; Hou, X.; Jia, C.; Li, J.; Miao, H.; Wang, Z.; Tie, W.; et al. An aquaporin gene *MaPIP2-7* is involved in tolerance to drought, cold and salt stresses in transgenic banana (*Musa acuminata* L.). *Plant Physiol. Biochem.* **2020**, *147*, 66–76. [CrossRef]
55. Verdoucq, L.; Rodrigues, O.; Martinière, A.; Luu, D.T.; Maurel, C. Plant aquaporins on the move: Reversible phosphorylation, lateral motion and cycling. *Curr. Opin. Plant Biol.* **2014**, *22*, 101–107. [CrossRef] [PubMed]
56. Ricardi, M.M.; González, R.M.; Zhong, S.; Domínguez, P.G.; Duffy, T.; Turjanski, P.G.; Salgado Salter, J.D.; Alleva, K.; Carrari, F.; Giovannoni, J.J.; et al. Genome-wide data (ChIP-seq) enabled identification of cell wall-related and aquaporin genes as targets of tomato ASR1, a drought stress-responsive transcription factor. *BMC Plant Biol.* **2014**, *14*, 29. [CrossRef]

forests

Article

Genomic Survey and Cold-Induced Expression Patterns of *bHLH* Transcription Factors in *Liriodendron chinense* (Hemsl) Sarg.

Rongxue Li [1], Baseer Ahmad [2], Delight Hwarari [1], Dong'ao Li [3], Ye Lu [4], Min Gao [1], Jinhui Chen [4,*] and Liming Yang [1,*]

1 College of Biology and the Environment, Nanjing Forestry University, Nanjing 210037, China; lirongxue2022@163.com (R.L.); tondehwarr@njfu.edu.cn (D.H.); gm1749486041@163.com (M.G.)
2 Faculty of Veterinary and Animal Sciences, Muhammad Nawaz Sharif University of Agriculture, Multan 25000, Punjab, Pakistan; dr.baseerahmadkhan@gmail.com
3 College of Plant Protection, Northwest Agriculture and Forestry University, Xianyang 712100, China; 2019050129@nwafu.edu.cn
4 College of Forestry, Nanjing Forestry University, Nanjing 210037, China; luye@njfu.edu.cn
* Correspondence: chenjh@njfu.edu.cn (J.C.); yangliming@njfu.edu.cn (L.Y.)

Abstract: *bHLH* transcription factors play an animated role in the plant kingdom during growth and development, and responses to various abiotic stress. In this current study, we conducted, the genome-wide survey of *bHLH* transcription factors in *Liriodendron chinense* (Hemsl) Sarg., 91 *LcbHLH* family members were identified. Identified *LcbHLH* gene family members were grouped into 19 different subfamilies based on the conserved motifs and phylogenetic analysis. Our results showed that *LcbHLH* genes clustered in the same subfamily exhibited a similar conservative exon-intron pattern. Hydrophilicity value analysis showed that all *Lc*bHLH proteins were hydrophilic. The Molecular weight (Mw) of *Lc*bHLH proteins ranged from 10.19 kD (*LcbHLH*15) to 88.40 kD (*LcbHLH*50). A greater proportion, ~63%, of *Lc*bHLH proteins had a theoretical isoelectric point (pI) less than seven. Additional analysis on the collinear relationships within species and among dissimilar species illustrated that tandem and fragment duplication are the foremost factors of amplification of this family in the evolution process, and they are all purified and selected. RNA-seq and real-time quantitative PCR analysis of *LcbHLH* members showed that the expression of *LcbHLH*35, 55, and 86 are up-regulated, and the expression of *LcbHLH*9, 20, 39, 54, 56, and 69 is down-regulated during cold stress treatments while the expression of *LcbHLH*24 was up-regulated in the short term and then later down-regulated. From our results, we concluded that *LcbHLH* genes might participate in cold-responsive processes of *L. chinense*. These findings provide the basic information of *bHLH* gene in *L. chinense* and their regulatory roles in plant development and cold stress response.

Keywords: *bHLH* transcription factor; cold stress; expression pattern; genome-wide identification; *Liriodendron chinense*

Citation: Li, R.; Ahmad, B.; Hwarari, D.; Li, D.; Lu, Y.; Gao, M.; Chen, J.; Yang, L. Genomic Survey and Cold-Induced Expression Patterns of *bHLH* Transcription Factors in *Liriodendron chinense* (Hemsl) Sarg. *Forests* **2022**, *13*, 518. https://doi.org/10.3390/f13040518

Academic Editor: Cristina Vettori

Received: 9 February 2022
Accepted: 24 March 2022
Published: 28 March 2022

1. Introduction

Globally out of all abiotic stress factors, cold, drought, and heat stresses are declared as the most complex ones affecting plant growth, survival, and crop productivity. Molecular regulation at the post-transcriptional level possesses a vital role for development, growth, nutrient allocation, and defensive mechanism in plants [1,2]. The *bHLH* family regulates growth and development, morphogenesis, and stress responses in plants [3–5], characterized by a helix-loop-helix (HLH) domain, with an approximated 15 amino acids N-terminal as the base region: known for recognizing and binding to specific DNA while, the C-terminal is the HLH region with about 50 amino acids [6–8]. The helix is also associated with DNA sequences that recognize protein-specific binding [9] and can form homodimer or heterodimer with other proteins [10]. On top of an α-helix near the N-terminal is another α- helix [11]. The two α-helices are connected with a ring formed by amino acid chains to form an HLH structure.

Generally, *bHLH* transcription factors are known to act as transcriptional activators or inhibitors for seed germination and flowering regulation [9]. However, a study in Arabidopsis mutant srl2, *AtPIF4* (*AtbHLH*09) spectacled a specific role in the signaling network in phytochrome B (phyB) and in light regulation [12]: *AtPRE1* (At*bHLH*136) and ILI1 were also identified to regulate cell elongation by interacting with IBH1 (At*bHLH*158) under the action of brassinosteroids (BR) and gibberellin signals [13]. Moreover, *AtPRE1* (*AtbHLH*136) and *IBH1* (*AtbHLH*158) form regulatory system with *AtACE*1/2/3 (*AtbHLH*049/074/077) that competitively regulate cell growth. IBH1 (*AtbHLH*158) has also been shown to negatively regulate cell growth by interacting with the positive regulatory gene *AtACE*1/2/3 (*AtbHLH*049/074/077) [13]. Certain members of the *bHLH* transcription factor family have also been shown to enhance resistance to harsh conditions when plants retort to abiotic stresses [14,15]. For instance, overexpression of *AtICE*1 (*AtbHLH*116) and *AtICE*2 (*AtbHLH*33) can augment the expression of *CBF* promoter at low temperature and mend the stress resistance of transgenic plants [16,17]. Feng et al. [18] has also demonstrated that *MdCIbHLH*1 protein binds to the *MdCBF*2 promoter and upregulates the expression of *CBF*2 through the C-repeat-binding factor (CBF) pathway and promote the cold tolerance of transgenic apple plants. A study in trifoliate orange has also shown *PtrbHLH* to increase cold resistance by activating *PtrCAT* [19].

To date, research on different plant genomes has concurred that the *bHLH* transcription factor family is incessantly distinguished, with the structural characteristics and response profiles to various environmental stresses [10,20–22]. Nonetheless, few studies on the *bHLH* gene family of the forest tree species have been conducted with less on the *L. chinense*. *L. chinense* is a kind of tall deciduous tree, which is of economic, ornamental, medicinal, and ecological value [23,24]. The recent release of the *L. chinense* genome provided the opportunity for its *LcbHLH* gene family (which will be referred to as *Lc* in this study) to be analyzed [23]. In this current study we identified 91 *LcbHLH* transcription factors, which were further analysed using Bioinformatic approach for evolution, conserved motif arrangement, exon-intron patterns, and other physiochemical proprieties. Additionally, each subfamily of the *LcbHLH* gene family was shown to play imperative biological functions in abiotic stress responses. The identification and distinctive analysis of the *bHLH* transcription factor of *L. chinense* will assist in comprehending the structural characteristics of gene families in *L. chinense* and preliminarily predict the function of *bHLH* members, which will provide the gene resources for the improvement of *L. chinense* germplasm by genetic engineering technology in the future.

2. Materials and Methods

2.1. Identification and Physicochemical Properties Analysis of bHLH Family Members of Liriodendron chinense

The nucleic acid and protein sequences of *L. chinense* were collected from the local protein database [23]. The protein sequences of the bHLH family of *Arabidopsis* and rice were retrieved and downloaded from the plant transcription factor database (http://planttfdb.cbi.pku.edu.cn (accessed on 12 November 2021)) [25]. The bHLH protein sequences of *Arabidopsis* and rice were used as query sequences, while the candidate protein-containing bHLH/HLH domain was screened from the *L. chinense* database by local blastp program. Then, the HMMER model downloaded from the Pfam database was used to identify the candidate bHLH protein of *L. chinense* in a local protein database. Finally, proteins with the bHLH/HLH domain were taken as the final bHLH family members of *L. chinense*. The physical and chemical properties (including molecular weight, isoelectric point, and hydrophilicity) of LcbHLH family members were analyzed using the Protparamin EXPASY database.

2.2. Phylogenetic Analysis of LcbHLHs

ClustalX2 was used for multiple sequence alignment of the *bHLH* domain. The *bHLH* proteins of three plants, rice, Arabidopsis, and poplar, have been downloaded from National Center for Biotechnology Information (NCBI). The phylogenetic tree was constructed using MEGA7.0 with the Neighbor-Joining method [26,27]. The evolutionary distance was obtained through the p-distance method, with the distances employed to estimate the number of amino acids at each locus. The reliability of each phylogenetic tree was guaranteed by 1000 bootstrap sampling iterations.

2.3. Chromosome Location and Gene Replication of LcbHLHs

The data of the chromosomal location of *LcbHLH* members were obtained from annotated files in the Liriodendron genomic database, while the distribution of *LcbHLH* members was plotted using the biological software TBtools [28]. The gene replication events were analyzed according to the following three standard definitions: (1) the length of one shorter sequence is greater than 70% of that of the other longer sequence; (2) the similarity between the two sequences is greater than 70%; (3) two genes separated by five or fewer genes in a 100 kb chromosome segment are considered as tandem repeat genes [29]. To analyze the collinearity correlation between *LcbHLH*s and *bHLH*s in other species, the genome data of Arabidopsis and rice were downloaded from Ensemble (http://plants.ensembl.org/index.html (accessed on 13 November 2021)). The multicollinearity scanning tool MCsanX was employed to compare the whole genome sequence of Liriodendron with that of Arabidopsis and rice, respectively [30]. The visualization of chromosome distribution was obtained through the Circos in TBtools. The ratio of Ka/Ks was calculated by using KaKs_calculator to acquire the natural purification selection between target gene pairs [31].

2.4. Analysis of Gene Structure, Conserved Motifs and Cis-Regulation Elements of LcbHLHs

TBtools software was adopted to map the gene structure of *LcbHLH* members onto a diagram. MEME was used to predict and analyze the conservative motif of the *bHLH* protein in *L. chinense*. Cis-regulation elements of *LcbHLH* members were predicted by the software Plantcare (http://bioinformatics.psb.ugent.be/webtools/plantcare/html/ (accessed on 25 November 2021)) and plotted by TBtools.

2.5. Analysis of Protein Interaction among LcbHLHs

The protein interaction network was generated using the STRING (www.string-db.org (accessed on 3 December 2021)) based on the high homology between *LcbHLH*s and *AtbHLH*s proteins. In addition, six *LcbHLH* proteins with high homology to *AtbHLH* were selected to map the extrafamilial protein interaction network using Cytoscape 3.8.2 [32].

2.6. Three-Dimensional Structure Modeling and Verification of bHLH Protein

The full-length atomic structures of *LcbHLH*24, *LcbHLH*72, and *At*ICE1 proteins were constructed based on the synthesis method on the Robetta online website. Homologous modeling was used for proteins with the sequence matching model, while the threading method was used for proteins with the sequence non-matching model. Then, the sequence was assembled to construct the protein structure. The reliability of their protein structures was further confirmed by ERRAT, PROVE, and Ramachandran on the online website Savesv6.0. VMD software was used for 3D modeling.

2.7. Expression Analysis of LcbHLHs in Response to Cold Stress by RNA-seq and qRT-PCR

The somatic embryo-regenerated seedlings of hybrid Liriodendron with consistent growth were cultured in an incubator (23 °C, 16 h light, and 8 h dark) and then treated at 4 °C. Seedling leaves were sampled at 0 h, 6 h, 1 day, and 3 days with three biological replicates. The collected leaves were quickly frozen in the liquid nitrogen and put in a −80 °C refrigerator for storage. Transcriptome sequencing was performed on the above samples. Transcriptome data of *LcbHLH* members were extracted from the sequencing results. The expression levels of each member at each period of cold stress treatment (the maximum expression value of each *LcbHLH* gene was set to 1, and then the expression values of the gene at other stress and growth stages were normalized to the maximum expression value) were normalized and displayed on the heatmap. The expression patterns of ten *LcbHLH* members were determined by quantitative RT-PCR analysis (qRT-PCR). The qRT-PCR was performed using SYBR-green in the Roche Light Cycler®480 real-time PCR system (Switzerland, Sweden). The relative expression abundance of *LcbHLH* was calculated with the ΔΔCT method. 18s rRNA was used as the internal reference. All qRT-PCR primers were designed by Primer5.0 and were listed in Table S1.

3. Results

3.1. Identification and Physiochemical Characteristics of LcbHLHs

Based on the search of the conserved bHLH domain (Pfam number: PF00010), 91 *LcbHLH* family members were recognized after further validation in the conserved domain database (CDD) and Pfam database. They were renamed as *Lc*bHLH 1~91 based on their chromosomal position. The physical and chemical properties of *Lc*bHLH members were computed. Analysis of the hydrophilicity value of all *Lc*bHLH proteins showed a negative total average value that ranged from −0.816 (*LcbHLH*47) to −0.143 (*LcbHLH*68), concluding that *Lc*bHLH proteins are hydrophilic. The Molecular weight (Mw) of *Lc*bHLH proteins ranged from 10.19 kD (*LcbHLH*15) to 88.40 kD (*LcbHLH*50), the majority (61%) were in the range of 21.41 kD to 48.85 kD, and the molecular weight of 24 members (about 26%) was in the range of 20 kD to 30 kD. Additionally, the theoretical isoelectric points (pI) of *Lc*bHLH proteins ranged from 4.59 (*LcbHLH*81) to 9.91 (*LcbHLH*53). Most *Lc*bHLH proteins (about 63%) were less than 7, and about 30% of *LcbHLH* proteins had a pI between 6 and 7 (Table S2).

3.2. Phylogenetic Characteristics of LcbHLHs

To fully comprehend the evolutionary relationship of the identified *Lc*bHLH protein sequences, *L. Chinense (Lc)*, *Arabidopsis thaliana* (*At*), *Oryza sativa* (*Os*), and *Populus. trichocarpa* (*Pt*), *bHLH* gene families were further compared and subjected in phylogenetic tree analysis (Figure 1A and Figure S1). A total of 581 bHLH protein sequences were obtained and divided into 31 groups, which were identified as evolutionary branches with high bootstrap values. Among the 31 subfamilies, 26 subfamilies were presented in all four species, signifying that the genes of these subfamilies had high homology in the four species and strong phylogenetic conservatism. Some *LcbHLH* genes in Arabidopsis and rice were clustered in the same subfamily. *LcbHLH* proteins were clustered in 29 subfamilies and an orphan sequence was observed. Subfamily 13 was clustered with *LcbHLH*14, *LcbHLH*15, *AtPRE*1/2/3/4/5, and *AtKDR*. Subfamily 17 was clustered with *LcbHLH*24, *LcbHLH*31, *LcbHLH*82, *LcbHLH*18, *At033SCRM*, and *At116ICE1* (Table S3). Subfamily 25 was clustered with *LcbHLH*16, *LcbHLH*69, *LcbHLH*78, *At*SPCH, and OsSPC1/2. Additionally, subfamily 6 was only found in Poplar, indicating individual evolution and functional diversity of Poplar (Figure 1B).

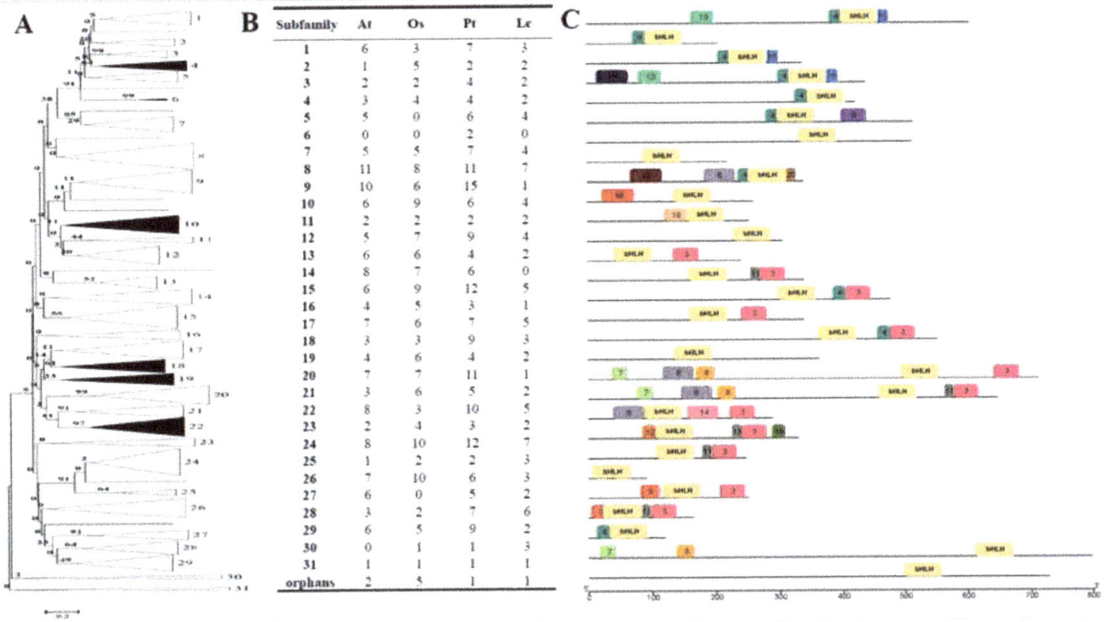

Subfamily	At	Os	Pt	Lc
1	6	3	7	3
2	1	5	2	2
3	2	2	4	2
4	3	4	4	2
5	5	0	6	4
6	0	0	2	0
7	5	5	7	4
8	11	8	11	7
9	10	6	15	1
10	6	9	6	4
11	2	2	2	2
12	5	7	9	4
13	6	6	4	2
14	8	7	6	0
15	6	9	12	5
16	4	5	3	1
17	7	6	7	5
18	3	3	9	3
19	4	6	4	2
20	7	7	11	1
21	3	6	5	2
22	8	3	10	5
23	2	4	3	2
24	8	10	12	7
25	1	2	2	3
26	7	10	6	3
27	6	0	5	2
28	3	2	7	6
29	6	5	9	2
30	0	1	1	3
31	1	1	1	1
orphans	2	5	1	1

Figure 1. Phylogenetic tree of four species proteins. (**A**) The phylogenetic tree of four species; *Liriodendron* (*Lc*), Rice (*Os*), *Arabidopsis* (*At*), *Poplar* (*Pt*). The branches with a bootstrap value greater than 50 were represented by black triangles, while those with a bootstrap value less than 50 were represented by white triangles, which are divided into 31 subfamilies. (**B**) Summary of each group plant-species member representation in phylogeny analysis, plant species, (*At*) Arabidopsis, (*Os*) Rice, (*Pt*) Poplar, and (*Lc*) Liriodendron, group presentation denoted relative to their group marked as subfamily. Orphan genes are shown in the bottom column denoted orphans. (**C**) The motif patterns of *Lc*bHLH subfamilies, showing the bHLH domain present in all protein sequence analysed and other motif.

3.3. Gene Structure and Conserved Motifs of LcbHLHs

Gene structure prediction plays an animated role in studying the evolution of gene family members. To further explore the phylogenetic relationships within the *LcbHLH* members, the intron/exon structures of the *LcbHLH* gene were analyzed based on the genomic annotation files of 91 *LcbHLH* members in combination with phylogenetic tree (Figure 2A). The number of introns in the *LcbHLH* gene ranged from 1 to 11. *LcbHLH* genes were clustered together by parallel exon/intron patterns in exon length and intron number (Figure 2B).

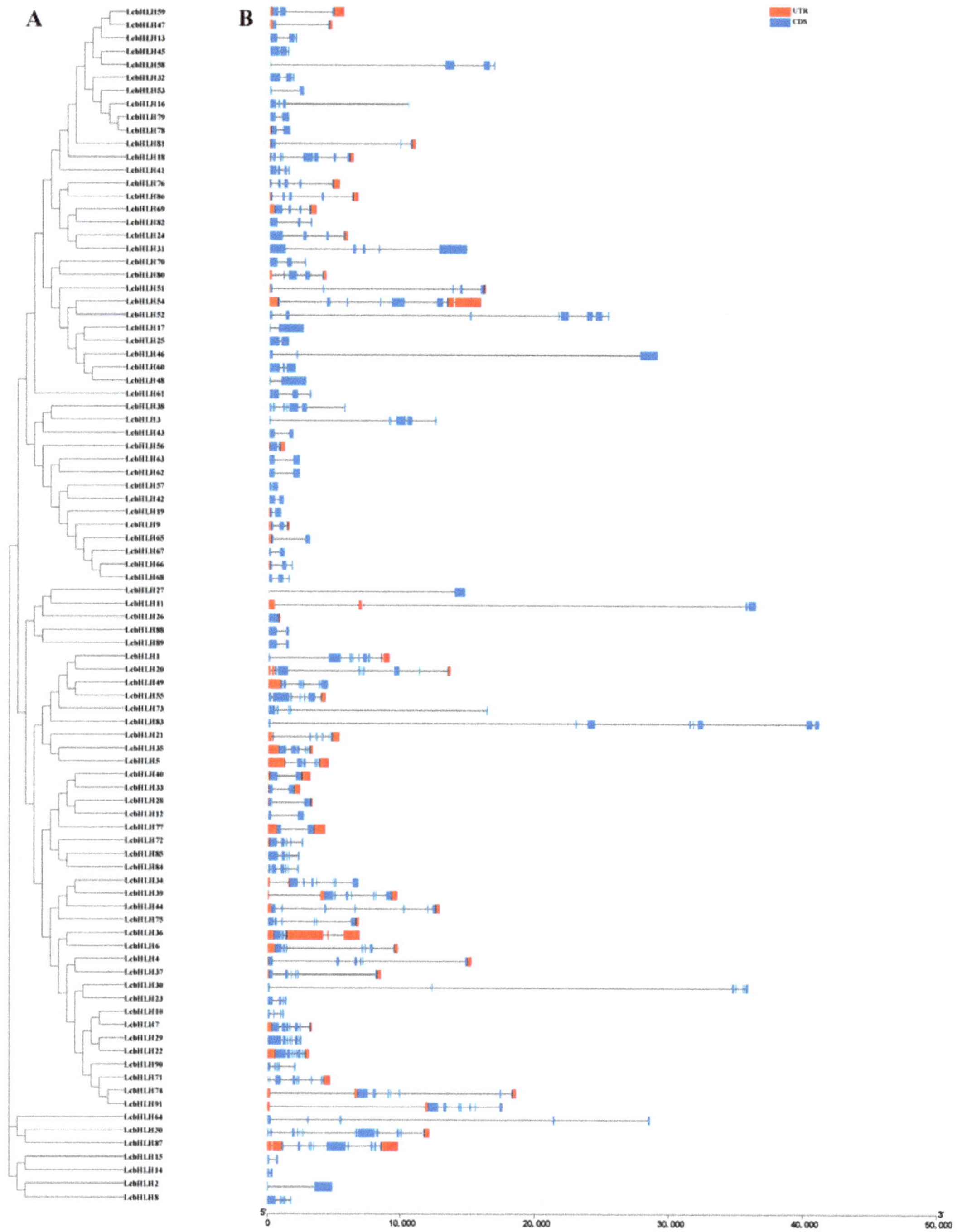

Figure 2. Phylogenetic relationships and exon/intron structures of LcbHLH protein. (**A**) The phylogenetic tree of LcbHLH protein. (**B**) Exon/intron structure analysis of LcbHLHs. Blue boxes represent CDS, red boxes represent UTR, and gray lines represent introns. The size of exons and introns can be estimated by the scale at the bottom.

In this study, the configuration of the *LcbHLH* conservative motif was discovered through the protein conservative theme sites predicted by online software MEME (Figure 1C and Table S4). *bHLH* conserved domain was constituted by motif 1 and motif 2 (Figure 1C). The meticulously connected *LcbHLH* proteins on immediate evolutionary branches of the phylogenetic tree had the same or comparable motif structures. Moreover, there were significant differences between dissimilar subfamilies, suggesting that members of the identical subfamily of *bHLHs* might play related roles in *L. chinense*. Seven subfamilies shared motif 11, eleven subfamilies shared motif 3, and nine subfamilies shared motif 4. Motif 19 only occurred in subfamily 4, motif 17 and motif 20 only occurred in subfamily 10, motif 16 only occurred in subfamily 11, motif 10 only occurred in subfamily 12.

3.4. *Cis-Regulation Elements of LcbHLHs*

The cis-regulatory element plays an imperative role in regulating the expression of stress response genes [33]. The presence of the cis-elements of the *LcbHLH* members in the promoter region (2000 bp upstream of the transcription initiation site) was predicted. Twenty-five typical elements with relatively robust functions were divided into three groups shown in Figure 3. Based on the functional annotations, cis-elements were categorized into three major classes: plant growth and development, phytohormone responsive, and abiotic and biotic stresses (Figure 3). Our findings showed that G-Box and ABRE were the most represented transcription factors in the *LcbHLH* gene family. Specifically, *LcbHLH7* had the most representation of G-Box and ABRE. 67 *LcbHLH* members had elements responsive to the methyl Jasmonate, including CGTCA-motif and TGACG-motif. Fifty-four members had gibberellin-responsive elements, including P-box and GARE-motif. Seventy-two members had salicylic acid responsiveness elements, TCA-element. Moreover, 45 members had auxin-responsive elements, including AuxRR-core and TGA-element. 52 *LcbHLHs* contained LTR elements that might be interrelated to the cold stress response of *L. chinense*.

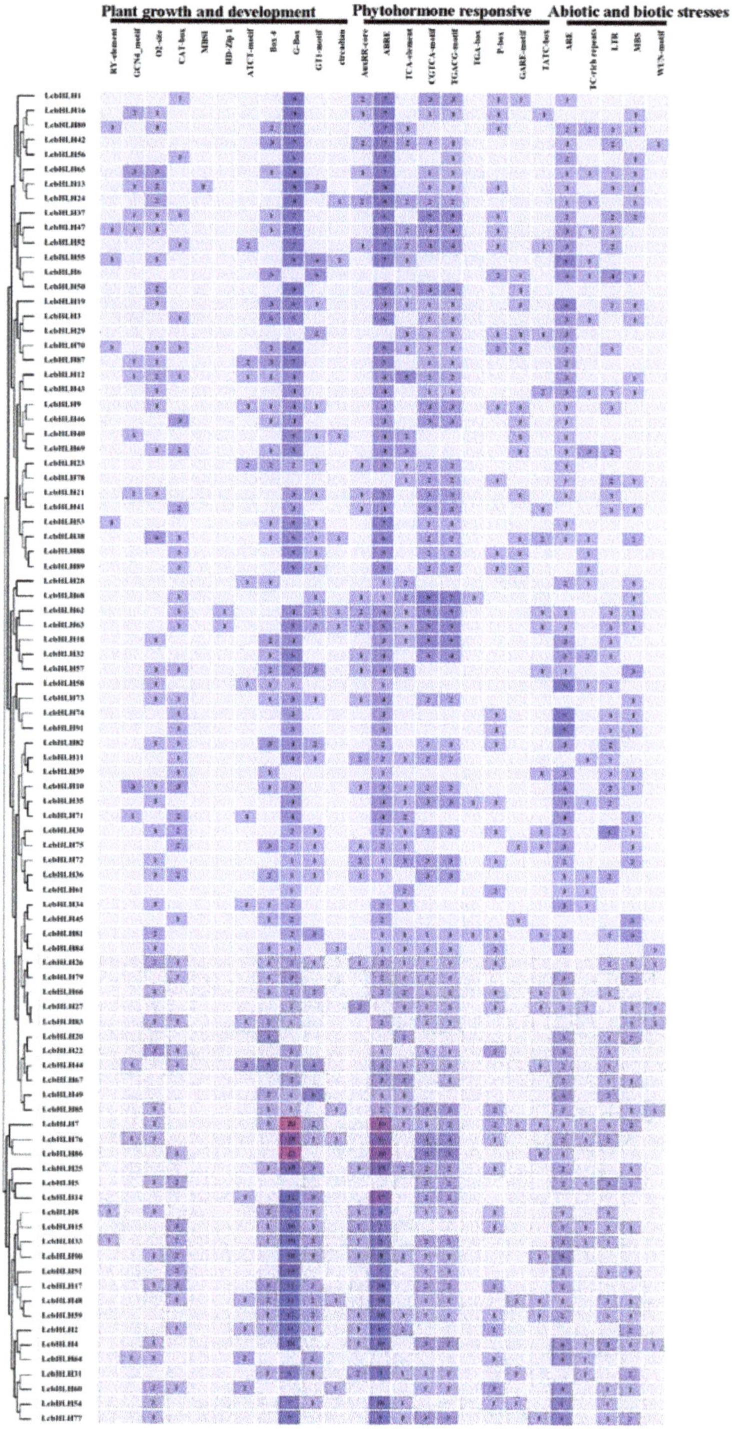

Figure 3. Cis-regulatory elements in the promoters of LcbHLHs.

3.5. Intergenomic Collinearity and Gene Replication of LcbHLHs

Amongst 91 *LcbHLH* genes, 89 were distributed on 19 chromosomes, and the other two were assigned to unassembled genomic contigs (Figure 4). The number of *LcbHLH* genes on each chromosome ranged from 1 to 9.

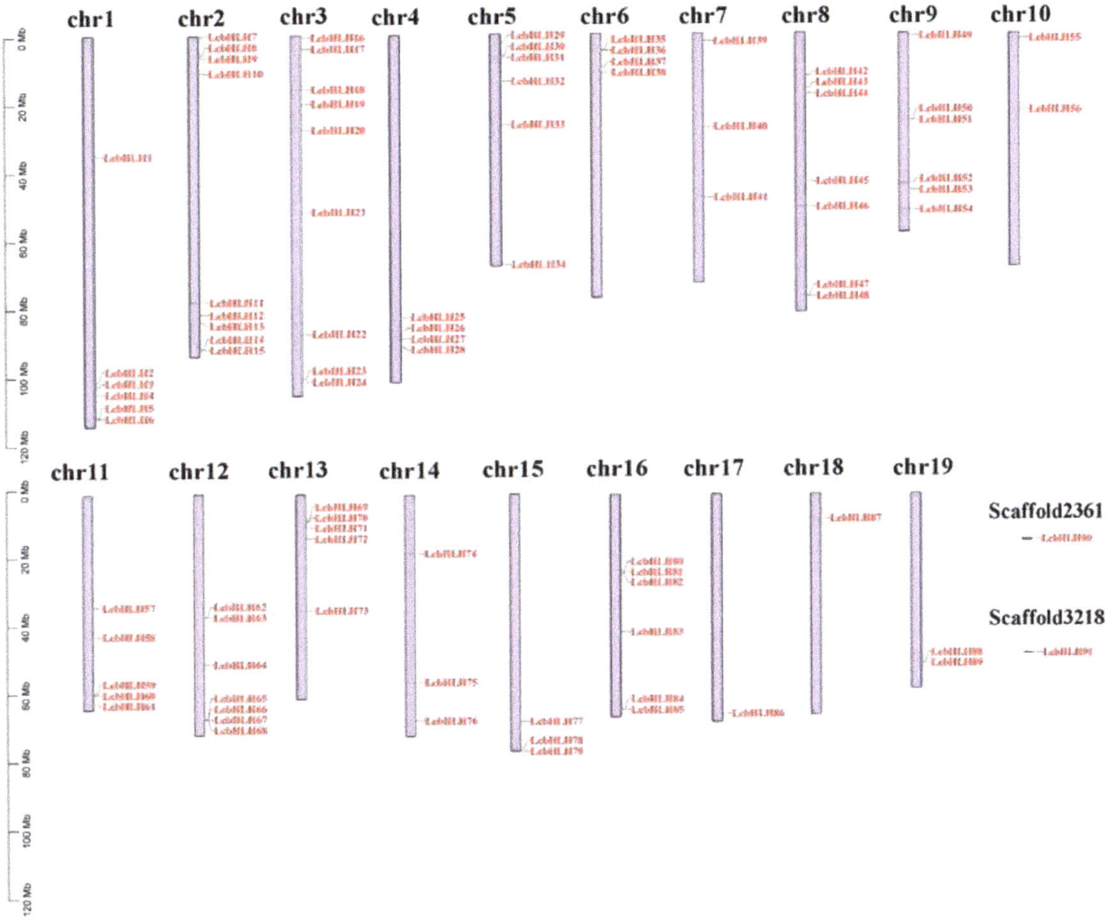

Figure 4. Chromosome distribution of *LcbHLH* gene. Ninety-one genes were labeled on 19 chromosomes and two scaffolds. Positional information for each *LcbHLH* gene is displayed on each chromosome (chr). The left scale represents the length of the chromosome.

The analysis of genome-wide replication, fragment replication, and tandem replication of gene family has a significant role in explaining the process of gene family expansion. In this analysis, intraspecies comparisons of *L. chinense* and *A. thaliana*, *L. chinense*, and rice were implemented at the genome-wide level (Figure 5). A total of 24 pairs of replication genes were found in the *LcbHLH* family, and 21 pairs of gene clusters with high similarity were institute in *LcbHLHs* (Figure 5A). For example, the protein sequences of *LcbHLH88* and *LcbHLH89* shared 99.23% resemblance. The similarity between *LcbHLH63* and *LcbHLH62* was 99.65%, respectively.

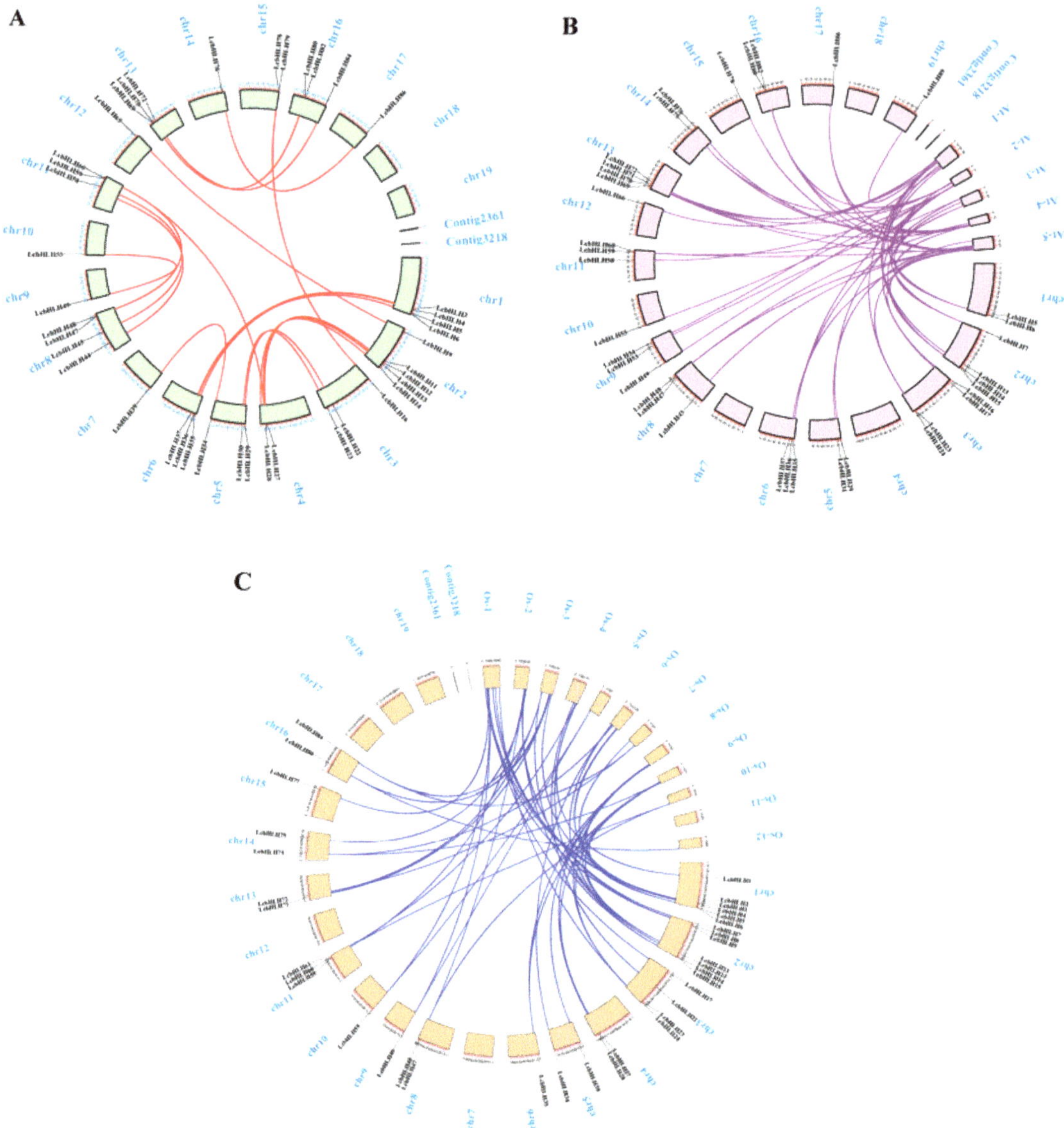

Figure 5. Fragment replication and chromosome distribution of *bHLH* genes in *Liriodendron chinense*. (**A**) Nineteen chromosomes were represented by green segments, red lines connected with homologous genes. (**B**) Collinearity analysis of *Liriodendron chinense* and *Arabidopsis thaliana*; (**C**) Collinearity analysis of Liriodendron chinense and Rice. The gene pairs between them are represented by purple lines and blue lines respectively.

Additionally, tandem repeat genes comprised the same number of exons due to closely related imitation associations. The tandem repeat genes *LcbHLH*14 and *LcbHLH*15 and *LcbHLH*62 and *LcbHLH*63 had a similar two exon and intron-exon structure pattern. Likewise, *LcbHLH*84 and *LcbHLH*85 had a similar intron structure pattern. Remarkably, as revealed in Figure 5A, there were four pairs of fragment-repetitive genes: *LcbHLH*3, *LcbHLH*4 and *LcbHLH*36, and *LcbHLH*37; *LcbHLH*12, *LcbHLH*13 and *LcbHLH*27, and *LcbHLH*28;

LcbHLH69, *LcbHLH70* and *LcbHLH80*, and *LcbHLH82*; *LcbHLH47*, *LcbHLH48* and *LcbHLH59*, and *LcbHLH60*. Together, these results show that the *LcbHLH* gene family was amplified by fragment replication and tandem replication of the *LcbHLH* genes.

The tandem repeated *bHLH* gene has a related gene structure, motif composition, and expression. The tandem repeated and intra-and inter-chromosome repeated regions of *LcbHLH* members were examined in the present study. Our results showed that greater than 38% (15 tandem and 22 fragment-repeat genes) of the *LcbHLHs* might have evolved from some genomic replication event. The substitution rate (Ka/Ks) between nonsynonymous and synonymous was an operative quantity of selection pressure after gene replication [34]. Consequently, the Ka/Ks of the *LcbHLH* repeat gene was premeditated (Table S5). For all tandem repeat pairs, the Ka/Ks values were well below one, which indicated that there were purification options during amplification. Besides, for gene pairs with fragment repeats, all Ka/Ks were less than one, indicating that there was strong purification selection pressure during evolution.

With genome-wide comparison and analysis of *L. chinense*, *A. thaliana*, and rice, it was established that most *LcbHLHs* were positively homologous in rice and *A. thaliana* (54% and 60%), respectively (Figure 5B,C, Tables S6 and S7). The Ka/Ks ratios of *L. chinense* to rice and *A. thaliana* were 0.175 and 0.186, respectively. These results indicate that *bHLH* gene pairs underwent strong purification selection and that there was a close correlation between them before. In brief, gene replication events, including tandem and fragment repeats, appeared to be essential for the expansion of the *bHLH* gene family in Liriodendron, as well as for the functional preservation and differentiation.

3.6. Protein Interaction Network of bHLHs

Diverse *bHLH* proteins bind to specific DNA and regulate the downstream target's transcription by forming homodimer or heterodimer mediated by their α-helix near the N-terminal [10]. Hence, protein interaction analysis is essential to fully review the function of *LcbHLH* proteins (Figure 6). It can be speculated that *LcbHLHs* might have played a role in forming protein complexes and attempted to construct an interaction network of *LcbHLHs*. In this current study, the interaction network within the *LcbHLH* gene family was constructed based on the orthogonal analysis of *AtbHLHs* (Figure 6A and Table S8). The protein interaction network indicated that most *LcbHLH* proteins could interact with more than one *bHLH* protein. More than a quarter of *LcbHLH* proteins can interact with four or more other *bHLH* proteins. Numerous imperious interactions were predicted, such as how *CIB1* (*LcbHLH7*) can participate in the regulation of flowering time [35]. *ICE1* (*LcbHLH24*, 31) interacts with *FMA* (*LcbHLH32*), *SPCH* (*LcbHLH78*, 79, 16) and *MUTE* (*LcbHLH53*) to regulate stomatal diversity [34]. *LRL1* (*LcbHLH75*) and *RDH6* (*LcbHLH8*) can interact with *RSL2* (*LcbHLH85* and 86) and contribute to the regulation of root hair development. These protein interaction networks further ascertained that the *LcbHLH* genes exerted their diverse biological functions through interaction and coordination with other members.

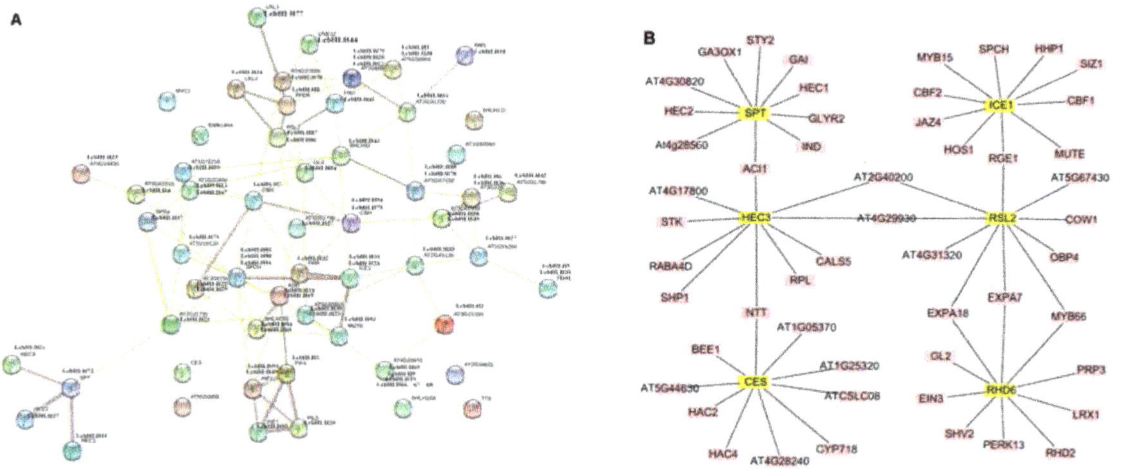

Figure 6. Prediction of LcbHLH protein interaction network based on Arabidopsis orthologs. (**A**) The protein interaction analysis in the LcbHLH family is predicted according to the homology with Arabidopsis thaliana by using a string online website, and the name of LcbHLH protein is marked next to Arabidopsis thaliana orthologous. (**B**) With Cytoscape software, six LcbHLH proteins with high gene homology with Arabidopsis thaliana were predicted and analyzed for extracellular protein interaction prediction according to String website.

3.7. Structural Modeling of LcbHLH Protein

The *bHLH* transcription factor family plays a vital role in plant response to abiotic stress by forming dimer and its helical structure [36]. *ICE*, one of the *bHLH* families, activates CBF via transcription and persuades its expression, playing a central role in cold response and signal transcription [16,37–41]. The amino acid sequence of *LcbHLH*24 in *L. chinense* is extremely homologous to that of *ICE*1 in *A. thaliana*. For that reason, this research predicted that these two protein structures, *LcbHLH*24 and *LcbHLH*72 (homologous gene of *AtRSL*2), interacted with *LcbHLH*24 through *RGE*1 in the protein network (Figure 7A). The structure of *LcbHLH*24 consisted of 14 α-helices and 19 loops (Figure 7A), and the model of *LcbHLH*72 had ten α-helices and eight loops (Figure 7B). The three-dimensional structure of *AtICE*1 protein consisted of 14 α-helices and 12 loops (Figure 7C).

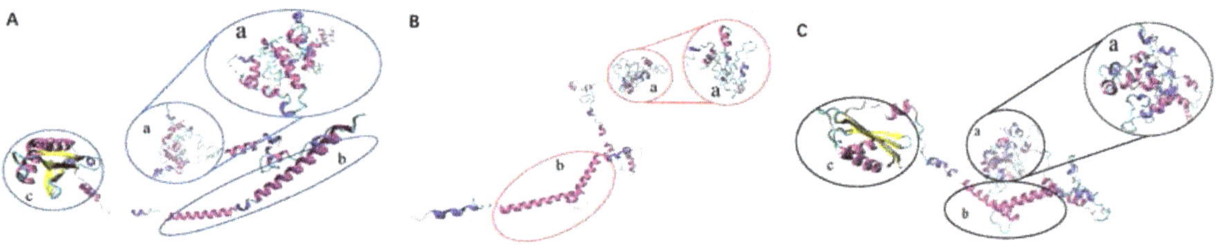

Figure 7. Three-dimensional structure of *bHLH* protein. a, b and c represent same protein regions in three different protein structures, respectively. (**A**) Three-dimensional structure of the protein of *LcbHLH*24; (**B**) Three-dimensional structure of the protein of *LcbHLH*72; (**C**) Three-dimensional structure of the protein of *At*ICE1.

In the 3D model of *LcbHLH*24, the structural model could be roughly divided into three regions, exposed as a, b, and c. *LcbHLH*72 could be divided into two regions, designated as a and b. Three structural regions could be found in *At*ICE1, in which region b was similar to the structure of the other two proteins. Nevertheless, region a of *LcbHLH*24 and *At*ICE1 is a little richer than that of *LcbHLH*72. According to the homology modeling of SWISS-MODEL and the prediction of the conserved domain of NCBI (CDD), region b is the *bHLH* conserved domain of three proteins. The conserved structural region b of *LcbHLH*24 and *AtICE*1 was predicted by SWISS-MODEL to have the domain characteristics of the MYC2 subfamily. Alternatively, region b of *LcbHLH*72 showed high consistency with MITF/CLEAR box structure. Interestingly, special structural region Berninger c was only identified in *LcbHLH*24 and *AtICE*1, and region c in *LcbHLH*24 was almost identical to *At*ICE1. In summary, comparative analysis of *LcbHLH*24 and *AtICE*1 protein sequences, region c is a highly conservative Zipper domain.

3.8. Cold Stress-Induced Expression Pattern of LcbHLHs

The expression patterns of *LcbHLHs* under cold stress in transcriptome data were analyzed (Figure 8) to understand the responses of *LcbHLHs* to cold stress, and 78 *LcbHLH* genes were examined to express in the seedling leaves of *L. chinense*. During the cold stress treatment, the expression patterns of *LcbHLH* members were coarsely defined by constant up-regulations and down-regulations (Figure 8). The expression patterns under the cold treatment of 20 *LcbHLH* genes (22.2%) showed a constant up-regulation trend, 15 *LcbHLH* genes (16.7%) were incessantly down-regulated; 28 of the total *LcbHLHs* (31.1%) were up-regulated and then subsequently down-regulated with the extension of cold treatment time, and only four genes (4%) showed the down and then increased trends.

To further verify the expression pattern of *LcbHLHs* under cold stress, ten *LcbHLHs* (*LcbHLH*9, 20, 24, 35, 39, 54, 55, 56, 69, 86) were chosen to quantify the expression abundance in *L. chinense* by qRT-PCR. As shown in Figure 8B, the expression trends of these ten genes were almost consistent with their transcriptomic patterns. Three *LcbHLH* genes (*LcbHLH*35, 55, 86) showed an up-regulation trend in response to cold stress, six *LcbHLH* genes (*LcbHLH*9, 20, 39, 54, 56, 69) displayed a down-regulation trend, and the expression profile of one *LcbHLH* gene (*LcbHLH*24) was up-regulated at 1d and then down-regulated at 3d.

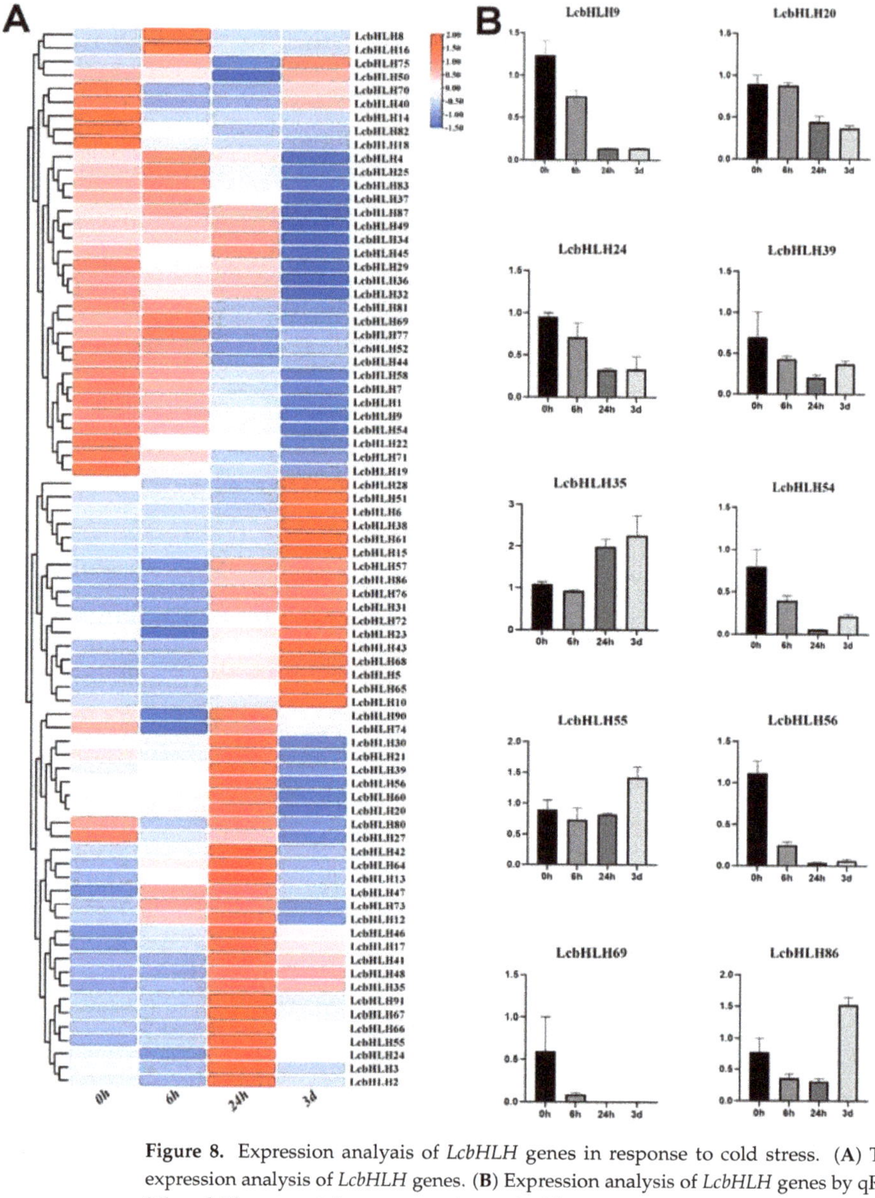

Figure 8. Expression analyais of *LcbHLH* genes in response to cold stress. (**A**) Transcriptomic expression analysis of *LcbHLH* genes. (**B**) Expression analysis of *LcbHLH* genes by qRT-PCR. 0h, 6h, 24h and 3d represent the treatment times of cold stress.

4. Discussion

Given the significant character and diverse functions in biological processes, the *bHLH* transcription factors have attracted more and more attention in recent years [21,42,43]. In this current study, members of the *bHLH* family identified from the genome of *L. chinense* had analogous structural characteristics to those of other species, especially the *bHLH* domain. That was highly conservative with 19 amino acid residues, of which five were base regions, five were distributed in the first helix, one in the loop, and eight in the second helix [44]. However, typical conserved sites were found in the domain of the *L. chinense bHLH* gene family, like the *AtbHLH* families. This indicated that *LcbHLH*s might have DNA-binding activity like that of *A. thaliana* [45].

We constructed a phylogenetic tree to better understand the evolutionary relationship of *bHLH* gene families between different species, *L. chinense, A. thaliana,* rice, and poplar. Interestingly, genes with the same functions were clustered into the same clade. For example, *LcbHLH78*, *LcbHLH79* and *AT5G53210* (*AtSPCH*), *Os02g15760* (*OsSPCH2*) and *Os02g33450* (*OsSPCH1*) were clustered into subfamily 25. We used this evolutionary clustering on the same branch to speculate the functional importance of identified *LcbHLH*s. Previous studies in *A. thaliana* have shown that *At*SPCH can regulate the formation of stomata together with *AtMUTE* and *AtFAMA* [46]. In rice, *SPCH* and *MUTE* have also been shown to exhibit the same functional importance in stomatal formation [47]. Hence, it is reasonable to speculate that *LcbHLH78* and *LcbHLH79* are imperative genes regulating the stomatal switch in *L. chinense*. Furthermore, the *LcbHLH24* and *LcbHLH31* were also clustered into the same subfamily (subfamily 17) as *AT1G12860* (*AtICE1*), *AT3G26744* (*AtICE2*), *Os11G32101* (*OsICE1*), and *Os01G0310* (*OsICE2*). *AtICE1* and *AtICE2* are the main transcription factors found in *A. thaliana* responding to low-temperature stress [17]. *OsICE1* can be phosphorylated by *OsMAPK3*, thus enhancing the activation of *OsbHLH* to its target gene *Os*TPP1 in response to low-temperature stress [48]. So, it is reasonable to speculate that *LcbHLH24* and *LcbHLH31* are most likely to participate in the signal transduction of *L. chinense* in response to low-temperature stress.

Similarly, exon-intron patterns and similar conservative motif arrangements are consistent with the subfamily classification. It is known that genes with few or no introns have low levels of expression in plants [49]. However, a gene structure with compact exons may facilitate rapid expression in response to both endogenous and exogenous stimuli [50]. We observed that the exon structures of *LcbHLH5* and *LcbHLH35* were relatively tight, and they both belonged to the subfamily 29. According to the transcriptomic data, the expression of these two genes under low-temperature stress was increasing in response to an increase in the duration of treatment exposure.

Genomic replication events occur throughout plant evolution, often leading to the expansion of gene families [51,52]. Tandem and fragment gene replication events are two major replication patterns common in the evolution of angiosperms [34,53] and play an essential role in gene family extension [51,54]. In the present study, several distinct gene clusters of *LcbHLH*s were distributed in the different chromosomes. Therefore, gene duplication might be an important reason for the large number of *LcbHLH*s. Gene replication is a common phenomenon in many organisms, which can regulate gene expression, improve genetic and environmental adaptability, and serve as a steppingstone in the evolution of new biological functions [55,56]. The relatively strong sequence diversity besides the *bHLH* domain suggests that the *bHLH* family has undergone extensive domain reorganization after gene replication [57]. More than 20 different conserved motifs with different arrangements were found in the *bHLH* family of *L. chinense*. Thus, extensive domain reorganization occurred in the protein structure of the *bHLH* members. This phenomenon implies that the evolutionary position of Liriodendron is difficult to determine accurately [23].

Time-specific expression patterns of genes in plant growth usually reflect variances in biological functions of gene family members and interactions among related pathways [58,59]. In transcriptional expression profiles, the diverse expression patterns of *LcbHLH* genes under cold stress inferred that each *LcbHLH* member might participate in the various cascades of signal transduction in *L. chinense* in response to cold stress. By predicting the cis-regulation elements of these *LcbHLH* genes, we observed regulatory elements responsive to temperature stress, including LTR, TCA, and AT-rich. The low-temperature responsive element LTR, with CCGAC as the core sequence, demonstrated diverse expression patterns under low-temperature stress, suggesting that LTR plays a key role in responding to low-temperature stresses [60,61]. CRT/DRE element is an important low-temperature response element in the *bHLH* family. *CBF* transcription factor can bind to CRT/DRE sequence and induce the expression of the *COR* gene to improve the cold resistance of plants [62,63].

Numerous proteins in the *bHLH* family are intricate in the tolerance to low-temperature stress, and *ICE*1 is a typical transcription factor that can regulate cold-responsive signal transduction in plants [37,64]. Two members (*LcbHLH*24 and *LcbHLH*31) were found to be highly homologous to *AtICE*1 and *AtICE*2 in the genome of *L. chinense*. The expression of *LcbHLH*31 was continuously up-regulated under cold stress, while the expression *LcbHLH*24 was continuously increased during one day but decreased after three-day treatment, but its abundance was still higher than that of the control. This indicated that two genes, *LcbHLH*31, and *LcbHLH*24, participated in the response of *L. chinense* to low-temperature stress. Over the comparative analysis of the protein sequences of *LcbHLH*24 and *AtICE1*, it can be inferred that *LcbHLH*24 has the characteristics of the typical *ICE* gene family, which contains an S-rich region and disulfide bonds. They can preserve the stability of its gene, but not in *LcbHLH*72. Consequently, it can be reasonably inferred that the stability of *LcbHLH*24 protein is stronger than that of *LcbHLH*72. Region c, which is found in the structure of *LcbHLH*24, shares the same characteristic with the structure of Zipper found in *ICE* of *A. thaliana* and other species. It can be expected and assumed that the special zipper protein structure of *LcbHLH*24 may be beneficial for further exploring and analysing the response of the *bHLH* family to low-temperature stress in *L. chinense*.

Protein-protein interaction analysis predicted interacted relationship among *LcbHLH*s, which of them were confirmed by previous reports. *ICE*1 [16], *ICE*2 [65], and *MYB*15 [66] have been recognized as regulatory factors that induce CBF expression. In response to low temperature, *ICE*1 can be sumoylated by SIZ1, thus promoting the binding of *ICE*1 and increasing *CBF*s expression [67]. In addition, *SCRM2* plays an important role in regulating the stomatal development of *SPCH*, *MUTE*, and *FAMA* [36]. Evidence suggests that there may be a relationship between transcriptional regulation of environmental adaptation and stomatal development in plants [68].

5. Conclusions

This comprehensive genome-wide study systematically identified and functionally analyzed the *bHLH* gene family in *L. chinense*. A total of 91 *LcbHLH* family members were identified and divided into 31 subfamilies, which were unevenly distributed on 19 chromosomes of *L. chinense*. The reported gene structures, conservative motifs, and phylogeny further supported the characteristics of the phylogenetic trees. The amplification of the *LcbHLH* gene was due to duplication during evolution, suggesting that this gene family may play an important role in polyploid plants. Cis-regulation elements responding to low temperature were found in the upstream region of the *LcbHLH* gene, which indicated that the *LcbHLH*s might play an important role in response to cold stress. RNA-seq and qRT-PCR analysis showed that members of the *LcbHLH* genes had various expression patterns during cold treatments. These results may contribute to further functional studies of *LcbHLH* genes and may provide gene resources for the genetic improvement of *L. chinense*.

Supplementary Materials: The following supporting information can be downloaded at: https://www.mdpi.com/article/10.3390/f13040518/s1, Figure S1. Members of the bHLH family from four species: Arabidopsis thaliana (blue triangle), rice (red quadrangle), poplar (green circle), and Liriodendron chinense (purple square). The number on the right indicates their grouping; Figure S2. Logo of 10 conservative motifs of LcbHLH. Table S1. Basic protein information of LcbHLH family members. Table S2. The primers used in the qRT-PCR. Table S3. The segmental and tandem duplication events of LcbHLHs. Table S4. The Ka/Ks ratios between *L. chinenese* and *Arabidopsis thaliana*. Table S5. The Ka/Ks ratios between *L. chinenese* and *Oryza sativa*. Table S6. LcbHLH cis-regulation elements. Table S7. Phylogenetic Analysis and Classification of LcbHLH TF Family. Table S8. Detailed information of interaction network of LcbHLHs. Table S9. Detailed information of interaction network of LcbHLHs with other genes.

Author Contributions: Conceptualization, L.Y. and R.L.; methodology and software, R.L., D.L. and M.G.; validation, B.A., D.H. and D.L.; formal analysis, resources, and data curation, Y.L., J.C. and L.Y.; writing—original draft preparation, R.L. and L.Y.; writing—review and editing, B.A., D.H. and D.L.;

visualization, supervision and funding acquisition, Y.L. and J.C. All authors have read and agreed to the published version of the manuscript.

Funding: This research was funded by the National Natural Science Foundation of China (Nos. 31971682, 32071784), the Research Startup Fund for High-Level and High-Educated Talents of Nanjing Forestry University, and the Priority Academic Program Development of Jiangsu Higher Education Institutions (PAPD).

Institutional Review Board Statement: Not applicable.

Informed Consent Statement: Not applicable.

Data Availability Statement: Data is contained within the article or Supplementary Materials. It is also available from the correspondence author (yangliming@njfu.edu.cn).

Conflicts of Interest: The authors declare no conflict of interest.

References

1. Gali, K.K.; Sackville, A.; Tafesse, E.G.; Lachagari, V.B.R.; McPhee, K.; Hybl, M.; Mikić, A.; Smýkal, P.; McGee, R.; Burstin, J.; et al. Genome-Wide Association Mapping for Agronomic and Seed Quality Traits of Field Pea (*Pisum sativum* L.). *Front. Plant Sci.* **2019**, *10*, 1538. [CrossRef] [PubMed]
2. Asim, A.; Gokce Ozturk, Z.N.; Bakhsh, A.; Cayli, I.T.; Aksoy, E.; Caliskan, S.; Caliskan, M.E.; Demirel, U. Individual and combined effect of drought and heat stresses in contrasting potato cultivars over expressing miR172b-3p. *Turk. J. Agric. For.* **2021**, *45*, 651–668. [CrossRef]
3. Ludwig, J.P.; Hull, C.N.; Ludwig, M.E.; Auman, H.J. Food habits and feeding ecology of nesting double-crested cormorants in the upper Great Lakes, 1986–1989. *Jack-Pine Warbler* **1989**, *67*, 115–126.
4. Goossens, J.; Mertens, J.; Goossens, A. Role and functioning of bHLH transcription factors in Jasmonate signalling. *J. Exp. Bot.* **2017**, *68*, 1333–1347. [CrossRef] [PubMed]
5. Buti, S.; Hayes, S.; Pierik, R. The bHLH network underlying plant shade-avoidance. *Physiol. Plant.* **2020**, *169*, 312–324. [CrossRef]
6. Murre, C.; McCaw, P.S.; Baltimore, D. A new DNA binding and dimerization motif in immunoglobulin enhancer binding, daughterless, MyoD, and myc proteins. *Cell* **1989**, *56*, 777–783. [CrossRef]
7. Ferré-D'Amaré, A.R.; Pognonec, P.; Roeder, R.G.; Burley, S.K. Structure and function of the b/HLH/Z domain of USF. *EMBO J.* **1994**, *13*, 180–189. [CrossRef]
8. Heim, M.A.; Jakoby, M.; Werber, M.; Martin, C.; Weisshaar, B.; Bailey, P.C. The basic helix-loop-helix transcription factor family in plants: A genome-wide study of protein structure and functional diversity. *Mol. Biol. Evol.* **2003**, *20*, 735–747. [CrossRef]
9. Feller, A.; Machemer, K.; Braun, E.L.; Grotewold, E. Evolutionary and comparative analysis of MYB and bHLH plant transcription factors. *Plant J.* **2011**, *66*, 94–116. [CrossRef]
10. Toledo-Ortiz, G.; Huq, E.; Quail, P.H. The Arabidopsis basic/helix-loop-helix transcription factor family. *Plant Cell* **2003**, *15*, 1749–1770. [CrossRef]
11. Aggarwal, P.; Das Gupta, M.; Joseph, A.P.; Chatterjee, N.; Srinivasan, N.; Nath, U. Identification of specific DNA binding residues in the TCP family of transcription factors in Arabidopsis. *Plant Cell* **2010**, *22*, 1174–1189. [CrossRef]
12. Huq, E.; Quail, P.H. PIF4, a phytochrome-interacting bHLH factor, functions as a negative regulator of phytochrome B signaling in Arabidopsis. *EMBO J.* **2002**, *21*, 2441–2450. [CrossRef] [PubMed]
13. Ikeda, M.; Fujiwara, S.; Mitsuda, N.; Ohme-Takagi, M. A triantagonistic basic helix-loop-helix system regulates cell elongation in Arabidopsis. *Plant Cell* **2012**, *24*, 4483–4497. [CrossRef] [PubMed]
14. Zhang, Y.; Zhang, M.; Hu, H.; Yang, J.; Cui, J.; Xu, J. Cloning and cold-resistance analyses of CfICE1 gene in Cryptomeria fortunei. *Plant Physiol. Biochem.* **2021**, *162*, 456–467. [CrossRef] [PubMed]
15. Zheng, P.F.; Yang, Y.Y.; Zhang, S.; You, C.X.; Zhang, Z.L.; Hao, Y.J. Identification and functional characterization of MdPIF3 in response to cold and drought stress in Malus domestica. *Plant Cell Tissue Organ Cult. (PCTOC)* **2021**, *144*, 435–447. [CrossRef]
16. Chinnusamy, V.; Ohta, M.; Kanrar, S.; Lee, B.H.; Hong, X.; Agarwal, M.; Zhu, J.K. ICE1: A regulator of cold-induced transcriptome and freezing tolerance in Arabidopsis. *Genes Dev.* **2003**, *17*, 1043–1054. [CrossRef] [PubMed]
17. Fursova, O.V.; Pogorelko, G.V.; Tarasov, V.A. Identification of ICE2, a gene involved in cold acclimation which determines freezing tolerance in Arabidopsis thaliana. *Gene* **2009**, *429*, 98–103. [CrossRef]
18. Feng, X.; Zhao, Q.; Zhao, L.L.; Qiao, Y.; Xie, X.B. The cold-induced basic helix-loop-helix transcription factor gene MdCIbHLH1 encodes an ICE-like protein in apple. *BMC Plant Biol.* **2012**, *12*, 22. [CrossRef]
19. Geng, J.; Wei, T.; Wang, Y.; Huang, X. Overexpression of PtrbHLH, a basic helix-loop-helix transcription factor from Poncirus trifoliata, confers enhanced cold tolerance in pummelo (Citrus grandis) by modulation of H_2O_2 level via regulating a CAT gene. *Tree Physiol.* **2019**, *39*, 2045–2054. [CrossRef]
20. Rushton, P.J.; Bokowiec, M.T.; Han, S.; Zhang, H. Tobacco transcription factors: Novel insights into transcriptional regulation in the Solanaceae. *Plant Physiol.* **2008**, *147*, 280–295. [CrossRef]

21. Carretero-Paulet, L.; Galstyan, A.; Roig-Villanova, I.; Martínez-García, J.F. Genome-wide classification and evolutionary analysis of the bHLH family of transcription factors in Arabidopsis, poplar, rice, moss, and algae. *Plant Physiol.* **2010**, *153*, 1398–1412. [CrossRef] [PubMed]
22. Bailey, P.C.; Martin, C.; Toledo-Ortiz, G.; Quail, P.H. Update on the basic helix-loop-helix transcription factor gene family in Arabidopsis thaliana. *Plant Cell* **2003**, *15*, 2497–2502. [CrossRef] [PubMed]
23. Chen, J.; Guang, X.; Zhao, C.; Wang, P. Liriodendron genome sheds light on angiosperm phylogeny and species-pair differentiation. *Nat. Plants* **2019**, *5*, 18–25. [CrossRef] [PubMed]
24. Hao, Z.; Liu, S.; Hu, L.; Shi, J.; Chen, J. Transcriptome analysis and metabolic profiling reveal the key role of carotenoids in the petal coloration of Liriodendron tulipifera. *Hortic. Res.* **2020**, *7*, 70. [CrossRef] [PubMed]
25. Jin, J.; Tian, F.; Yang, D. PlantTFDB 4.0: Toward a central hub for transcription factors and regulatory interactions in plants. *Nucleic Acids Res.* **2017**, *45*, D1040–D1045. [CrossRef]
26. Saitou, N.; Nei, M. The neighbor-joining method: A new method for reconstructing phylogenetic trees. *Mol. Biol. Evol.* **1987**, *4*, 406–425.
27. Hall, B.G. Building phylogenetic trees from molecular data with MEGA. *Mol. Biol. Evol.* **2013**, *30*, 1229–1235. [CrossRef]
28. Chen, C.; Chen, H.; Zhang, Y. TBtools: An Integrative Toolkit Developed for Interactive Analyses of Big Biological Data. *Mol. Plant* **2020**, *13*, 1194–1202. [CrossRef]
29. Zhu, Y.; Wu, N.; Song, W.; Yin, G. Soybean (Glycine max) expansion gene superfamily origins: Segmental and tandem duplication events followed by divergent selection among subfamilies. *BMC Plant Biol.* **2014**, *14*, 93. [CrossRef]
30. Wang, Y.; Tang, H.; Debarry, J.D. MCScanX: A toolkit for detection and evolutionary analysis of gene synteny and collinearity. *Nucleic Acids Res.* **2012**, *40*, e49. [CrossRef]
31. Zhang, Z.; Li, J.; Zhao, X.Q. KaKs_Calculator: Calculating Ka and Ks through model selection and model averaging. *Genom. Proteom. Bioinform.* **2006**, *4*, 259–263. [CrossRef]
32. Shannon, P.; Markiel, A.; Ozier, O.; Baliga, N.S. Cytoscape: A software environment for integrated models of biomolecular interaction networks. *Genome Res.* **2003**, *13*, 2498–2504. [CrossRef] [PubMed]
33. Wittkopp, P.J.; Kalay, G. Cis-regulatory elements: Molecular mechanisms and evolutionary processes underlying divergence. *Nat. Rev. Genet.* **2011**, *13*, 59–69. [CrossRef] [PubMed]
34. Lynch, M.; Conery, J.S. The evolutionary fate and consequences of duplicate genes. *Science* **2000**, *290*, 1151–1155. [CrossRef]
35. Liu, Y.; Li, X.; Li, K.; Liu, H. Multiple bHLH proteins form heterodimers to mediate CRY2-dependent regulation of flowering-time in Arabidopsis. *PLoS Genet.* **2013**, *9*, e1003861. [CrossRef]
36. Qi, X.; Torii, K.U. Hormonal and environmental signals guiding stomatal development. *BMC Biol.* **2018**, *16*, 21. [CrossRef]
37. Kashyap, P.; Deswal, R. Two ICE isoforms showing differential transcriptional regulation by cold and hormones participate in *Brassica juncea* cold stress signaling. *Gene* **2019**, *695*, 32–41. [CrossRef]
38. Zhou, L.; He, Y.J.; Li, J.; Li, L.Z. An eggplant SmICE1a gene encoding MYC-type ICE1-like transcription factor enhances freezing tolerance in transgenic Arabidopsis thaliana. *Plant Biol.* **2020**, *22*, 450–458. [CrossRef]
39. Verma, R.K.; Kumar, V.; Yadav, S.K.; Kumar, T.S. Overexpression of Arabidopsis ICE1 enhances yield and multiple abiotic stress tolerance in indica rice. *Plant Signal. Behav.* **2020**, *15*, 1814547. [CrossRef]
40. Liu, L.; Duan, L.S.; Zhang, J.C.; Zhang, Z.X. Cucumber (*Cucumis sativus* L.) over-expressing cold-induced transcriptome regulator ICE1 exhibits changed morphological characters and enhances chilling tolerance. *Sci. Hortic.* **2010**, *124*, 29–33. [CrossRef]
41. Lin, Y.; Guo, H.; Liu, C.X. Molecular cloning and expression analysis of cold-resistant transcription factor ICE1 from *Eucalyptus camaldulensis* L. *Plant Physiol. Commun.* **2011**, *47*, 488–494.
42. Song, X.M.; Huang, Z.N.; Duan, W.K.; Ren, J. Genome-wide analysis of the bHLH transcription factor family in Chinese cabbage (*Brassica rapa* ssp. *pekinensis*). *Mol. Genet. Genom.* **2014**, *289*, 77–91. [CrossRef] [PubMed]
43. Niu, X.; Guan, Y.; Chen, S.; Li, H. Genome-wide analysis of basic helix-loop-helix (bHLH) transcription factors in *Brachypodium distachyon*. *BMC Genom.* **2017**, *18*, 619. [CrossRef] [PubMed]
44. Atchley, W.R.; Fitch, W.M. A natural classification of the basic helix-loop-helix class of transcription factors. *Proc. Natl. Acad. Sci. USA* **1997**, *94*, 5172–5176. [CrossRef]
45. Shimizu, T.; Toumoto, A.; Ihara, K.; Shimizu, M. Crystal structure of PHO4 bHLH domain-DNA complex: Flanking base recognition. *EMBO J.* **1997**, *16*, 4689–4697. [CrossRef]
46. Chater, C.; Caine, R.S.; Fleming, A.J.; Gray, J.E. Origins and Evolution of Stomatal Development. *Plant Physiol.* **2017**, *174*, 624–638. [CrossRef]
47. Wu, Z.; Chen, L.; Yu, Q. Multiple transcriptional factors control stomata development in rice. *New Phytol.* **2019**, *223*, 220–232. [CrossRef]
48. Zhang, Z.; Li, J.; Li, F. *OsMAPK3* Phosphorylates *OsbHLH002/OsICE1* and Inhibits Its Ubiquitination to Activate *OsTPP1* and Enhances Rice Chilling Tolerance. *Dev. Cell* **2017**, *43*, 731–743.e5. [CrossRef]
49. Ren, X.Y.; Vorst, O.; Fiers, M. In plants, highly expressed genes are the least compact. *Trends Genet.* **2006**, *22*, 528–532. [CrossRef]
50. Jeffares, D.C.; Penkett, C.J.; Bähler, J. Rapidly regulated genes are intron poor. *Trends Genet.* **2008**, *24*, 375–378. [CrossRef]
51. Cannon, S.B.; Mitra, A.; Baumgarten, A.; Young, N.D. The roles of segmental and tandem gene duplication in the evolution of large gene families in *Arabidopsis thaliana*. *BMC Plant Biol.* **2004**, *4*, 10. [CrossRef] [PubMed]

52. Mehan, M.R.; Freimer, N.B.; Ophoff, R.A. A genome-wide survey of segmental duplications that mediate common human genetic variation of chromosomal architecture. *Hum. Genom.* **2004**, *1*, 335–344. [CrossRef] [PubMed]
53. Vision, T.J.; Brown, D.G.; Tanksley, S.D. The origins of genomic duplications in Arabidopsis. *Science* **2000**, *290*, 2114–2117. [CrossRef] [PubMed]
54. Xie, T.; Chen, C.; Li, C.; Liu, J. Genome-wide investigation of WRKY gene family in pineapple: Evolution and expression profiles during development and stress. *BMC Genom.* **2018**, *19*, 490. [CrossRef] [PubMed]
55. Andersson, D.I.; Hughes, D. Gene amplification and adaptive evolution in bacteria. *Annu. Rev. Genet.* **2009**, *43*, 167–195. [CrossRef] [PubMed]
56. Zhang, J. Evolution by gene duplication: An update. *Trends Ecol. Evol.* **2003**, *18*, 292–298. [CrossRef]
57. Morgenstern, B.; Atchley, W.R. Evolution of bHLH transcription factors: Modular evolution by domain shuffling? *Mol. Biol. Evol.* **1999**, *16*, 1654–1663. [CrossRef]
58. Hu, C.H.; Wei, X.Y.; Yuan, B.; Yao, L.B. Genome-Wide Identification and Functional Analysis of NADPH Oxidase Family Genes in Wheat During Development and Environmental Stress Responses. *Front. Plant Sci.* **2018**, *9*, 906. [CrossRef]
59. Zhao, Y.; Zhou, M.; Xu, K.; Li, J. Integrated transcriptomics and metabolomics analyses provide insights into cold stress response in wheat. *Crop J.* **2019**, *7*, 857–866. [CrossRef]
60. Brown, A.P.; Dunn, M.A.; Goddard, N.J.; Hughes, M.A. Identification of a novel low-temperature-response element in the promoter of the barley (*Hordeum vulgare* L.) gene *blt101.1*. *Planta* **2001**, *213*, 770–780. [CrossRef]
61. Maruyama, K.; Todaka, D.; Mizoi, J.; Yoshida, T.; Kidokoro, S. Identification of cis-acting promoter elements in cold- and dehydration-induced transcriptional pathways in Arabidopsis, rice, and soybean. *DNA Res.* **2012**, *19*, 37–49. [CrossRef] [PubMed]
62. Liu, Q.; Kasuga, M.; Sakuma, Y.; Abe, H. Two transcription factors, DREB1 and DREB2, with an EREBP/AP2 DNA binding domain separate two cellular signal transduction pathways in drought- and low-temperature-responsive gene expression, respectively, in Arabidopsis. *Plant Cell* **1998**, *10*, 1391–1406. [CrossRef] [PubMed]
63. Kasuga, M.; Liu, Q.; Miura, S.; Yamaguchi-Shinozaki, K. Improving plant drought, salt, and freezing tolerance by gene transfer of a single stress-inducible transcription factor. *Nat. Biotechnol.* **1999**, *17*, 287–291. [CrossRef] [PubMed]
64. Hwarari, D.; Guan, Y.; Ahmad, B. ICE-CBF-COR Signaling Cascade and Its Regulation in Plants Responding to Cold Stress. *Int. J. Mol. Sci.* **2022**, *23*, 1549. [CrossRef] [PubMed]
65. Kim, Y.S.; Lee, M.; Lee, H.J.; Park, C.M. The unified ICE-CBF pathway provides a transcriptional feedback control of freezing tolerance during cold acclimation in Arabidopsis. *Plant Mol. Biol.* **2015**, *89*, 187–201. [CrossRef] [PubMed]
66. Agarwal, M.; Hao, Y.; Kapoor, A.; Dong, C.H. A R2R3 type MYB transcription factor is involved in the cold regulation of CBF genes and in acquired freezing tolerance. *J. Biol. Chem.* **2006**, *281*, 37636–37645. [CrossRef] [PubMed]
67. Miura, K.; Jin, J.B.; Lee, J.; Yoo, C.Y. *SIZ1*-mediated sumoylation of *ICE1* controls *CBF3/DREB1A* expression and freezing tolerance in Arabidopsis. *Plant Cell* **2007**, *19*, 1403–1414. [CrossRef]
68. Kanaoka, M.M.; Pillitteri, L.J.; Fujii, H.; Yoshida, Y. *SCREAM/ICE1* and *SCREAM2* specify three cell-state transitional steps leading to arabidopsis stomatal differentiation. *Plant Cell* **2008**, *20*, 1775–1785. [CrossRef]

forests

MDPI

Article

Identification of AP2/ERF Transcription Factor Family Genes and Expression Patterns in Response to Drought Stress in *Pinus massoniana*

Shuang Sun [1,2,†], Xingxing Liang [1,3,4,†], Hu Chen [1,2,3,4], La Hu [1,3] and Zhangqi Yang [1,2,3,4,*]

1 Guangxi Key Laboratory of Superior Timber Trees Resource Cultivation, Guangxi Forestry Research Institute, Nanning 530002, China
2 Masson Pine Engineering Research Center, The State Forestry Administration, Nanning 530002, China
3 Masson Pine Engineering Research Center of Guangxi, Nanning 530002, China
4 Key Laboratory of Central South Fast-Growing Timber Cultivation, Forestry Ministry of China, Nanning 530002, China
* Correspondence: yangzhangqi@163.com
† These authors contributed equally to this work.

Abstract: *Pinus massoniana* Lamb. is found in 17 Chinese provinces and is an important timber tree species in southern China. The current seasonal drought climate is becoming increasingly severe, threatening *P. massoniana* growth and limiting the development of the *P. massoniana* industry. Plant growth, development, and stress were all regulated by AP2/ERF. We identified 124 AP2/ERF transcription factor family members in this study and discovered that all the genes had their own conserved structural domains and that PmAP2/ERFs were divided into 12 subfamilies with high conservation and similarity in gene structure and evolutionary level. Nine *PmAP2/ERF* genes were constitutively expressed under drought treatment, and it was hypothesized that the *PmAP2/ERF96* gene negatively regulated drought stress, *PmAP2/ERF46* and *PmAP2/ERF49* genes showed a positive or negative response to drought in different tissues, while the remaining six genes were positively regulated. The *PmAP2/ERF* genes responded to drought stress following treatment with the exogenous hormones SA, ABA, and MeJA, but the expression patterns differed, with each gene responding to at least one exogenous hormone to induce up-regulation of expression under drought stress, with *PmAP2/ERF11*, *PmAP2/ERF44*, *PmAP2/ERF77*, and *PmAP2/ERF80* genes significantly induced by three hormones. The genes mentioned above may be involved in hormone signaling pathways in response to drought stress. The results indicate that the *PmAP2/ERF* genes may positively or negatively regulate the corresponding signaling pathways in *P. massoniana* to improve drought resistance.

Keywords: *Pinus massoniana* Lamb.; AP2/ERF transcription factor; bioinformatics; drought stress; exogenous hormone; expression pattern

Citation: Sun, S.; Liang, X.; Chen, H.; Hu, L.; Yang, Z. Identification of AP2/ERF Transcription Factor Family Genes and Expression Patterns in Response to Drought Stress in *Pinus massoniana*. *Forests* **2022**, *13*, 1430. https://doi.org/10.3390/f13091430

Academic Editor: Yuepeng Song

Received: 9 July 2022
Accepted: 2 September 2022
Published: 6 September 2022

1. Introduction

Drought, cold, salt, and other abiotic stressors have a significant impact on plant growth and development [1–3]. When plants are subjected to drought stress, they respond with a comprehensive series of physiological and molecular regulatory mechanisms [4,5]. To reduce plant damage, plants regulated osmoregulatory substances, antioxidant defense systems, and endogenous hormone levels to maintain cell morphology and scavenge excess oxygen radicals [4,6,7]. Drought induced the expression of plant-related genes, and the main regulatory genes that responded to drought were transcription factors. Such genes can respond quickly after the plant becomes stressed and form their own regulatory network by regulating downstream genes or collaborating with one another to resist drought stress [8,9].

AP2/ERF (APETALA2/ethylene -responsive factor) is one of the largest families of transcription factors in plants. AP2/ERF genes were firstly identified in *Arabidopsis thaliana* and were associated with flower development [10]. AP2/ERF gene families have been discovered in an increasing number of plant species, and AP2/ERF family genes are more numerous and functionally diverse, involved in physiological and biochemical processes such as growth and development, hormone signaling, and the response to biotic and abiotic stresses in plants [11,12]. The AP2 functional structural domain specific to the AP2/ERF gene consists of 60–70 conserved amino acid residues, which include the YRG and RAYD structural domains; the YRG structure is located at the N-terminus of the AP2 structural domain and consists of about 20 amino acid residues; the YRG structure's role is to allow the AP2/ERF gene to contact DNA and recognize cis-acting elements [13]. A structural domain at the C-terminus with about 40 amino acid residues may participate in transcription factor interactions [13]. The AP2/ERF gene family is divided into four subfamilies based on sequence similarity and the number of AP2/ERF functional structural domains [14]. The AP2 subfamily contains two AP2/ERF structural domains that have been linked to plant flower development [15,16]; the RAV subfamily contains one AP2/ERF structural domain and one B3 structural domain; and the ERF and Soloist subfamilies have one AP2/ERF structural domain each.

Genes of the *AP2/ERF* family are considered to be plant-specific transcription factors. In *A. thaliana* [11] and *Oryza sativa* L. [17], a large number of AP2/ERF genes have been discovered; for instance, *AP2/ERF* genes can improve drought tolerance by specifically binding to downstream genes, for example, *SpERF1* activated and regulated downstream genes Meanwhile, the function of *AP2/ERF* genes in model plants such as *Arabidopsis* and rice has been studied more frequently and intensively, and it has been shown that *AP2/ERF* genes play an important role in the molecular regulation mechanism of drought in transgenic plants to improve drought stress tolerance by binding to DRE/CRT elements in the promoters of drought-related genes *HSP101*, *RD29A*, *P5CS*, and others [18]. By regulating hormone signaling pathways such as abscisic acid (ABA) and jasmonic acid (JA), *AP2/ERF* genes play an important role in plant signaling and improve plant drought resistance [19,20]. Plant drought resistance was also improved by *AP2/ERF* genes, which regulate transpiration, photosynthesis, plant development, and endogenous hormone content [21,22]. In contrast, *AP2/ERF* genes have been relatively little studied in forest trees, due to the lack of genomic and related expression data, etc., in many tree species, and functional studies of *AP2/ERF* genes have also been carried out in recent years in forest trees, and the results showed that *AP2/ERF* genes are involved in the process of phellogen activity/phellem differentiation [23], in the early stage of leaf primordium development [24], in signal transduction such as ethylene [23] and gibberellin [25], in phosphorus stress and drought stress [7,26], etc.

Pinus massoniana Lamb. is distributed in 17 Chinese provinces and is an important timber species with significant economic value in southern China [27,28], as well as a pioneer tree species for afforestation [29]. Seasonal drought is common in southern China, which has a negative impact on *P. massoniana* growth and limited the development of the *P. massoniana* industry [6]. The mechanism of *P. massoniana AP2/ERF* genes in response to drought stress is not clear at the moment, and the mechanism of *P. massoniana*-related transcription factors involved in drought resistance from the molecular level study has rarely been investigated [5,30]. Since genome-wide data are not yet available for *P. massoniana*, the identification of the *AP2/ERF* gene family in *P. massoniana* is a feasible approach. In this study, we conducted gene family identification by identifying *PmAP2/ERF* genes in *P. massoniana*, explored related drought resistance genes, and explored the signaling pathways that the genes may be involved in, to provide a reference for revealing the function of *PmAP2/ERF* genes and drought response mechanism studies in *P. massoniana*.

2. Materials and Methods

2.1. Plant Materials and Treatments

Drought-tolerant line 19–220 seedlings and drought-sensitive line 19–214 seedlings were chosen as experimental materials, and seedlings with good development and consistency were chosen and put in pots (18 cm in diameter and 25 cm in height) with substrate (yellow clay soil:coconut coir = 3:1), one plant per pot, and the experiment was carried out after 1 month of normal cultivation in a greenhouse.

Five treatments were established based on the results of the previous pre-experiment: CK1 normal watering; CK2 drought stress; 50 mg/L salicylic acid (SA), SA + drought stress; 0.5 mmol/L methyl jasmonate (MeJA) + drought stress; and 25 mg/L ABA + drought stress, with three replications of each treatment and ten plants in each replicate. Following three days of continuous appropriate watering, the above concentrations of SA, MeJA, ABA, and distilled water (CK1 and CK2) were sprayed for four days, with 10 mL on the above-ground part (stems and needles) and 10 mL on the below-ground part (roots). Following that, all experimental seedlings were rehydrated and recorded as day 0 of drought stress for ongoing natural drought stress, with CK1 serving as the regular watering control group, which was watered once every two days. At soil drought levels [9] of mild drought (55%–70%), moderate drought (45%–55%), severe drought (30%–45%), and 48 h after rehydration, the needles (middle part of the area with needles), stems (3 cm long in the middle), and roots (2 cm in the apical part of the main and lateral roots) of *P. massoniana* were sampled and stored in a refrigerator at −80 °C.

2.2. Identification of PmAP2/ERFs Gene Family

The *AP2/ERF* genes in *P. massoniana* were identified based on the full-length transcriptome, insect resistance transcriptome [31], lateral branch differentiation transcriptome [25], and drought resistance transcriptome (unpublished) of the previous research group. Sequences from the three transcriptome databases were removed from redundant sequences and annotated by NR, SwissProt, and Pfam databases, and genes annotated as AP2/ERF were extracted, and their nucleic acid sequences and protein sequences were extracted using Perl language. Hidden Markov models of the AP2 structural domain were downloaded from the Pfam database [32] (https://pfam.xfam.org/, accessed on 7 June 2021) and the AP2/ERF genes were identified using HMMER software, while the protein sequences of the *A. thaliana AP2/ERF* genes were downloaded from the PlantTFBD database [33] (http://planttfdb.gao-lab.org/index_ext.php, accessed on 7 June 2021) to download the protein sequences of *A. thaliana AP2/ERF* genes, and the AP2/ERF protein sequences of *A. thaliana* and *P. massoniana* were compared using BLAST software, and the protein sequences obtained by the above two methods were taken as a concatenation. After removing the redundant sequences, NCBI CD Search [34] (https://www.ncbi.nlm.nih.gov/Structure/cdd/wrpsb.cgi, accessed on 8 June 2021), SMART [35] (http://smart.embl.de/, accessed on 8 June 2021), and Pfam were used to determine whether the candidate protein sequences contained AP2/ERF structural domains, and sequences lacking structural domains or containing incomplete structural domains were removed.

2.3. Analysis of Physicochemical Properties of PmAP2/ERFs Proteins

ExPASy [36] (https://web.expasy.org/, accessed on 14 December 2021) was used to predict the physicochemical properties of PmAP2/ERF molecular mass, isoelectric point, etc.; Plant-mPLoc [37] (http://www.csbio.sjtu.edu.cn/bioinf/plant-multi/, accessed on 14 December 2021) and WoLF PSORT [38] (https://psort.hgc.jp/, accessed on 14 December 2021) online sites for subcellular localization prediction analysis; TMHMM Server v. 2.0 (http://www.cbs.dtu.dk/services/TMHMM/, accessed on 14 December 2021) for protein transmembrane structure analysis.

2.4. PmAP2/ERFs Protein Phylogenetic Analysis and Multiple Sequence Alignment

The *A. thaliana* AP2/ERF protein sequences were obtained from the TAIR database [39] and the PlantTFBD database, and the phylogenetic evolutionary trees of *P. massoniana* and *A. thaliana* were generated using MEGA7 software [40] with the following parameters: Using the Neighbor-Joining (NJ) approach, the P-distance model was chosen, and the evolutionary tree was decorated with iTOL [41].

2.5. PmAP2/ERFs Protein Conserved Motif Analysis

The conserved motifs of 124 *P. massoniana* PmAP2/ERFs proteins were evaluated online using the MEME online tool [42] (https://meme-suite.org/meme/, accessed on 26 July 2021), with the following parameters: The expected motif count was ten, and the motif length ranged from 6 to 60 AA.

2.6. Prediction of PmAP2/ERFs Protein Interactions

Using the STRING website [43] (https://string-db.org/cgi, accessed on 27 March 2022) and Cytoscape software [44], potential interaction networks and biological functions between PmAP2/ERFs proteins of *P. massoniana* were predicted based on the AP2/ERF protein analysis of *A. thaliana*.

2.7. RNA-seq Data Analysis of PmAP2/ERF Genes

The expression heat map of *PmAP2/ERF* genes under drought stress was drawn based on the pre-drought transcriptome data. The treatment groups were continuous natural drought stress (D) and the control group (C) was normal watering. The root systems (main and lateral root tips were 2 cm) of seedlings in the treatment and control groups corresponding to the three time points were taken at the 7th d (1), the 8th d before rehydration for 7 h (2), and the 8th d (3) of the drought stress treatment, respectively. Based on the obtained transcriptome data, the expression heat map of *PmAP2/ERFs* gene family under drought stress was drawn using TBtools.

2.8. Expression Analysis of PmAP2/ERFs Gene

RNA extraction was performed using the polyphenol polysaccharide plant RNA extraction kit from Tiangen (Beijing, China), cDNA synthesis was performed using M-MLV reverse transcriptase from Takara (Shanghai, China), and finally, the concentration of all cDNA samples was adjusted to 50 ng/μL. Fluorescent quantitative PCR was performed using the TB Green® *Premix Ex Taq*™ II (Tli RNaseH Plus) kit (Shanghai, China) from Takara Bio, with the reaction system configured according to the instructions; the internal reference genes were *PmUBI4* (tissue-specific internal reference gene) and *PmCYP* (drought stress-treated internal reference gene) [45], and primer information is provided in (Table S1); real-time fluorescence quantitative PCR was performed using a Bio-Rad CFX96 quantitative PCR instrument (San Diego, CA, USA), and each sample was technically repeated three times, and the relative expression of genes was calculated using the $2^{-\Delta\Delta Ct}$ method [46], and the data were analyzed for significant differences using SPSS software (IBM, New York, NY, USA), and finally plotted using GraphPad Prism 8 software (GraphPad Software, San Diego, CA, USA).

3. Results

3.1. Identification and Naming of PmAP2/ERFs Gene Family

Non-redundant sequences were obtained based on the previous full-length transcriptome and annotated as *AP2/ERF* genes, and 453 sequences were identified by HMMER software and local Blast. After removing the redundant sequences, the candidate protein sequence structural domains were analyzed using NCBI CD Search, SMART, and Pfam, and 124 PmAP2/ERF genes of *P. massoniana* were finally identified and obtained (Table S2).

3.2. Analysis of Physicochemical Properties of PmAP2/ERFs Proteins

PmAP2/ERFs physicochemical properties and functional structure analysis showed that PmAP2/ERFs protein encoded 101 (PmAP2/ERF58) to 683 (PmAP2/ERF43) amino acids. The molecular weight of PmAP2/ERFs protein ranged from 11.34 (PmAP2/ERF58) to 76.18 (PmAP2/ERF43) kDa. The theoretical isoelectric points of PmAP2/ERFs proteins ranged from 4.48 (PmAP2/ERF27) to 11.65 (PmAP2/ERF89) with an average pI of 7.50; 61 PmAP2/ERFs proteins had isoelectric points < 7, which were acidic, and 63 PmAP2/ERFs proteins had isoelectric points >7, which were alkaline. A total of 9 PmAP2/ERFs proteins

(PmAP2/ERF35\49\71\74\ 101\110\113\121\123) with instability coefficients less than 40 were stable proteins, while the remaining 115 PmAP2/ERFs proteins with instability coefficients greater than 40 were unstable proteins. The average hydrophobic value of 23 PmAP2/ERFs proteins was greater than −0.5, which were hydrophilic proteins, while the average hydrophobic value of the remaining 101 proteins was less than −0.5, which were hydrophobic. The subcellular localization showed that most of the proteins were localized in the nucleus, and some of them were also distributed in the cytoplasm. It was speculated that the *PmAP2/ERF* genes might play different regulatory roles in different organelles, and none of the *PmAP2/ERFs* had transmembrane structures (Table S2).

3.3. *PmAP2/ERFs Protein Phylogeny and Multiple Sequence Alignment Analysis*

The phylogenetic tree showed that the *AP2/ERF* gene family of *P. massoniana* can be divided into three subfamilies, AP2, RAV, and ERF, and did not contain the Soloist subfamily [11]. Including 8 members of the AP2 subfamily and 9 members of the RAV subfamily, 107 belong to the ERF subfamily. The ERF subfamily was further divided into two subfamilies, DREB and ERF, and in this study the DREB subfamily was divided into subgroups I, II, III, and IV, containing 6, 38, 14, and 4 genes, respectively, for a total of 62 genes; while the ERF subfamily was divided into subgroups V, VI, VII, VIII, IX, and X subfamilies, containing 1, 4, 7, 15, 13, and 5 member genes, respectively, for a total of 45 genes. The AP2 subfamily and the RAV subfamily clustered on one major branch and later on two different minor branches; the DREB and ERF subfamilies clustered on two different major branches, respectively (Figure 1).

The AP2 subfamily genes included two AP2 structural domains (AP2-1 and AP2-2), both containing relatively conserved YRG and RAYD structural domains, with a C-terminal motif deletion in PmAP2/ERF102 in the second AP2 structural domain. The main differences between the ERF and DREB subfamilies were 14 and 19 amino acids of the ERF subfamily were alanine (A) and aspartic acid (D) and 14 and 19 amino acids of the DREB subfamily are valine (V) and glutamic acid (E) [14]. Both the ERF and DREB subfamily genes in the present study also contained the YRG and RAYD structural domains, and most genes were highly conserved in these two structural domains, while a few genes had amino acid residue variants or deletions at positions 14 and 19 (Figure 2).

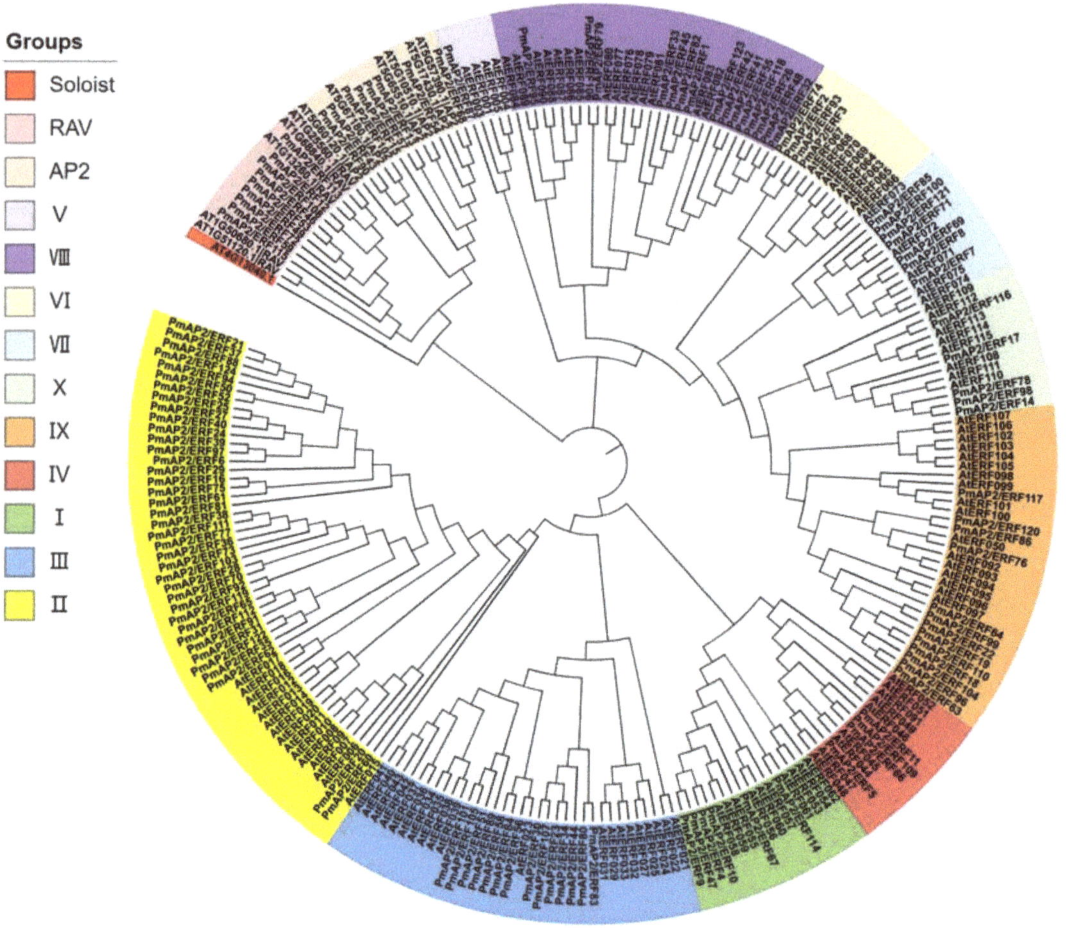

Figure 1. Phylogenetic analysis of PmAP2/ERFs proteins in *P. massoniana*. *A. thaliana* AP2/ERF protein sequences were downloaded from TAIR database and PlantTFBD database, and the phylogenetic evolutionary trees of PmAP2/ERFs of *P. massoniana* and AtERFs of *A. thaliana* were constructed using MEGA7 software and iTOL, and different colors in the figure represent different groupings.

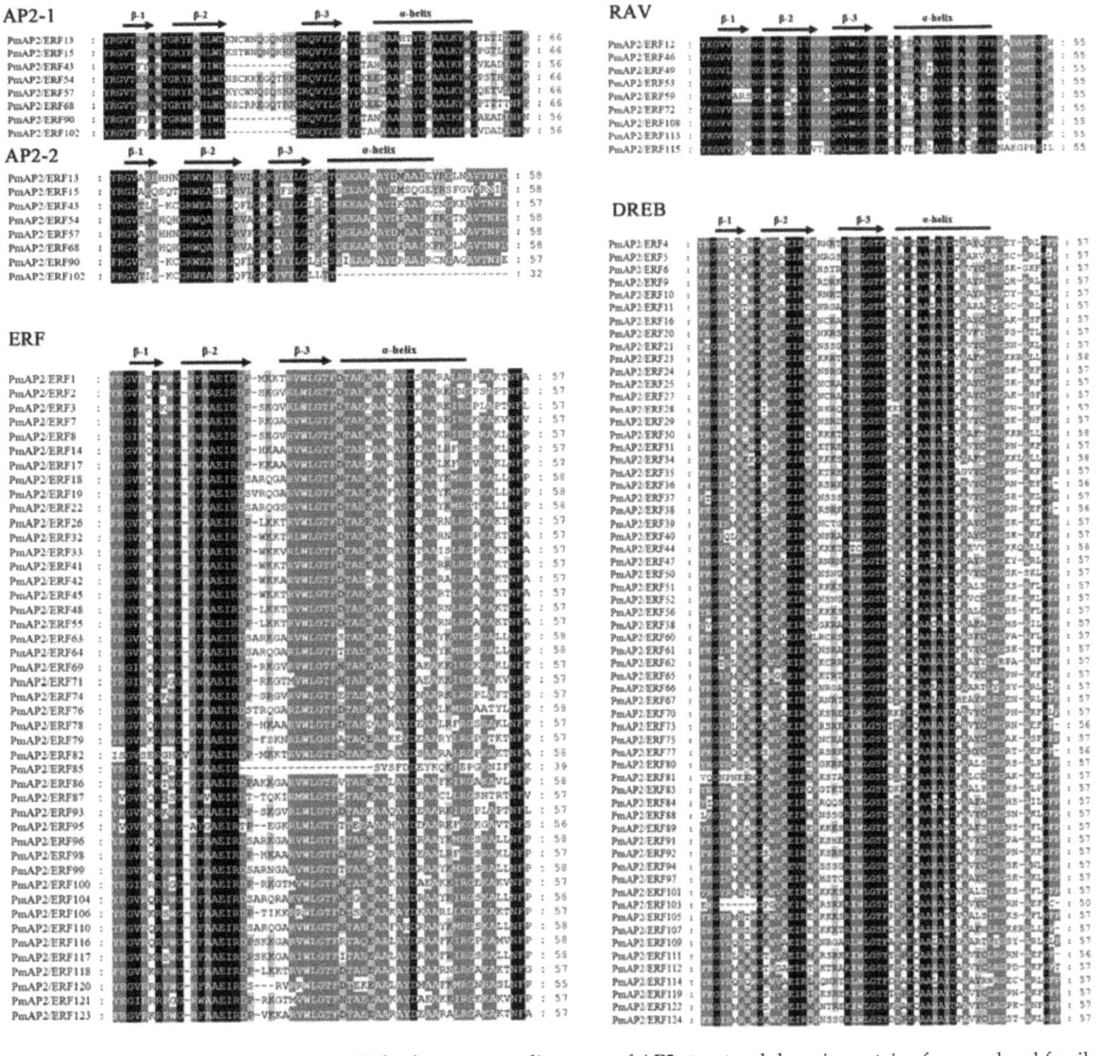

Figure 2. Multiple sequence alignment of AP2 structural domain proteins from each subfamily of PmAP2/ERFs. The sequence comparison results of AP2 structural domain proteins of AP2, RAV, ERF, and DREB subfamilies are shown in the figure, where AP2 subfamily contains two AP2 structural domains, AP2-1 and AP2-2, respectively. Arrows represent β-sheets and horizontal lines represent α-helix.

3.4. Conserved Motif Analysis of PmAP2/ERFs Proteins

The conserved motif analysis of PmAP2/ERFs protein showed that 10 conserved Motifs were obtained, ranging from 29 to 58 amino acids in length. Among them, Motif 8, Motif 1, and Motif 10 formed the first AP2 structural domain of AP2 subfamily, and Motif 6 and Motif 1 formed the second AP2 structural domain. Motifs 6, 1, and 10 formed the AP2 structural domain of RAV subfamily, and Motif 5 and Motif 4 formed the B3 structural domain of RAV subfamily. Motif 2, Motif 1, and Motif 3 formed the AP2 domain of the ERF subfamily. The results showed that the genes of the same subfamily contained basically the same motifs, but there were a few differences, for example, members of the RAV subfamily contained Motif 6, Motif 1, Motif 10, Motif 5, and Motif 4, while PmAP2/ERF115 of the same RAV subfamily contained one less Motif 10 and one more Motif 4. This phenomenon also exists in other subfamilies, which may be due to mutations during protein evolution (Figure 3A,B).

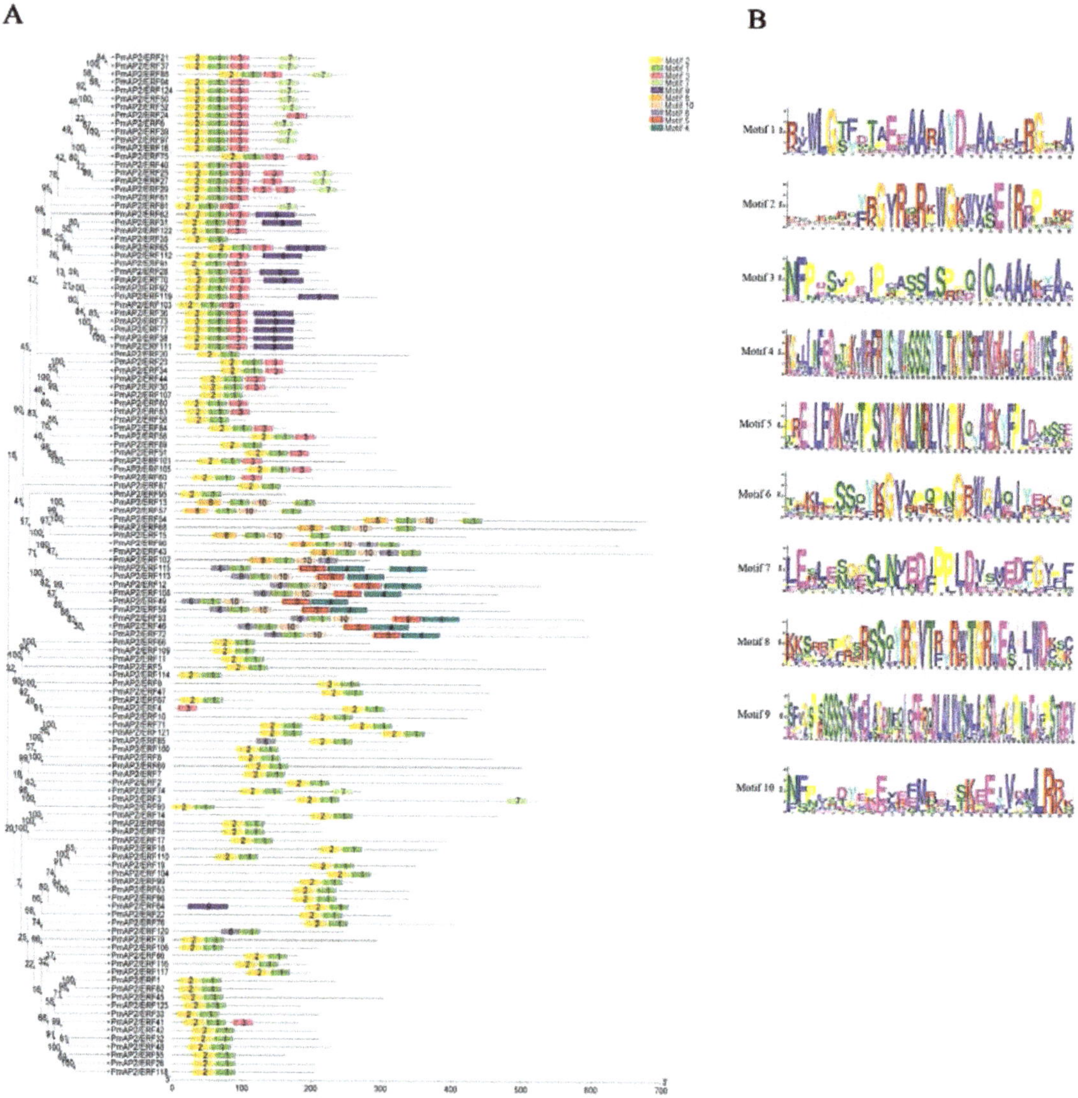

Figure 3. Conserved motif analysis of PmAP2/ERFs proteins of *P. massoniana* and multiple sequence alignment of AP2 structural domain proteins: (**A**): Distribution of conserved motifs of PmAP2/ERFs proteins of *P. massoniana*. A total of 10 conserved Motifs were obtained, indicated by different numbers and colors, and arranged in order. (**B**): Conserved motifs of PmAP2/ERFs proteins of *P. massoniana*. Corresponds to Motif in A.

3.5. Protein Interaction Analysis of PmAP2/ERFs

The results of the protein interaction network map (Figure 4) showed that most of the PmAP2/ERFs could interact with more than one protein, among which there were interactions between PmAP2/ERF102 and PmAP2/ERF109. PmAP2/ERF102 (AT4G36920) and PmAP2/ERF109 (AT5G05410) played a vital role in the overall reciprocal network.

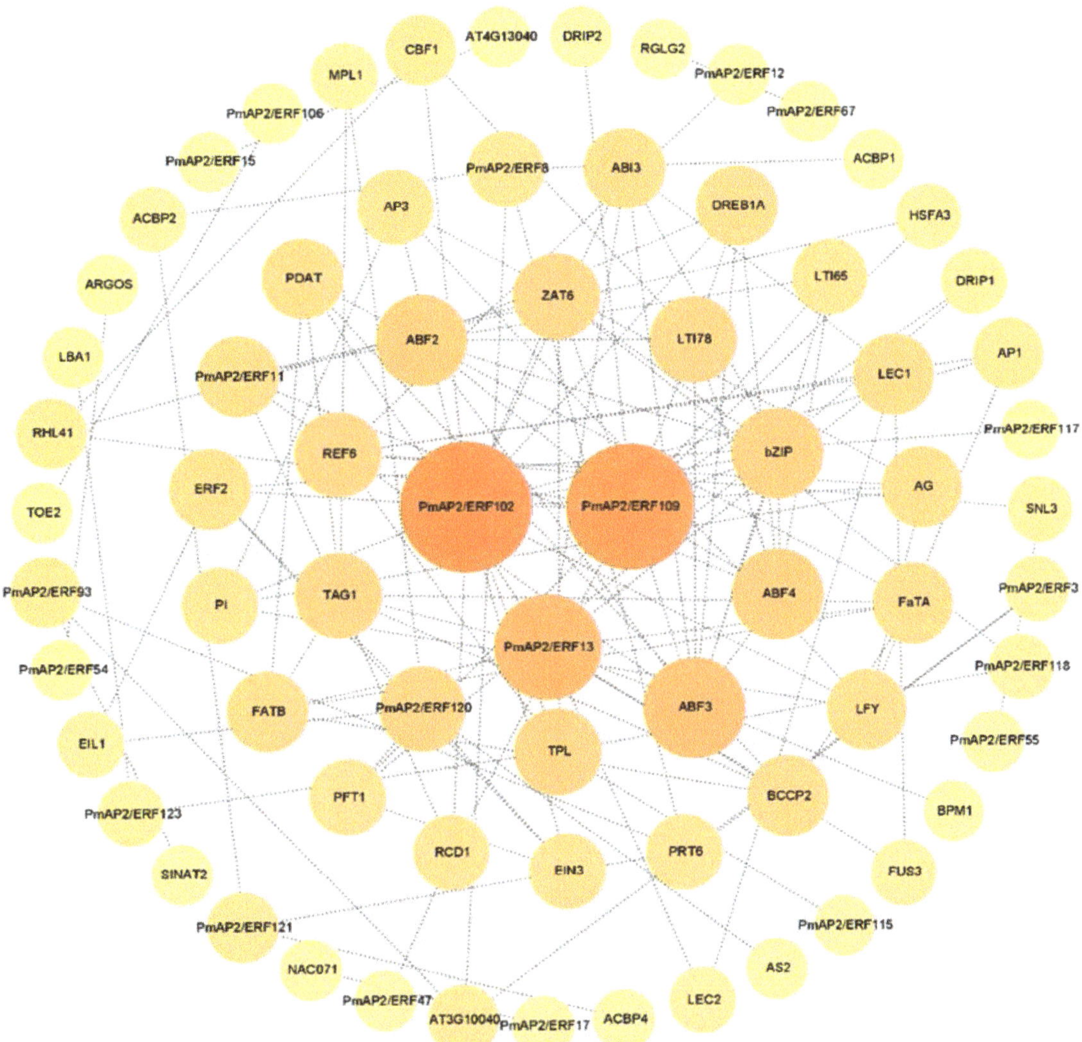

Figure 4. Prediction of the protein interaction network between PmAP2/ERFs of *P. massoniana* and AP2/ERFs of *A. thaliana*. The protein interaction network map of PmAP2/ERFs was constructed by STRING and Cytoscape software, which was based on the AP2/ERFs proteins of *A. thaliana* for analysis and prediction. The circles of different colors and sizes represent the importance of different proteins in the whole interaction network, and the dashed lines represent the possible interactions between the proteins.

3.6. Expression Analysis of PmAP2/ERF Genes in RNA-seq

The expression heat map of PmAP2/ERF genes under drought based on transcriptome data (Figure 5) revealed that 118 genes expressed during drought stress, while 6 genes did not express. A total of 13 genes peaked at D1, 14 genes peaked at D2, 29 genes peaked at D3, and the rest of the genes peaked at CK1. Further, FDR ≤ 0.001 and $|\log2FC| \geq 2$ were used as the screening criteria for significantly different genes, and *PmAP2/ERF11*, *PmAP2/ERF14*, *PmAP2/ERF44*, *PmAP2/ERF46*, *PmAP2/ERF49*, *PmAP2/ERF77*, *PmAP2/ERF80*, *PmAP2/ERF96*, and *PmAP2/ERF109* genes were studied for their expression patterns under hormonal and drought stresses.

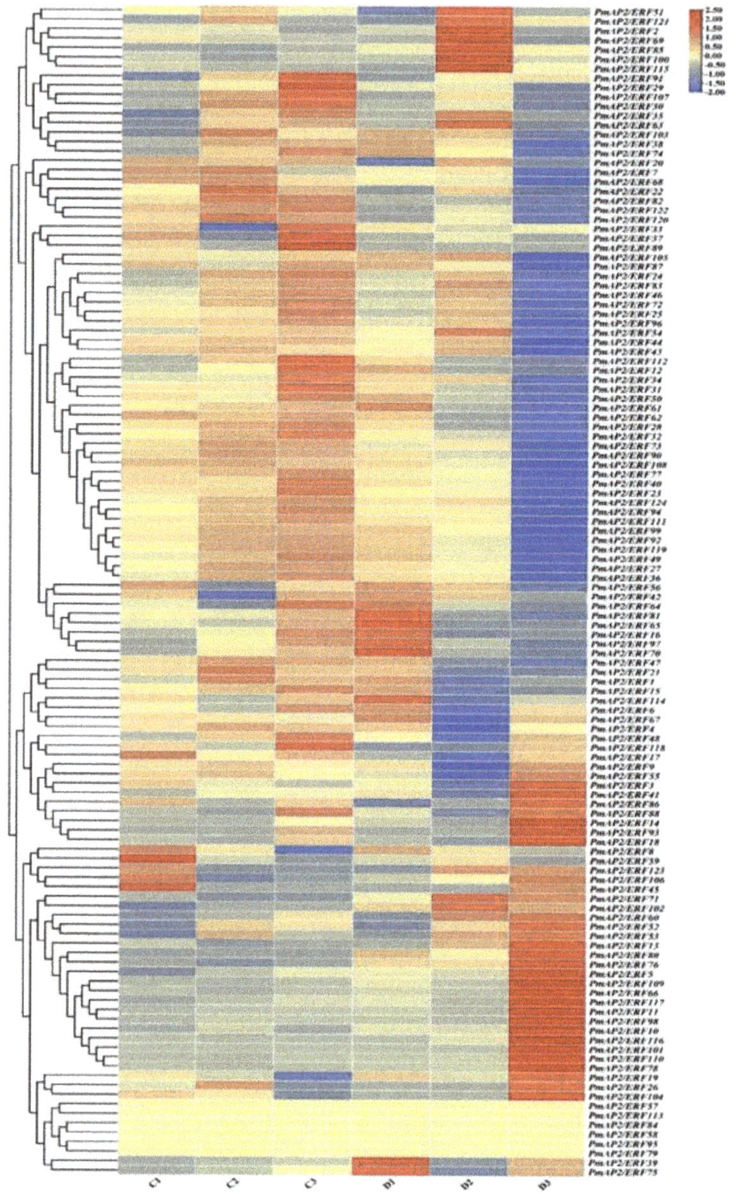

Figure 5. Expression heat map of *PmAP2/ERFs* gene expression in RNA-seq data. Blue represents low expression levels and red represents high expression levels. C and D indicate control (normal growth) and treatment (drought stress) groups, respectively; 1, 2, and 3 indicate control and drought treatment groups on the 7th day, 8th day rehydration at 7 h, and 8th day, respectively. Gene expression heat map was plotted using TBtools.

3.7. Tissue-Specific Analysis of PmAP2/ERF Genes

The tissue-specific results showed that *PmAP2/ERF* genes were expressed in needles, stems, and roots, but the expression levels differed (Figure 6). The genes *PmAP2/ERF14*, *PmAP2/ERF44*, *PmAP2/ERF49*, and *PmAP2/ERF109* were expressed in the roots of different drought-resistant materials. The highest expression was found in needle leaves. These genes may play an important role in regulating the growth and development of roots or needles (Figure 6A,B).

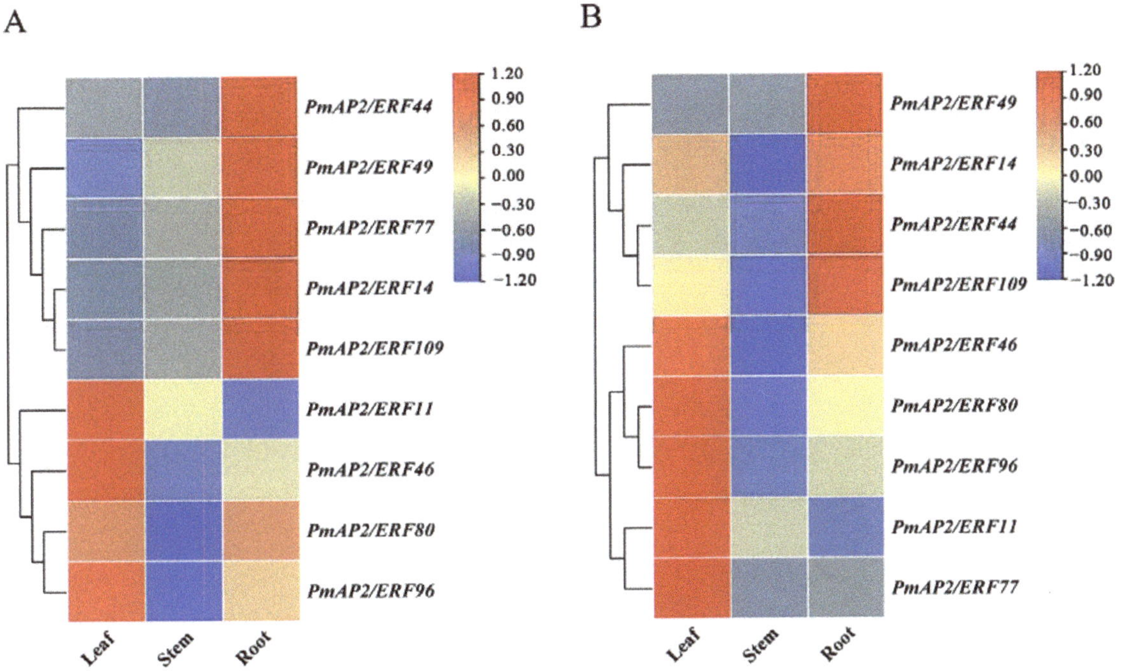

Figure 6. Tissue-specific analysis of *PmAP2/ERF* genes in *P. massoniana*: (**A**): Heat map of *PmAP2/ERFs* gene expression in needle leaves, stems, and roots of drought-sensitive lines. (**B**): Heat map of *PmAP2/ERFs* gene expression in needles, stems, and roots of drought-resistant lines. The expression levels of *PmAP2/ERF* genes in needles, stems, and roots of *P. massoniana* were analyzed by real-time fluorescence quantitative PCR, and the expression of each gene in leaves was used as a control for quantification in stems and roots, respectively. Blue color represents low expression levels and red color represents high expression levels. Plotting was performed using TBtools software.

3.8. Expression Pattern Analysis of PmAP2/ERF Genes under Hormone Treatment and Drought Stress

Expression pattern studies showed that *PmAP2/ERF* genes expressed in all tissues as constitutive expression, but there were differences in expression patterns. In different drought-tolerant lines, drought stress (CK2) induced up-regulated expression of *PmAP2/ERF11*, *PmAP2/ERF14*, *PmAP2/ERF44*, *PmAP2/ERF77*, *PmAP2/ERF80*, and *PmAP2/ERF109* genes, and down-regulated expression of *PmAP2/ERF46*, *PmAP2/ERF49*, and *PmAP2/ERF96*. Expression patterns of PmAP2/ERF genes induced by hormones differed in different tissues of different families (Figures 7 and 8).

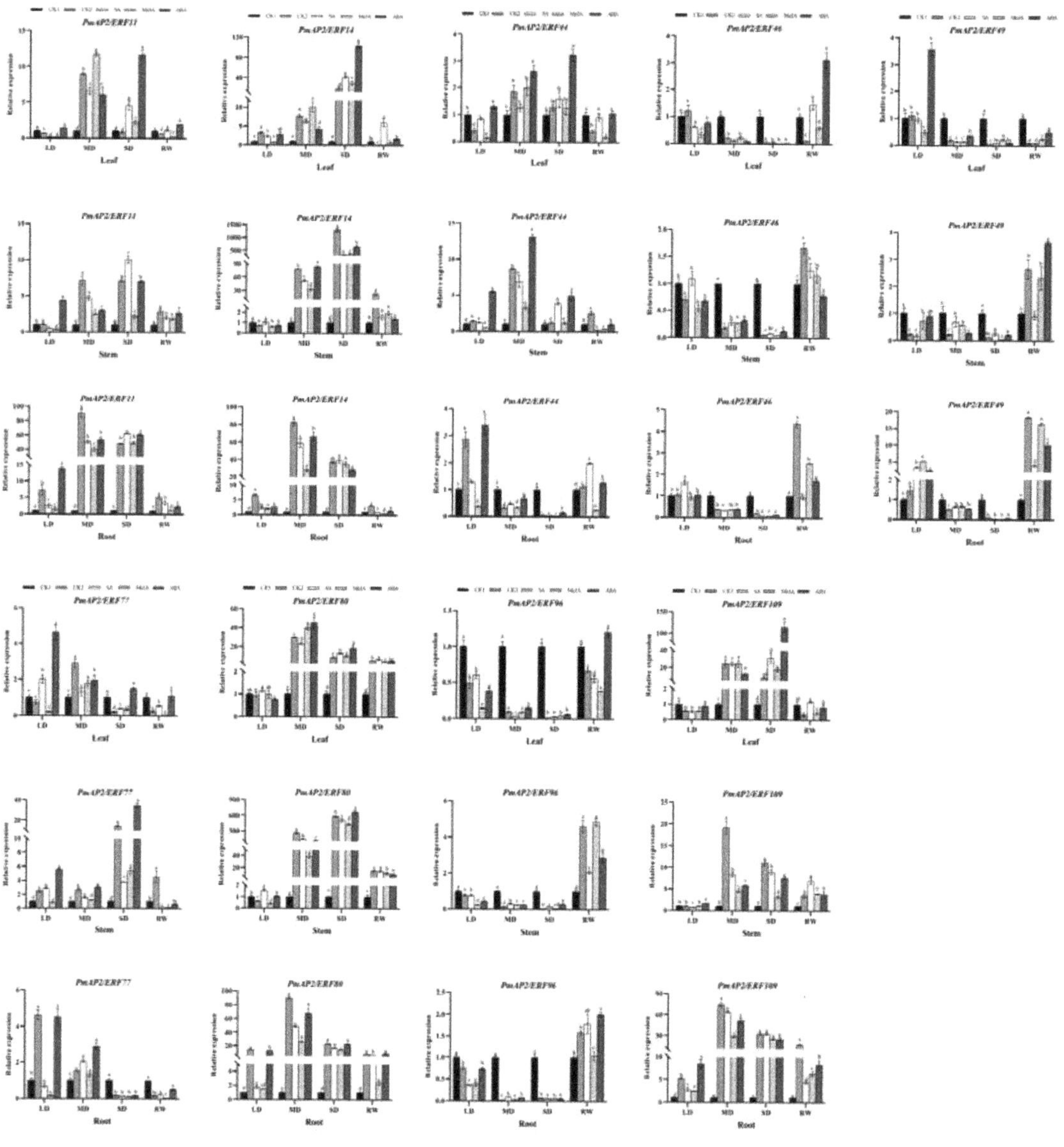

Figure 7. Expression patterns of *PmAP2/ERF* genes in drought-sensitive lines during drought stress. LD indicates light drought, MD indicates moderate drought, SD indicates severe drought, and RW indicates rehydration. Different lowercase letters indicate differences in gene expression at the $p < 0.05$ level between treatments at each sampling site. The standard error of the mean for three biological replicates is represented by the error bars.

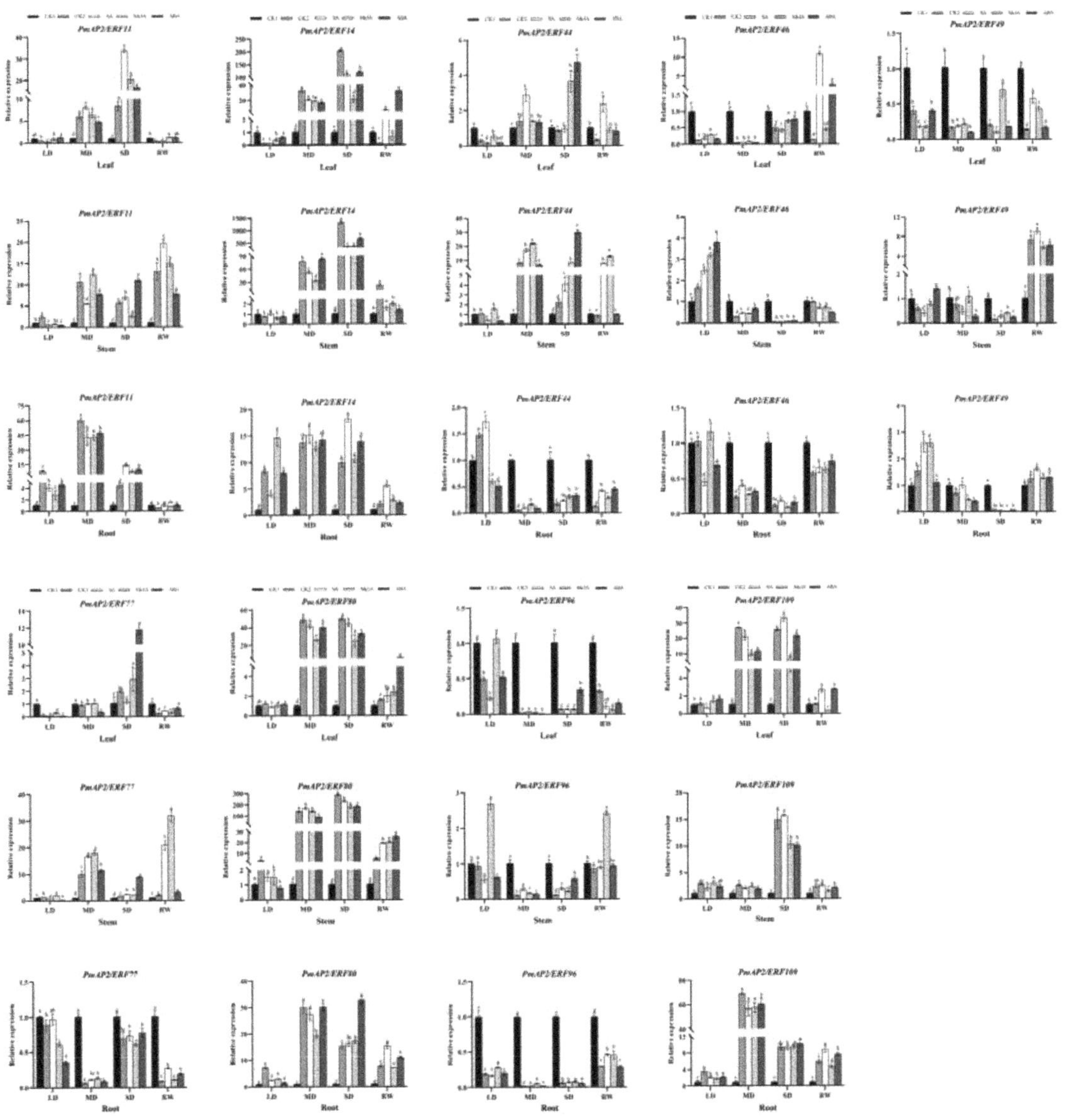

Figure 8. Expression patterns of *PmAP2/ERF* genes in drought-resistant lines during drought stress. LD indicates mild drought, MD indicates moderate drought, SD indicates severe drought, and RW indicates rehydration. Different lowercase letters indicate differences in gene expression at the $p < 0.05$ level between treatments at each sampling site. The standard error of the mean for three biological replicates is represented by the error bars.

In needle leaves, 6 *PmAP2/ERF* genes were up-regulated in leaves of two lines, with *PmAP2/ERF11* and *PmAP2/ERF14* genes expressing more in drought-resistant lines than in drought-sensitive lines. The expression patterns of *PmAP2/ERF* genes induced by hormones under drought stress were similar and different in drought-sensitive and drought-resistant lines. Compared with CK2, *PmAP2/ERF11* and *PmAP2/ERF109* genes were significantly up-regulated when induced by SA, MeJA, and ABA in both lines, while *PmAP2/ERF11* gene expression was higher in drought-resistant lines than in drought-sensitive lines.

PmAP2/ERF14 and *PmAP2/ERF80* genes were up-regulated by three hormones in drought-sensitive lines, while they were not significantly up-regulated by hormones in drought-resistant lines and both were smaller than CK2. *PmAP2/ERF44* and *PmAP2/ERF77* gene expression was induced by three hormones in drought-resistant lines, whereas *PmAP2/ERF44* was induced by ABA at a higher expression level than CK2 in drought-sensitive lines; the *PmAP2/ERF77* gene was significantly expressed when induced by SA and ABA during mild drought. The remaining three genes did not significantly express when induced by hormones. After rehydration, *PmAP2/ERF* genes were expressed at higher levels in drought-resistant lines (Figures 7 and 8).

In the stems, there were similarities and also differences in the expression patterns of *PmAP2/ERF* genes induced by hormones in different lines. Compared with CK2, *PmAP2/ERF11*, *PmAP2/ERF77*, and *PmAP2/ERF80* genes were significantly up-regulated when induced by SA, MeJA, and ABA in drought-sensitive lines, and *PmAP2/ERF11*, *PmAP2/ERF44*, *PmAP2/ERF46*, and *PmAP2/ERF77* genes were significantly expressed when induced by three exogenous hormones in drought-resistant lines. The *PmAP2/ERF14* gene was expressed when induced by SA and ABA in drought-sensitive lines, but only by ABA in drought-tolerant lines. The *PmAP2/ERF49* gene was expressed when induced by ABA, and *PmAP2/ERF96* gene was expressed when induced by MeJA. The expression of the *PmAP2/ERF109* gene in drought-sensitive lines significantly increased by ABA treatment at mild drought and then decreased, which was always lower than CK2. In drought-resistant lines, the expression of the *PmAP2/ERF109* gene induced by hormones was less than CK2 or not significantly different from CK2. After rehydration, the expression of *PmAP2/ERF* genes induced by different hormones was significantly higher than CK1 and CK2 after rehydration (Figures 7 and 8).

In the roots, there were significant differences in expression levels of *PmAP2/ERF* genes in the two lines under hormone treatments. Compared with CK2, the *PmAP2/ERF49* gene was significantly up-regulated by three hormones in drought-sensitive lines, whereas *PmAP2/ERF49* genes were expressed when induced by SA and MeJA in the drought-resistant line. The *PmAP2/ERF11*, *PmAP2/ERF14*, and *PmAP2/ERF80* genes were significantly expressed when induced by the three hormones in the drought-resistant lines, while the *PmAP2/ERF11* gene was significantly induced by MeJA and ABA in the drought-sensitive lines, and the *PmAP2/ERF14* and *PmAP2/ERF80* genes were not significantly induced by the hormones. Both were basically smaller than CK2. In drought-sensitive lines, the *PmAP2/ERF44* and *PmAP2/ERF109* genes were expressed when induced by ABA, and the *PmAP2/ERF46* gene was expressed when induced by SA; while in the drought-resistant lines, the *PmAP2/ERF44* and *PmAP2/ERF46* genes were induced by SA and MeJA, respectively, with the highest expression at mild drought, and *PmAP2/ERF109* did not express. The *PmAP2/ERF77* and *PmAP2/ERF96* genes were less or not significantly different from CK2 in both lines, while the *PmAP2/ERF44* and *PmAP2/ERF46* genes were significantly expressed when induced by SA and MeJA, respectively, and the *PmAP2/ERF109* gene was not significantly expressed in the drought-sensitive lines after rehydration (Figures 7 and 8).

4. Discussion

AP2/ERF is a major transcription factor in plants that is involved in plant growth, development, biotic, and abiotic stresses [47–49]. Based on transcriptome data identification, 124 *PmAP2/ERF* genes were identified in this study, which was comparable to 122 in *A. thaliana* [11], less than *Oryza sativa* L. (163) [17], *Zea mays* L. (292) [3], and more than *Taxus wallichiana* var. *chinensis* (49) [20], implying that the number of *AP2/ERF* gene family may not be directly related to species and genome size. The current analysis was more detailed than previous investigations, which revealed 88 *AP2/ERF* genes in *P. massoniana* [50]. There were eight AP2 subfamily genes, nine RAV subfamily genes, and 107 ERF subfamily genes in *PmAP2/ERFs*. The ERF subfamily was divided into two subfamilies, DREB and ERF [11], which included 62 and 45 genes, respectively, and it has been shown that DREB genes bound to DRE elements in the promoter regions of downstream genes to regulate the expression of related genes, and ERF genes activated downstream gene expression by binding to GCC-box elements in the promoter regions of downstream target genes [51]. ERF subfamily genes were implicated in both plant hormone response and abiotic stress response [48,52]. The result showed that around half of the proteins were acidic, the majority were hydrophobic, and the protein structures were unstable. PmAP2/ERFs may perform various regulatory roles in different organelles based on their subcellular location in the nucleus or cytoplasm. Conservative motifs revealed that each cluster had its own distinct distribution pattern, and the motifs contained in each cluster branch and the same subfamily of genes were basically the same, implying that their roles may be similar, which was consistent with earlier research [51,53]. The current findings indicated that *AP2/ERF* genes were highly conserved and structurally and evolutionary comparable.

Protein interactions revealed that PmAP2/ERF102 (AT4G36920) and PmAP2/ERF109 (AT5G05410) played important roles in the overall interplay network. PmAP2/ERF10 may have a similar function to the AT4G36920 (*APETALA 2*) gene, which was discovered to be important in the development of the floral meristem, embryo, endosperm, and seed coat [54,55]. AT5G05410 (*DREB2A*) was associated with drought and high temperature [56,57], and *A. thaliana* plants with overexpressing *DREB2A* had significant drought tolerance [56], implying that *PmAP2/ERF109* may have similar function with AT5G05410 (*DREB2A*), and the *PmAP2/ERF109* gene was up-regulated under drought stress in our study, which indicated the result was accuracy.

Tissue-specific analysis of nine *PmAP2/ERF* genes revealed that *PmAP2/ERF* genes were expressed in all tissues, and there were differences in the expression levels of genes in different tissues of different lines, while the expression levels of some genes in the same tissues of different lines also differed significantly. For example, in drought-sensitive lines, *PmAP2/ERF77* gene expression was highest in roots and lowest in needle leaves, whereas in drought-tolerant lines, *PmAP2/ERF77* gene expression was the highest in needle leaves and the lowest in stems, and the tissue-specific expression of *PmAP2/ERF77* gene was completely different in the two lines, which could be related to the difference in drought resistance.

The results of this study showed that nine *PmAP2/ERF* genes constitutively expressed under drought stress, *PmAP2/ERF46*, *PmAP2/ERF49*, and *PmAP2/ERF96* genes were negatively regulated by drought stress, and the remaining six *PmAP2/ERF* genes were positively regulated to respond to drought stress. It was shown that the DREB subfamily of *TtAP2/ERF* genes, *TtAP2/ERF-176*, *TtAP2/ERF-206*, and *TtAP2/ERF-227*, were significantly up-regulated under drought stress [47]. Five DREB subfamily genes in this study (*PmAP2/ERF11*, *PmAP2/ERF44*, *PmAP2/ERF77*, *PmAP2/ERF80*, and *PmAP2/ERF109*) were up-regulated under drought stress and positively regulated drought stress. *GmDREB1* regulated the expression of downstream stress-related genes by interacting with *GmERF008* and *GmERF106* to form a heterodimer, which significantly improved drought tolerance and increases yield in transgenic soybean [58]. The *NtERF172* gene is directly bound to the promoter region of the downstream *NtCAT* gene and positively regulates *NtCAT* gene expression, resulting in higher catalase activity and less H_2O_2 accumulation in transgenic plants, indicating that the *NtERF172* gene significantly improved the drought tolerance of the plants [59]. The *PalERF2* gene directly regulates drought response genes *PalRD20* and *PalSAG113* to improve drought resistance in poplar. The above studies suggest that *AP2/ERF* genes can activate and regulate downstream gene expression in response to drought stress through intergenic interactions. Whether there are interactions between the nine PmAP2/ERF genes in this study and the regulatory mechanisms in response to drought needs to be further investigated.

Hormones played a crucial role in plant response to abiotic stresses such as drought. SA, ABA, and JA act as hormone signaling molecules in plant drought resistance [60–62]. *AP2/ERF* genes can improve plant resistance by participating in hormone signaling networks, e.g., the *ZmEREB160* gene increased survival and proline accumulation in transgenic *A. thaliana* under drought stress. The expression levels of the ABA signaling pathway and drought-related genes such as *ABI2*, *ABI5*, and *DREB2A* were also found to be significantly up-regulated, indicating that the *ZmEREB160* gene was involved in the ABA signaling pathway to improve drought resistance [63]. In this study, *P. massoniana* was pretreated with exogenous hormones SA, ABA, and MeJA, and the expression pattern of the *PmAP2/ERF* genes in *P. massoniana* was found to be different in different tissues of different lines induced by hormones under drought stress, in which the *PmAP2/ERF96* gene significantly expressed when induced by MeJA only in the stem of the drought-resistant line at mild stress, and was not significantly affected by hormones in other tissues. *PmAP2/ERF* gene expression was induced by at least one exogenous hormone in response to drought stress in both lines and its expression was higher than that of CK2. It was hypothesized that *PmAP2/ERF* genes may be involved in the corresponding hormone signaling pathways.

The *MdDREB2* gene directly bonded to the DRE motif in the promoters of *MdNCED6* and *MdNCED9* genes, activating the transcription of ABA biosynthesis genes to promote ABA synthesis, and the *MdDREB2* gene interacted with the *MdCoL* gene to more effectively promote the expression of *MdNCED6/9* for ABA synthesis [64]. *A. thaliana* overexpressing the sweet potato *IbRAP2-12* gene showed up-regulated expression of genes related to ABA and JA signaling pathways under drought stress, while the *IbRAP2-12* gene improved *A. thaliana* tolerance during abiotic stress [23]. *PmAP2/ERF11*, *PmAP2/ERF44*, *PmAP2/ERF77*, and *PmAP2/ERF80* genes were significantly up-regulated when induced by three hormones, and it was hypothesized that these four *PmAP2/ERF* genes enhance drought tolerance in hormone signaling in *P. massoniana*, but the specific regulatory mechanisms need to be further investigated. The *PmAP2/ERF46* and *PmAP2/ERF49* genes were members of the RAV subfamily of the *AP2/ERF* gene family, and *PmAP2/ERF11* and *PmAP2/ERF109* were members of subfamily IV. In this study, we found similarities in expression patterns between the above two groups of genes under drought stress and hormone induction, further demonstrating that genes of the same subfamily may have similarities. The difference in expression of *PmAP2/ERF* genes induced by hormones in two different families was similar to the expression of *CsPRX* genes in two different *Camellia sinensis* varieties [65], which may be caused by the genetic background of two families with different drought tolerance levels. We also suggest that the high expression of some *PmAP2/ERF* genes in drought-resistant lineages increased the drought resistance of *P. massoniana*, and caused differences in drought resistance phenotypes.

5. Conclusions

In this study, we successfully identified 124 *PmAP2/ERF* genes and analyzed the physicochemical properties, phylogeny, and conserved motifs of the gene family members. The expression patterns of nine *PmAP2/ERF* genes were also analyzed under drought treatment, and it was found that all nine genes underwent expression changes in response to drought stress, but the expression patterns were different. The expression patterns showed that *PmAP2/ERF11*, *PmAP2/ERF14*, *PmAP2/ERF44*, *PmAP2/ERF77*, *PmAP2/ERF80*, and *PmAP2/ERF109* genes were up-regulated in response to drought stress and were positively regulated; *PmAP2/ERF96* gene was negatively regulated; *PmAP2/ERF46* and *PmAP2/ERF49* genes are up-regulated and down-regulated. The expression pattern of *PmAP2/ERF* genes induced by hormones differs in different tissues of different families, and the *PmAP2/ERF* genes responded to at least one hormone signal, suggesting that the *PmAP2/ERF* genes may positively or negatively regulate the response to hormones to improve drought tolerance in *P. massoniana*. Our study will provide a theoretical basis for the functional study of the *AP2/ERF* gene family and help to further investigate the molecular mechanism of *PmAP2/ERF* gene regulation in response to drought in *P. massoniana*. At the same time, this study also provides a reference for exploring the molecular mechanism of *AP2/ERF* genes in response to drought in other gymnosperms because of the close kinship between gymnosperms and the similarity in phylogeny and gene functions.

Supplementary Materials: The following supporting information can be downloaded at: https://www.mdpi.com/article/10.3390/f13091430/s1, Table S1: qRT-PCR Primer sequences of the genes; Table S2: Physical and chemical analysis of AP2/ERF in Pinus massoniana.

Author Contributions: S.S. and H.C. designed and conducted the experiments and wrote the manuscript; X.L. and H.C. contributed to manuscript writing and editing; L.H. executed the bioinformatics tools; and Z.Y. contributed to the experimental design and editing. All authors have read and agreed to the published version of the manuscript.

Funding: The Natural Science Foundation of China (32060348, 32160382), The Guangxi Natural Science Foundation (2019GXNSFDA245033, 2019GXNSFBA245064), the Special Fund for Bagui Scholar (2019A26) and Bagui Young Scholar, and the Guangxi Science and Technology and Talents Special Project (AD19254004).

Institutional Review Board Statement: Not applicable.

Informed Consent Statement: Not applicable.

Data Availability Statement: Data are contained within the article and Supplementary Materials. It is also available from the correspondence author (yangzhangqi@163.com).

Conflicts of Interest: The authors declare no conflict of interest.

References

1. Peleg, Z.; Blumwald, E. Hormone balance and abiotic stress tolerance in crop plants. *Curr. Opin. Plant Biol.* **2011**, *3*, 290–295. [CrossRef] [PubMed]
2. Sharma, P.; Dubey, R. Drought induces oxidative stress and enhances the activities of antioxidant enzymes in growing Rice seedlings. *Plant Growth Regul.* **2005**, *3*, 209–221. [CrossRef]
3. Zhou, M.; Tang, Y.; Wu, Y. Genome-Wide analysis of AP2/ERF transcription factor family in *Zea mays*. *Curr. Bioinform.* **2012**, *7*, 324–332. [CrossRef]
4. Fang, Y.; Xiong, L. General mechanisms of drought response and their application in drought resistance improvement in plants. *Cell. Mol. Life Sci.* **2015**, *72*, 673–689. [CrossRef]
5. Xu, H.; Gao, X.; Yu, C. Physiological and transcriptomic analysis of *Pinus massoniana* seedling response to osmotic stress. *Biol. Plant.* **2021**, *65*, 145–156. [CrossRef]
6. Li, M.; Wang, H.; Zhao, X.; Lu, Z.; Sun, X.; Ding, G. Role of *Suillus placidus* in improving the drought tolerance of Masson Pine (*Pinus massoniana* Lamb.) seedlings. *Forests* **2021**, *12*, 332. [CrossRef]
7. Wang, X.; Han, H.; Yan, J.; Chen, F.; Wei, W. A New AP2/ERF transcription factor from the oil plant *Jatropha curcas* confers salt and drought tolerance to transgenic Tobacco. *Appl. Biochem. Biotechnol.* **2015**, *176*, 582–597. [CrossRef]
8. Yang, G.; Zhang, W.; Sun, Y.; Zhang, T.; Zhai, M. Two novel WRKY genes from *Juglans regia*, *JrWRKY6* and *JrWRKY53*, are involved in abscisic acid-dependent stress responses. *Biol. Plant.* **2017**, *61*, 611–621. [CrossRef]
9. Du, M.; Ding, G.; Cai, Q. The transcriptomic responses of *Pinus massoniana* to drought stress. *Forests* **2018**, *9*, 326. [CrossRef]
10. Jofuku, K.D.; Boer, B.G.W.D.; Montagu, M.V.; Okamuro, J.K. Control of *Arabidopsis* flower and seed development by the homeotic gene APETALA2. *Plant Cell* **1994**, *9*, 1211–1225.
11. Nakano, T.; Suzuki, K.; Fujimura, T.; Shinshi, H. Genome-Wide analysis of the ERF gene family in *Arabidopsis* and Rice. *Plant Physiol.* **2006**, *2*, 411–432. [CrossRef] [PubMed]
12. Lixia, Z.; Rajesh, Y. Genome-Wide identification and characterization of AP2/ERF transcription factor family genes in oil palm under abiotic stress conditions. *Int. J. Mol. Sci.* **2021**, *22*, 2821–2835.
13. Okamuro, J.K.; Caster, B.; Villarroel, R.; Van Montagu, M.; Jofuku, K.D. The AP2 domain of APETALA2 defines a large new family of DNA binding proteins in *Arabidopsis*. *Proc. Natl. Acad. Sci. USA* **1997**, *13*, 7076–7081. [CrossRef]
14. Yoh, S.; Qiang, L.; Joseph, G.D.; Hiroshi, A.; Kazuo, S.; Kazuko, Y.S. DNA-Binding specificity of the ERF/AP2 domain of *Arabidopsis* DREBs, transcription factors involved in dehydration- and cold-Inducible gene expression. *Biochem. Biophys. Res. Commun.* **2002**, *3*, 998–1009.
15. Kunst, L.; Klenz, J.E.; Haughn, M.Z.W. AP2 gene determines the identity of perianth organs in flowers of *Arabidopsis thaliana*. *Plant Cell* **1989**, *1*, 1195–1208. [CrossRef] [PubMed]
16. Zhao, L.; Xu, S.; Chai, T.; Wang, T. OsAP2-1, an AP2-like gene from *Oryza sativa*, is required for flower development and male fertility. *Sex. Plant Reprod.* **2006**, *19*, 197–206. [CrossRef]
17. Sharoni, A.M.; Nuruzzaman, M.; Satoh, K.; Shimizu, T.; Kondoh, H.; Sasaya, T.; Choi, I.R.; Omura, T.; Kikuchi, S. Gene structures, classification and expression models of the AP2/EREBP transcription factor family in Rice. *Plant Cell Physiol.* **2010**, *2*, 344–360. [CrossRef]
18. Yang, Y.; Dong, C.; Li, X.; Du, J.; Qian, M.; Sun, X.; Yang, Y. A novel Ap2/ERF transcription factor from *Stipa purpurea* leads to enhanced drought tolerance in *Arabidopsis thaliana*. *Plant Cell Rep.* **2016**, *35*, 2227–2239. [CrossRef]

19. Li, Y.; Zhang, H.; Zhang, Q.; Liu, Q.; Zhao, H.; Zhao, N.; He, S. An AP2/ERF gene, IbRAP2-12, from sweetpotato is involved in salt and drought tolerance in transgenic *Arabidopsis*. *Plant Sci.* **2019**, *281*, 19–30. [CrossRef]
20. Zhang, G.; Chen, M.; Li, L.; Xu, Z.; Chen, X.; Guo, J.; Ma, Y. Overexpression of the soybean *GmERF3* gene, an AP2/ERF type transcription factor for increased tolerances to salt, drought, and diseases in transgenic tobacco. *J. Exp. Bot.* **2009**, *60*, 3781–3796. [CrossRef]
21. Abogadallah, G.M.; Nada, R.M.; Malinowski, R.; Quick, P. Overexpression of HARDY, an AP2/ERF gene from *Arabidopsis*, improves drought and salt tolerance by reducing transpiration and sodium uptake in transgenic *Trifolium alexandrinum* L. *Planta* **2011**, *233*, 1265–1276. [CrossRef]
22. Mawlong, I.; Ali, K.; Srinivasan, R.; Rai, R.D.; Tyagi, A. Functional validation of a drought-responsive AP2/ERF family transcription factor-encoding gene from rice in *Arabidopsis*. *Mol. Breed.* **2015**, *35*, 163. [CrossRef]
23. Lopes, S.T.; Sobral, D.; Costa, B.; Perdiguero, P.; Chaves, I.; Costa, A.; Miguel, C.M. Phellem *versus* xylem: Genome-wide transcriptomic analysis reveals novel regulators of cork formation in cork oak. *Tree Physiol.* **2020**, *40*, 129–141. [CrossRef]
24. Zong, Y.; Hao, Z.; Tu, Z.; Shen, Y.; Zhang, C.; Wen, S.; Yang, L.; Ma, J.; Li, H. Genome-wide survey and identification of AP2/ERF genes involved in shoot and leaf development in Liriodendron Chinense. *BMC Genom.* **2021**, *22*, 807. [CrossRef] [PubMed]
25. Zhang, K.; Jiang, L.; Wang, X.; Han, H.; Chen, D.; Qiu, D.; Yang, Y. Transcriptome-wide analysis of AP2/ERF transcription factors involved in regulating Taxol biosynthesis in *Taxus × media*. *Ind. Crops Prod.* **2021**, *171*, 113972. [CrossRef]
26. Chen, N.; Qin, J.; Tong, S.; Wang, W.; Jiang, Y. One AP2/ERF transcription factor positively regulates Pi uptake and drought tolerance in Poplar. *Int. J. Mol. Sci.* **2022**, *23*, 5241. [CrossRef] [PubMed]
27. Liu, Q.; Zhou, Z.; Wei, Y.; Shen, D.; Feng, Z.; Hong, S. Genome-Wide identification of differentially expressed genes associated with the high yielding of oleoresin in secondary xylem of Masson pine (*Pinus massoniana* Lamb.) by transcriptomic analysis. *PLoS ONE* **2015**, *7*, e132624. [CrossRef]
28. Liu, Q.; Wei, Y.; Xu, L. Transcriptomic profiling reveals differentially expressed genes associated with pine wood nematode resistance in Masson pine (*Pinus massoniana* Lamb.). *Sci. Rep.* **2017**, *1*, 4693–4706. [CrossRef] [PubMed]
29. Chen, H.; Tan, J.; Liang, X.; Tang, S.; Jia, J.; Yang, Z. Molecular mechanism of lateral bud differentiation of *Pinus massoniana* based on high-throughput sequencing. *Sci. Rep.* **2021**, *11*, 9033. [CrossRef]
30. Fan, F.; Wang, Q.; Li, H.; Ding, G.; Wen, X. Transcriptome-wide identification and expression profiles of Masson Pine WRKY transcription factors in response to low phosphorus stress. *Plant Mol. Biol. Report.* **2021**, *39*, 1–9. [CrossRef]
31. Yang, Z.; Chen, H.; Jia, J.; Luo, Q.; Tang, S.; Li, K.; Wu, D.; Feng, Y. De novo assembly and discovery of metabolic pathways and genes that are involved in defense against pests in Songyun *Pinus massoniana* Lamb. *Bangladesh J. Bot.* **2016**, *45*, 855–863.
32. Mistry, J.; Chuguransky, S.; Williams, L.; Qureshi, M.; Salazar, G.A.; Sonnhammer, E.L.L.; Tosatto, S.; Paladin, L.; Raj, S.; Richardson, L.J.; et al. Pfam: The protein families database in 2021. *Nucleic Acids Res.* **2021**, *49*, 412–419. [CrossRef] [PubMed]
33. Tian, F.; Yang, D.; Meng, Y.; Jin, J.; Gaom, G. PlantRegMap: Charting functional regulatory maps in plants. *Nucleic Acids Res.* **2020**, *48*, D1104–D1113. [CrossRef]
34. Lu, S.; Wang, J.; Chitsaz, F.; Derbyshire, M.K.; Geer, R.C.; Gonzales, N.R.; Marc, G.; Hurwitz, D.I.; Marchler, G.H.; Song, J.; et al. CDD/SPARCLE: The conserved domain database in 2020. *Nucleic Acids Res.* **2020**, *48*, 265–268. [CrossRef] [PubMed]
35. Letunic, I.; Khedkar, S.; Bork, P. SMART: Recent updates, new developments and status in 2020. *Nucleic Acids Res.* **2021**, *49*, 458–460. [CrossRef] [PubMed]
36. Duvaud, S.; Gabella, C.; Lisacek, F.; Stockinger, H.; Durinx, C. Expasy, the swiss bioinformatics resource portal, as designed by its users. *Nucleic Acids Res.* **2021**, *49*, 216–227. [CrossRef]
37. Chou, K.C.; Shen, H.B. Plant-mPLoc: A top-down strategy to augment the power for predicting plant protein subcellular localization. *PLoS ONE* **2010**, *5*, e11335. [CrossRef]
38. Horton, P.; Park, K.J.; Obayashi, T.; Fujita, N.; Harada, H.; Adams-Collier, C.J.; Nakai, K. WoLF PSORT: Protein localization predictor. *Nucleic Acids Res.* **2007**, *35*, 585–587. [CrossRef]
39. Swarbreck, D.; Wilks, C.; Lamesch, P.; Berardini, T.Z.; Garcia-Hernandez, M.; Foerster, H.; Li, D.; Meyer, T.; Muller, R.; Ploetz, L.; et al. The *Arabidopsis* Information Resource (TAIR): Gene structure and function annotation. *Nucleic Acids Res.* **2007**, *36*, 1009–1014. [CrossRef]
40. Kumar, S.; Stecher, G.; Tamura, K. MEGA7: Molecular evolutionary genetics analysis version 7.0 for bigger datasets. *Mol. Biol. Evol.* **2016**, *33*, 1870–1874. [CrossRef]
41. Letunic, I.; Bork, P. Interactive Tree of Life (iTOL) v5: An online tool for phylogenetic tree display and annotation. *Nucleic Acids Res.* **2021**, *49*, 293–296. [CrossRef] [PubMed]
42. Bailey, T.L.; Johnson, J.; Grant, C.E.; Noble, W.S. The MEME Suite. *Nucleic Acids Res.* **2015**, *43*, 39–49. [CrossRef] [PubMed]
43. Szklarczyk, D.; Gable, A.L.; Nastou, K.; Lyon, D.; Kirsch, R.; Pyysalo, S.; Doncheva, N.T.; Legeay, M.; Fang, T.; Bork, P.; et al. The STRING database in 2021: Customizable protein–protein networks, and functional characterization of user-uploaded gene/measurement sets. *Nucleic Acids Res.* **2021**, *49*, 605–612. [CrossRef] [PubMed]
44. Shannon, P.; Markiel, A.; Ozier, O.; Baliga, N.; Wang, J.T.; Ramage, D.; Amin, N.; Schwikowski, B.; Ideker, T. Cytoscape: A software environment for integrated models of biomolecular interaction networks. *Genome Res.* **2003**, *13*, 2498–2504. [CrossRef]
45. Chen, H.; Yang, Z.; Hu, Y.; Tan, J.; Jia, J.; Xu, H.; Chen, X. Reference genes selection for quantitative gene expression studies in *Pinus massoniana* L. *Trees* **2016**, *30*, 685–696. [CrossRef]

46. Livak, K.J.; Schmittgen, T.D. Analysis of relative gene expression data using Real-time quantitative PCR and the $2^{-\Delta\Delta CT}$ Method. *Methods* **2001**, *25*, 402–408. [CrossRef]
47. Faraji, S.; Filiz, E.; Kazemitabar, S.K.; Vannozzi, A.; Palumbo, F.; Barcaccia, G.; Heidari, P. The AP2/ERF gene family in *Triticum durum*: Genome-wide identification and expression analysis under drought and salinity stresses. *Genes* **2020**, *11*, 1464. [CrossRef]
48. Sharma, S.; Anuraj, C.; Heerendra, P.; Abhishek, W.; Raj, K.; Sneha, D. Identification, phylogeny and transcript profiling of ERF family genes during temperature stress treatment in Pea (*Pisum sativum* L.). *J. Plant Biochem. Biotechnol.* **2022**, *31*, 561–572. [CrossRef]
49. Lv, K.; Li, J.; Zhao, K.; Chen, S.; Wei, H. Overexpression of an AP2/ERF family gene, *BpERF13*, in birch enhances cold tolerance through upregulating CBF genes and mitigating reactive oxygen species. *Plant Sci.* **2019**, *292*, 110375. [CrossRef]
50. Zhu, P.; Chen, Y.; Zhang, J.; Wu, F.; Wang, X.; Pan, T.; Wei, Q.; Hao, Y.; Chen, X.; Jiang, C.; et al. Identification, classification, and characterization of AP2/ERF superfamily genes in Masson pine (*Pinus massoniana* Lamb.). *Sci. Rep.* **2021**, *11*, 5441. [CrossRef]
51. Ohme-Takagi, M.; Shinshi, H. Ethylene-inducible DNA binding proteins that interact with an ethylene-responsive element. *Plant Cell* **1995**, *2*, 173–182.
52. Han, D.; Han, J.; Xu, T.; Li, X.; Yao, C.; Li, T.; Sun, X.; Wang, X.; Yang, G. Overexpression of MbERF12, an ERF gene from *Malus baccata* (L.) Borkh, increases cold and salt tolerance in *Arabidopsis thaliana* associated with ROS scavenging through ethylene signal transduction. *Vitr. Cell. Dev. Biol.-Plant* **2021**, *57*, 760–770. [CrossRef]
53. Yu, Y.; Yu, M.; Zhang, S.; Zhang, S.; Song, T.; Zhang, M.; Zhou, H.; Wang, Y.; Xiang, J.; Zhang, X. Transcriptomic identification of wheat AP2/ERF transcription factors and functional characterization of *TaERF-6-3A* in response to drought and salinity stresses. *Int. J. Mol. Sci.* **2022**, *23*, 3272. [CrossRef] [PubMed]
54. Sharma, P.; Kumar, V.; Singh, S.K.; Thakur, S.; Siwach, P.; Sreenivasulu, Y.; Srinivasan, R.; Bhat, S.R. Promoter trapping and deletion analysis show *Arabidopsis thaliana* APETALA2 gene promoter is bidirectional and functions as a Pollen- and Ovule-specific promoter in the reverse orientation. *Appl. Biochem. Biotechnol.* **2017**, *182*, 1591–1604. [CrossRef]
55. Ohto, M.; Floyd, S.K.; Fischer, R.L.; Goldberg, R.B.; Harada, J.J. Effects of APETALA2 on embryo, endosperm, and seed coat development determine seed size in *Arabidopsis*. *Sex. Plant Reprod.* **2009**, *22*, 277–289. [CrossRef] [PubMed]
56. Qin, F.; Sakuma, Y.; Tran, L.; Maruyama, K.; Kidokoro, S.; Fujita, Y.; Fujita, M.; Umezawa, T.; Sawano, Y.; Miyazono, K.; et al. Arabidopsis DREB2A-Interacting proteins function as RING E3 ligases and negatively regulate plant drought Stress–responsive gene expression. *Plant Cell* **2008**, *20*, 1693–1707. [CrossRef] [PubMed]
57. Mizoi, J.; Kanazawa, N.; Kidokoro, S.; Takahashi, F.; Qin, F.; Morimoto, K.; Shinozaki, K.; Yamaguchi-Shinozaki, K. Heat-induced inhibition of phosphorylation of the stress-protective transcription factor DREB2A promotes thermotolerance of *Arabidopsis thaliana*. *J. Biol. Chem.* **2019**, *294*, 902–917. [CrossRef]
58. Chen, K.; Tang, W.; Zhou, Y.; Chen, J.; Xu, Z.; Ma, R.; Dong, Y.; Ma, Y.; Chen, M. AP2/ERF transcription factor *GmDREB1* confers drought tolerance in transgenic soybean by interacting with *GmERFs*. *Plant Physiol. Biochem.* **2022**, *170*, 287–295. [CrossRef]
59. Zhao, Q.; Hu, R.S.; Liu, D.; Liu, X.; Wang, J.; Xiang, X.; Li, Y. The AP2 transcription factor *NtERF172* confers drought resistance by modifying *NtCAT*. *Plant Biotechnol. J.* **2020**, *18*, 2444–2455. [CrossRef]
60. Li, S.; Zhou, X.; Chen, L.; Huang, W.; Yu, D. Functional characterization of *Arabidopsis thaliana WRKY39* in heat stress. *Mol. Cells* **2010**, *5*, 475–483. [CrossRef]
61. Niu, F.; Cui, X.; Zhao, P.; Sun, M.; Jiang, Y. *WRKY42* transcription factor positively regulates leaf senescence through modulating SA and ROS synthesis in *Arabidopsis thaliana*. *Plant J.* **2020**, *104*, 171–184. [CrossRef] [PubMed]
62. Gulzar, F.; Fu, J.; Zhu, C.; Yan, J.; Li, X.; Meraj, T.A.; Shen, Q.; Hassan, B.; Wang, Q. Maize WRKY transcription factor *ZmWRKY79* positively regulates drought tolerance through elevating ABA biosynthesis. *Int. J. Mol. Sci.* **2021**, *22*, 10080. [CrossRef] [PubMed]
63. Liu, W.; Zhao, B.; Chao, Q.; Wang, B.; Li, X. The Maize AP2/EREBP transcription factor *ZmEREB160* enhances drought tolerance in *Arabidopsis*. *Trop. Plant Biol.* **2020**, *13*, 251–261. [CrossRef]
64. Sun, X.; Wen, C.; Xu, J.; Wang, Y.; Zhu, J.; Zhang, Y. The apple columnar gene candidate *MdCoL* and the AP2/ERF factor *MdDREB2* positively regulate ABA biosynthesis by activating the expression of *MdNCED6/9*. *Tree Physiol.* **2021**, *41*, 1065–1076. [CrossRef]
65. Li, H.J.; Wang, H.B.; Chen, Y.; Ma, Q.P.; Chen, X. Isolation and expression profiles of class III PRX gene family under drought stress in *Camellia sinensis*. *Biol. Plant.* **2020**, *64*, 280–288. [CrossRef]

forests

MDPI

Article

Comprehensive Analysis of GRAS Gene Family and Their Expression under GA3, Drought Stress and ABA Treatment in *Larix kaempferi*

Miaomiao Ma, Lu Li, Xuhui Wang, Chunyan Zhang, Solme Pak and Chenghao Li *

State Key Laboratory of Tree Genetics and Breeding, Northeast Forestry University, Harbin 150040, China
* Correspondence: chli@nefu.edu.cn; Tel.: +86-451-82191556

Abstract: The *GRAS* family transcription factors play important roles in regulating plant growth and responses to abiotic stress, which can be utilized to breed novel plants with improved abiotic stress resistance. However, the GRAS gene family has been largely unexplored for tree species, particularly for *Larix kaempferi*, which has high economic and ecological values, challenging practices for breeding abiotic stress-resistant *L. kaempferi*. In order to improve the stress resistance by regulating the transcription factors in *L. kaempferi*, we identified 11 *GRAS* genes in *L. kaempferi* and preliminarily characterized them through comprehensive analyses of phylogenetic relationships, conserved motifs, promoter *cis*-elements, and expression patterns, as well as protein interaction network prediction. The phylogenetic analysis showed that the *LkGRAS* family proteins were classified into four subfamilies, including DELLA, HAM, SCL, and PAT1, among which the SCL subfamily was the largest one. Conserved motif analysis revealed many putative motifs such as LHRI-VHIID-LHRII-PFYRE-SAW at C-terminals of the *LkGRAS* proteins; we discovered a unique motif of the *LkGRAS* genes. Promoter *cis*-acting element analysis exhibited several putative elements associated with abiotic stresses and phytohormones; the abscisic acid-responsive elements (ABRE) and G-box are the most enriched elements in the promoters. Through expression profiles of *LkGRAS* genes in different tissues and under drought-stress and phytohormones (GA_3 and ABA) treatments, it was demonstrated that *LkGRAS* genes are most active in the needles, and they rapidly respond to environmental cues such as drought-stress and phytohormone treatments within 24 h. Protein interaction network prediction analysis revealed that *LkGRAS* proteins interact with various proteins, among which examples are the typical GA, ABA, and drought-stress signaling factors. Taken together, our work identifies the novel *LkGRAS* gene family in *L. kaempferi* and provides preliminary information for further in-depth functional characterization studies and practices of breeding stress-resistant *L. kaempferi*.

Keywords: *Larix kaempferi*; GRAS family; genome-wide analysis; phytohormone; drought stress; qRT-PCR

Citation: Ma, M.; Li, L.; Wang, X.; Zhang, C.; Pak, S.; Li, C. Comprehensive Analysis of GRAS Gene Family and Their Expression under GA3, Drought Stress and ABA Treatment in *Larix kaempferi*. *Forests* **2022**, *13*, 1424. https://doi.org/10.3390/f13091424

Academic Editor: Yuepeng Song

Received: 28 July 2022
Accepted: 2 September 2022
Published: 5 September 2022

1. Introduction

The GRAS gene family encodes a large transcription factor (TF) family crucial for plant growth, development, and responses to environmental stresses. Its name "GRAS" was derived from three TFs including GAI (Gibberellic Acid Insensitive), RGA (Repressor of GAI), and SCR (Scarecrow) which are the typical members of GRAS TFs [1]. The GRAS domain is conserved throughout the GRAS TFs at the carboxyl (C)-terminus, which mainly includes the five motifs, namely, LHR I (Leucine Heptapide Repeat I), LHR II, VHIID, PFYRE, and SAW [2], while they have a high degree of variability at the amino (N)-terminus [3]. It is currently known that the GRAS gene family consists of seven to 16 subfamilies, and the number depends on the plant species; seven in *Arabidopsis thaliana* [4], eight in *Oryza sativa* [3], 11 in *Citrus sinensis* [5], 13 in *Ricinus communis* [6], and 16 in *Medicago truncatula* [7].

The *GRAS* genes play significant roles in plant growth, development, and defense responses to various biotic and abiotic stresses, as well as phytohormone signaling and symbiosis formation. Their expression has been observed in various plant organs and tissues, including needle, stem, root, fruit, coleoptile, radicle, anther, and silk [8], and vary according to developmental stages and environmental conditions [9,10], suggesting their roles in plant development and response to environmental cues. DELLA, DLT, HAM, PAT1, LAS, LISCL, SCR, SCL3, SHR, and SCL4/7 are typical subfamilies of GRAS proteins [11] that have been implicated in plant development as follows. In *A. thaliana*, DELLA is a central regulator that plays a major role in regulating GA signal [12], and HAM is involved in chlorophyll synthesis, the proliferation of meristem cells, and polar organization [4,13,14]. PAT1 is a putative component of the phytochrome A signaling pathway [4], while the LAS subfamily increases inflorescence number [15], shortens flowering time [6], and promotes flowering induction [16] and lateral bud growth [17,18]. LISCL regulates the pre-meiotic phase of anthers and promotes microspore genesis [19], and SCL3 integrates the gibberellin acid (GA) pathway [12]. The SHR and SCR complex participates in controlling plant organ development [20,21]. In addition, several *GRAS* genes are known to be associated with plant responses to abiotic stresses. In tobacco, *GRAS1* was induced by various stresses, which then increasing the level of reactive oxygen species [15]. Overexpression of *PAT1* enhanced tolerance to abiotic stress in *Arabidopsis* [22]. The SCL4/7 subfamily members in rapeseed enhanced tolerance against drought and salt stresses [23]. *GRAS6*-silenced tomato plants showed increased sensitivity to drought stress [20]. By regulating the expression of the stress-related gene, *GRAS23* has been demonstrated to enhance resistance against drought and oxidative stress in rice by regulating several stress-related genes [24]. In tomato, the *GRAS40* gene is essential to regulate the activation of abiotic stress-inducible promoters and auxin and gibberellin signaling [25].

L. kaempferi is an important fast-growing native tree species in northern China that has high economic and ecological value. *L. kaempferi* belongs to a conifer species, generally called larch trees, with great value for wood production and ecological afforestation. Larch trees constitute forests in large areas of China, Eastern Europe, and Western North America. Among larch trees, *L. kaempferi* has several superiorities over others; it grows faster at the juvenile stage, has longer, fibrous, denser wood, and can adapt more easily to the environment than other larch trees. Thus, *L. kaempferi* is now recognized as an important tree species for various economical uses, such as timber and pulp production and papermaking, as well as afforestation and ornamental purposes. The problem is that recent climate-change-derived abiotic stresses such as drought are severely challenging afforestation practices of *L. kaempferi*, which calls for breeding novel *L. kaempferi* varieties with improved abiotic stress resistance. The *GRAS* gene family is a candidate gene family that can be utilized to breed novel *L. kaempferi* varieties with improved abiotic stress resistance. However, the *GRAS* gene family has not yet been largely explored in *L. kaempferi*, probably due to the unavailability of *L. kaempferi* genome information. The whole genome of *L. kaempferi* was recently sequenced [26] and it is, therefore, possible to perform genome-wide identification analysis for important TFs such as the GRAS TFs.

In this study, we, for the first time, identified the GRAS gene family in the *L. kaempferi* whole genome and then performed comprehensive analyses. In total, we identified 11 *GRAS* genes from the *L. kaempferi* whole genome and analyzed the evolutionary relationship, conserved motifs, and promoter *cis*-elements. We further analyzed the expression pattern of *LkGRAS* genes in different organs and tissues, including the root, stem, and needles in *L. kaempferi*. We also analyzed the expression of the *GRAS* genes under GA3, ABA, and drought treatments. Finally, we predicted the protein interaction network of *LkGRAS* proteins. This study provides a comprehensive overview of the *L. kaempferi* GRAS gene family as well as a preliminary basis for further in-depth research on the roles of *LkGRAS* factors in regulating *L. kaempferi* responses to phytohormone and abiotic stresses. More importantly, this study provides valuable information for further studies of *L. kaempferi* to improve stress resistance by regulating transcription factors.

2. Materials and Methods

2.1. Genome-Wide Identification and Phylogenetic Analysis of LkGRAS Genes

The genomic DNA, CDS, and protein sequences of *L. kaempferi* were obtained from NCBI (http://www.ncbi.nlm.nih.gov/) (accessed on 11 September 2021). Whole GRAS family members were searched in *L. kaempferi* using profile hidden Markov models (HMM); the GRAS binding domain (PF03514) was queried in the Pfam database (http://pfam.xfam.org/) (accessed on 11 September 2021) and then used to search all putative *L. kaempferi* GRAS protein members with the HMMER3 package. Redundant sequences were manually detected and eliminated, and then the remaining sequences were examined to confirm whether the GRAS binding domain is conserved throughout the sequences using the online programs CDD (https://www.ncbi.nlm.nih.gov/cdd) (accessed on 11 September 2021), Pfam (http://pfam.xfam.org/) (accessed on 11 September 2021), and SMART (http://smart.embl-heidelberg.de/) (accessed on 11 September 2021). The *L. kaempferi* GRAS protein sequences were aligned and visualized using EMBL-EBI (https://www.ebi.ac.uk/Tools/services/web/tool/) (accessed on 23 December 2021) and Jalview. We set the basic options, including "annotations, format, and color". The physical and chemical properties of the *L. kaempferi* GRAS proteins were analyzed using the ExPASy proteomics server (http://web.expasy.org/protparam/) (accessed on 23 December 2021) to analyze the characteristics of the GRAS proteins.

The amino acid sequences of GRAS proteins in *A. thaliana* and *O. sativa* were downloaded from Phytozome (Phytozome v12.1: Home) and then aligned using Clustal X (version 2.0) and Bioedit (version 7.2.5) with a gap opening penalty and gap extension penalty of 10 and 0.1, respectively. Molecular features and phylogenetic relationships between the *GRAS* genes of *L. kaempferi*, *A. thaliana*, and *O. sativa* were analyzed using MEGA software (v7.0) with the maximum likelihood method parameters as the Poisson model, partial deletion (95%), and 500 bootstrap replications [27].

2.2. Conserved Motif and Promoter Cis-Element Analysis of LkGRAS Genes

Conserved motifs in the *LkGRAS* genes were investigated using MEME (Multiple Em for Motif Elicitation program 5.1.1; http://meme-suite.org/tools/meme) (accessed on 23 December 2021) with the following parameters: the maximum number of motifs was set to 15, and the optimum motif width was set to 6 to 50 residues [28]. The Pfam and SMART tools were used to perform each structural motif annotation.

The sequences of *LkGRAS* genes were downloaded from the *L. kaempferi* genome database in NCBI, and their promoters, 2000 bp upstream of the translation start site, were identified. Then, putative *cis*-elements were searched throughout the promoters using the online database PlantCARE [29].

2.3. Plant Materials and Treatments

Mature seeds of *L. kaempferi* were collected from 60-year-old trees in Qing Shan national Larch seed orchard in Heilongjiang province (the geographical coordinates are 133°53′28″–133°58′05″ E and 46°38′56″–46°44′20″ N) and stored at −20 °C. The seeds were sown in plastic pots (11 × 11 cm) containing a grit/soil mixture (1:3 ratio), and 30 days later, seedlings were transferred to 15 cm pots (one plant per pot) containing a grit/soil mixture (1:1 ratio). The seedlings were cultured for five months under a 16 h/8 h light/dark photoperiod, 150 μmol m^{-2} s^{-1} light intensity, 70% relative humidity [30], and the soil water content was kept at ≥70% field capacity [31].

We sampled roots, stem, and needles, respectively, before treatments to determine the tissue-specific expression pattern. For the ABA and GA treatment, the solution containing 100 μM ABA or 100 μM GA3 was prepared and sprayed on needles of the *L. kaempferi* seedlings [32]. The needles were then collected at 0, 6, 12, and 24 h after treatment [32] for further RNA extraction. In addition, for drought-stress treatment, watering was stopped, and soil moisture contents were temporally measured by the gravimetric method [26]. The degree of drought stress was determined by the soil moisture contents as follows: 70%–80%

(CK, non-drought), 50%–60% (mild drought, MD), and 20%–35% (severe drought, SD) of the maximum field water capacity [33]. Temporal change in soil water contents is shown in Figure S1 and the needles were sampled at 6, 9, and 12 d for further RNA extraction. The plants at 0 days were used as control. All the treatments were sampled with three biological repeats for each seedling. The needles were carefully sampled and frozen immediately in liquid nitrogen, stored at −80 °C until RNA extraction.

2.4. RNA Extraction and Gene Expression Analysis by qRT-PCR

Total RNA was extracted from the needles using the CTAB (cetyltrimethylammonium bromide) method [34] and then reverse-transcribed to cDNA by Hi Script® II Q Select RT Super Mix for qRT-PCR. The genome DNA was eliminated by gDNA Wiper Mix. The primers were designed and checked for *LkGRAS* genes using the NCBI primer designing tool (https://www.ncbi.nlm.nih.gov/tools) (accessed on 23 December 2021). The qRT-PCR was performed to determine transcript levels of *LkGRAS* genes using SYBR Premix Ex Taq II (TaKaRa, Dalian, China) according to the manufacturer's instructions. The whole-genome sequencing of *GRAS* genes in the *L. kaempferi* gene (Whole Genome Shotgun (WGS): INSDC: WOXR00000000.2) was used as target genes. The glyceraldehyde 3-phosphate dehydrogenase (GAPDH) gene was used as an internal control gene [30]. The $2^{-\Delta\Delta Ct}$ method was used to calculate the relative gene expression levels. All *LkGRAS* gene-specific primers used for qRT-PCR are listed in Table S1.

The qRT-PCR data were tabulated and loaded by HEML to generate a heat-map. We set "canvas" and "space" to resize the heat map. We also determined the position of the X and Y axes, meanwhile selecting "column and row" to generate the branch network. We set "note" to adjust the basic setting of the font, including size and color. In the end, we set "logarithmic 2" in the option of "statistics" and exported the image.

2.5. Protein Interaction Network Analysis

The STRING (version 11.0; https://string-db.org/cgi/input.pl) (accessed on 23 December 2021) database was employed to predict the protein interaction network of *LkGRAS* proteins; prediction was performed using amino acid sequence of *LkGRAS* proteins as query and *Arabidopsis thaliana* as the "organism". The basic settings included "evidence" and "text-mining, experiments, databases, co-expression, neighborhood, gene fusion and co-occurrence". The minimum required interaction score was set as medium confidence of 0.4.

2.6. Statistical Analysis of Data

The experimental data were analyzed by one-way analysis of variance (ANOVA) method using SPSS software (version 20, IBM, Chicago, IL, USA) to evaluate significant differences between the control and each treatment. Significant differences were defined as * $p < 0.05$ and ** $p < 0.01$.

3. Results

3.1. Identification of GRAS Genes Family in L. kaempferi

To determine the information of the *GRAS* family member in *L. kaempferi*, we identified 11 GRAS genes in *L. kaempferi* genome using HMM profile of the GRAS binding domain (PF03514) as a query and then analyzed their basic information as follows. Domain search analysis using SMART and Pfam databases demonstrated that all encoded *LkGRAS* proteins possess GRAS domains. We named these genes from *LkGRAS1* to *LkGRAS11* (Table 1). The number of protein lengths, molecular weight, grand average of hydrophilicity (GRAVY), and isoelectric points are shown in Table 1. The length of GRAS proteins in *L. kaempferi* is between 223 and 730 amino acids, and the molecular weights are from 25.25 kDa to 86.22 kDa. The predicted theoretical point (pI) value varies from 5.12 to 7.07. GRAVY values of all *LkGRAS* proteins are below zero, ranging from −0.533 to −0.075, suggesting that *LkGRAS* proteins belong to the hydrophilic protein group. The instability index for most *LkGRAS* proteins is greater than 40, indicating that most *LkGRAS* proteins are unstable. Only three *LkGRAS* proteins have a stable index from 37.84 to 39.67. The aliphatic index of all *LkGRAS* proteins ranged from 71.97 to 91.82. The research showed that the aliphatic index usually shows the domination of aliphatic side chains to indicate thermal stability [35].

Table 1. Basic information of *L. kaempferi GRAS* family members.

Name	Gene ID	Length	Molecular Weight (kDa)	Theoretical pI	GRAVY Value
LkGRAS1	Lk_f2p60_2509	619	68.86	5.12	−0.336
LkGRAS2	Lk_f2p57_2714	721	80.35	5.16	−0.533
LkGRAS3	Lk_f2p39_2015	594	64.40	5.65	−0.075
LkGRAS4	Lk_f4p60_3081	696	77.89	6.31	−0.423
LkGRAS5	Lk_f2p60_2987	730	82.16	5.67	−0.459
LkGRAS6	Lk_f2p49_1552	447	50.46	6.10	−0.331
LkGRAS7	Lk_f2p39_2775	781	86.22	5.19	−0.358
LkGRAS8	Lk_f2p16_2684	634	71.62	5.58	−0.291
LkGRAS9	Lk_f2p7_2221	476	51.89	7.07	−0.233
LkGRAS10	Lk_f2p60_2999	228	25.77	6.23	−0.258
LkGRAS11	Lk_f2p49_1141	223	25.25	5.66	−0.238

Figure 1 shows the multiple sequence alignments of the GRAS gene family members of *L. kaempferi*. In the multiple sequence alignments outcome, the blue color and its intensity represent conserved domains and their homology degrees; darker color means a higher homology level. There are four conserved domains, including LHR (C1), PFYRE (C2), VHIID (C3), and SAW (C4).

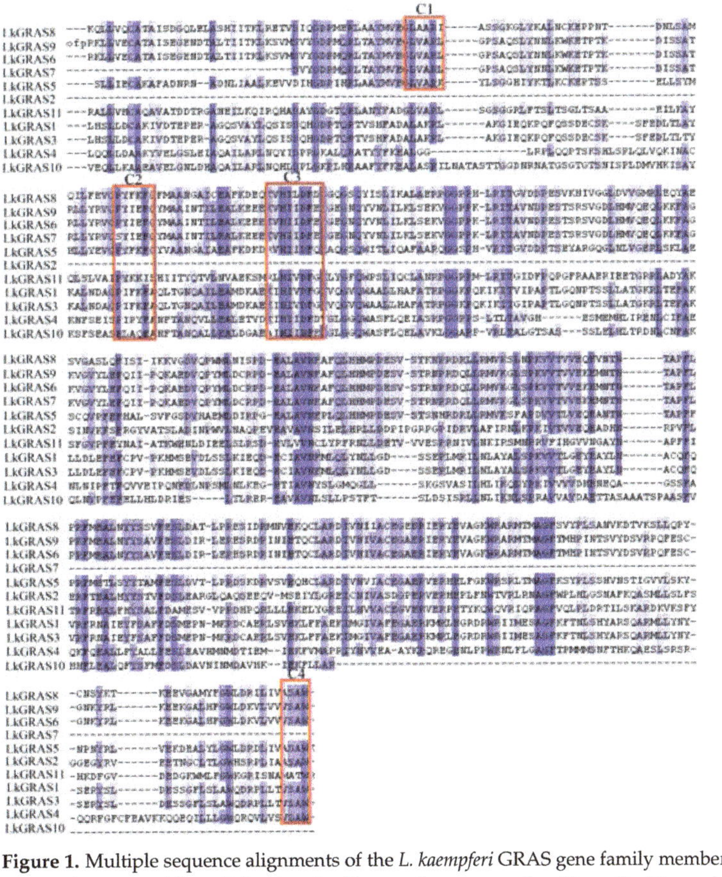

Figure 1. Multiple sequence alignments of the *L. kaempferi* GRAS gene family members. Blue shading marks identical residues, light blue shading marks conserved residues. Positions of the basic region of the GRAS domain and conserved domains (**C1–C4**) are demarcated by lines above sequences.

3.2. Phylogenetic Analysis of *L. kaempferi* GRAS Proteins

To investigate the evolutionary relationships and classification of the GRAS family in *L. kaempferi*, 37 *A. thaliana*, 63 *O. sativa*, and 11 *LkGRAS* proteins were used to construct a phylogenetic tree with the neighbor-joining (NJ) method in MEGA7.0 (Figure 2). According to the two clusterings and the relationship with *A. thaliana* and *O. sativa*, the GRAS proteins were classified into eight subfamilies (LISCL, RGL, PAT1, SCR, HAM, SCL3, SCL4/7, and DELLA). There are eight *LkGRAS* proteins belonging to the SCL (4) and PAT (4) subfamilies, while the other three proteins belong to the DELLA (1) and HAM (2). LkGRAS2, −5, −6, −7, −8, −9 proteins were clustered with the OsGRAS proteins, whereas LkGRAS1, −3, −4, −10, −11 were clustered with AtGRAS proteins. This indicates that the function of OsGRAS and AtGRAS proteins may provide a reliable reference to *LKGRAS* proteins.

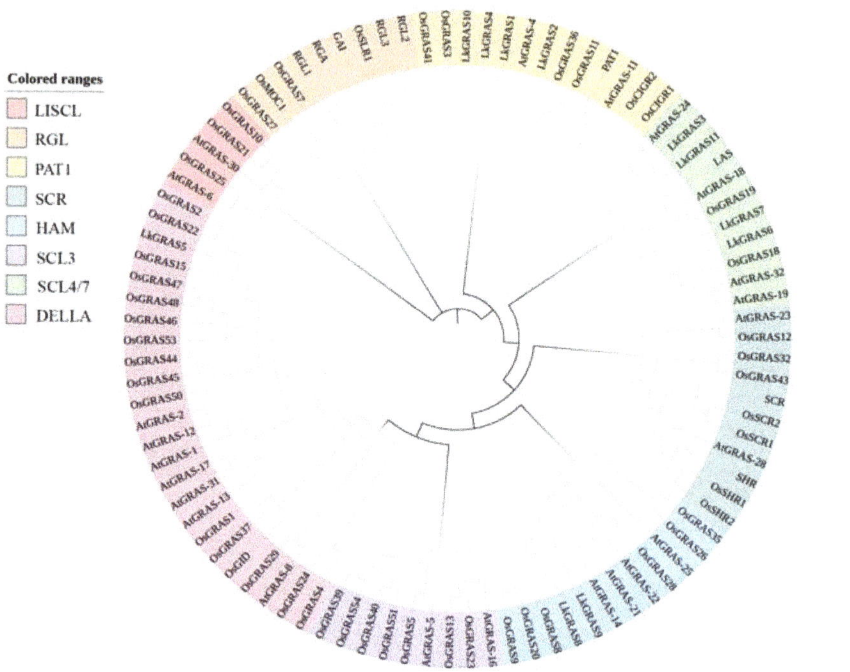

Figure 2. Phylogenetic analysis of the GRAS gene family members from *L. kaempferi*, *O. sativa*, and *A. thaliana*. Branches with less than 50% bootstrap support were collapsed. The phylogenetic tree was constructed using the maximum likelihood (ML) method of MEGA 7.0 with 500 bootstrap replicates.

3.3. Conserved Motifs of *LkGRAS* Proteins

The motifs analysis contributes to comprehensively understand the conserved characteristics of *LkGRAS* proteins and analyze structure in their conserved domain. We further confirmed the conserved motifs of *LkGRAS* proteins using MEME. In total, 15 distinct motifs were detected and named motif 1 to motif 15 (Figure 3). Since the structures and functions of the *LkGRAS* are not recognized completely, the motifs were defined based on sequence conservation. As per the previous research in GRAS domains characterization analysis, the LHRI-VHIID-LHRII-PFYRE-SAW structure domain determined the arrangements of motifs [1]. Motif 5 was highly conserved at the outermost part of C-terminal regions except for LkGRAS9 and LkGRAS10. The motifs were distributed mostly in the C-terminal. There were 10 motifs (motifs 1, 2, 3, 4, 5, 7, 8, 10, 12, and 14) in the C-terminal, while the remaining motifs (including motifs 6, 9, 11, 13, and 15) were at the N-terminal. Our results showed that conserved GRAS domains, including LHRI, VHIID, LHRII, PFYRE, and SAW domains (previously discovered by Pysh et al., 1999), included motif 1 (in VHIID domain), motif 2 (in PRYRE and SAW domains), motif 4 (in LHRII domain), motifs 5 and 6 (in LHRI domain), motif 5 (in SAW domain), and motif 7 (in PRYRE domain) (Figure S2). The motif 3 and motif 8 to motif 15 were not found to form a structure in certain domains in *LkGRAS* proteins, but they were still an indispensable part of the conserved structure domain [36].

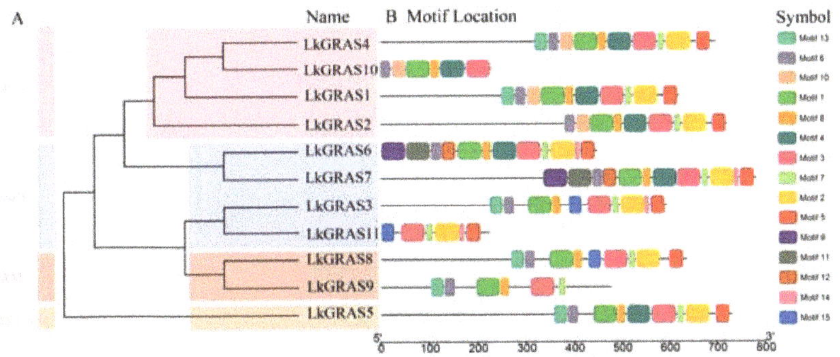

Figure 3. Phylogenetic relationships and conserved motifs of *LkGRAS* proteins. Phylogenetic tree (**A**) of *LkGRAS* proteins was constructed by using the neighbor-joining (NJ) method with 1000 bootstrap replicates in MEGA 7.0, and conserved motifs (**B**) were obtained using MEME.

3.4. Promoter Cis-Element Analysis

To understand possible regulation mechanisms of the *LkGRAS* genes, we analyzed the promoters of *LkGRAS* genes using PlantCARE and identified nine putative stress-related and phytohormone-related *cis*-elements (Figure S3). They include drought-inducibility elements (MBS) and low-temperature responsive elements (LTR), stress- and defense-responsive elements (TC-rich repeats elements), CGTCA/TGACG (MeJA-responsive elements), TCA-element (salicylic-acid-responsive elements), TGA-element (auxin-responsive elements), ABRE elements (abscisic-acid-responsive elements), and TA-rich repeats TC-box (gibberellin-responsive elements), as well as Box4 and G-box (light-responsive elements) (Table 2). The presence of these various stress- and phytohormone- responsive *cis*-elements suggested putative roles of *LkGRAS* genes in plant growth, development, and responses to abiotic stresses.

Table 2. *Cis*-element analysis of promoter regions of *LkGRAS* genes.

Name	MRE	MBS	LTR	ABRE	TGA	TCA	Box4	G-Box	CGTCA	TGACG
LkGRAS1	0	0	0	1	0	0	1	0	0	0
LkGRAS2	2	0	4	0	0	2	3	1	0	0
LkGRAS3	0	0	0	3	1	1	3	2	4	4
LkGRAS4	1	1	0	0	0	2	0	0	1	1
LkGRAS5	0	1	0	2	0	2	0	2	1	1
LkGRAS6	0	0	0	5	0	0	0	9	5	5
LkGRAS7	0	0	1	5	0	1	0	6	4	4
LkGRAS8	0	1	0	0	1	0	0	5	0	0
LkGRAS9	0	2	1	3	1	0	4	0	0	0
LkGRAS10	0	2	1	0	1	3	0	0	1	1
LkGRAS11	1	0	0	3	2	0	1	1	3	3

3.5. Tissue-Specific Expression Pattern of LkGRAS Genes

Tissue-specific expression profile for the genes belonging to a plant gene family reflects their tissue-specific functions. To determine tissue-specific expression profile of *LkGRAS* genes, we performed qPCR to analyze *LkGRAS* gene expression patterns in roots, stems, and needles at the same developmental stages and then generated a heat map (Figure 4) using the qPCR data. Expression levels of 11 *LkGRAS* genes were different to each other in the same tissue. In addition, different tissues exhibited different expression levels of *LkGRAS* genes. Most *LkGRAS* genes were weakly expressed in root tissues except the *LkGRAS10*, while they showed much higher expression levels in needle and stem tissues. In addition, the *LkGRAS10* showed the highest expression level among the *LkGRAS* genes in roots and needles, as well as high expression level in stem tissue. Taken together, we demonstrated that *LkGRAS* genes are expressed in mostly needle, and among them, *LkGRAS10* showed relatively high expression levels in all kinds of tissues tested here.

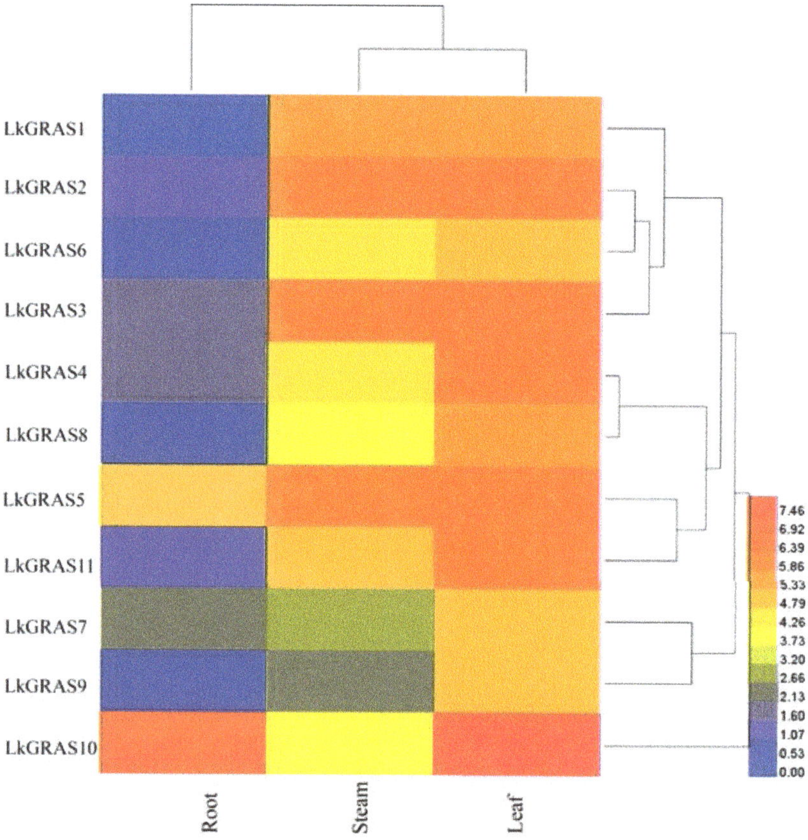

Figure 4. Tissue-specific expression pattern of *LkGRAS* genes. The relative gene expression levels were calculated using the $2^{-\Delta\Delta Ct}$ method. Different colors represent different expression levels: blue, green, and red colors represent low, mild, and high expression levels, respectively.

3.6. Expression Analysis of LkGRAS Genes under GA_3, ABA Treatment, and Drought Stress

The presence of various stress- and phytohormone-responsive *cis*-elements suggested involvement of *LkGRAS* genes in plant growth, development, and responses to abiotic stresses. To examine whether the *LkGRAS* genes take part in the abiotic stress and phytohormone response, we performed qPCR to analyze the expression level of *LkGRAS* genes in needles of *L. kaempferi* plants subjected to GA_3 (100 μM), ABA (100 μM) treatment, and drought stress. Fold change > 2 was considered as significantly differentially expressed genes. Firstly, we analyzed the *LkGRAS* gene expression under GA_3 treatment. As shown in Figure 5, all *LkGRAS* genes showed responses to exogenous GA_3 treatment with diverse expression profiles; nine *LkGRAS* genes were upregulated, among which the expression levels of *LkGRAS4*, *5*, and *7* shown were very significant. *LkGRAS6* and *10* did not show a significant response to GA_3 treatment (no more than twofold). Duration of GA_3 treatment also differentially influenced the expression pattern of the *LkGRAS* genes. *LkGRAS1*, *3*, and *8* were upregulated and reached a peak at 6 h, and *LkGRAS2*, *4*, *5*, and *7* at 12 h. The expression levels of *LkGRAS9* and *11* consistently increased for 24 h. *LkGRAS4*, *5*, and *7* had the highest expression levels among 11 *LkGRAS* genes in response to GA_3 treatment. Then, we analyzed the *LkGRAS* gene expression under drought stress. Except for *LkGRAS1*, *3*, *8*, and *9*, the other *LkGRAS* genes showed significant response to drought stress (Figure 6). *LkGRAS5*, *6*, and *10* were initially upregulated (at 6 d after drought treatment), and then declined gradually later. *LkGRAS2*, *4*, *7*, and *11* showed upregulation and reached a peak at 9 d after treatment. Finally, we analyzed the *LkGRAS* genes expression under ABA treatment. The *LkGRAS* genes were sensitive to ABA treatment except for *LkGRAS3*, *6*, *8*, and *9* (Figure 7). Though the *LkGRAS* genes showed different expression levels, they had a similar expression tendency under ABA treatment. Notably, the *LkGRAS* genes were significantly induced at various points in time under ABA treatment. The expression level was upregulated and reached a peak at 6 h, then downregulated later. Nearly all genes were in line with this trend, but the *LkGRAS3*, *6*, *8*, and *9* always showed dramatically downregulated expression levels. The expression level of *LkGRAS3*, *6*, and *8* were upregulated no more than twofold and showed lower expression levels together with *LkGRAS9*, while the expression levels of the other *LkGRAS* genes (*LkGRAS2*, *4*, *5*, *7*, *10*, and *11*) compared to them were significant, and the expression levels of *LkGRAS4*, *5*, *7*, and *11* were very significant.

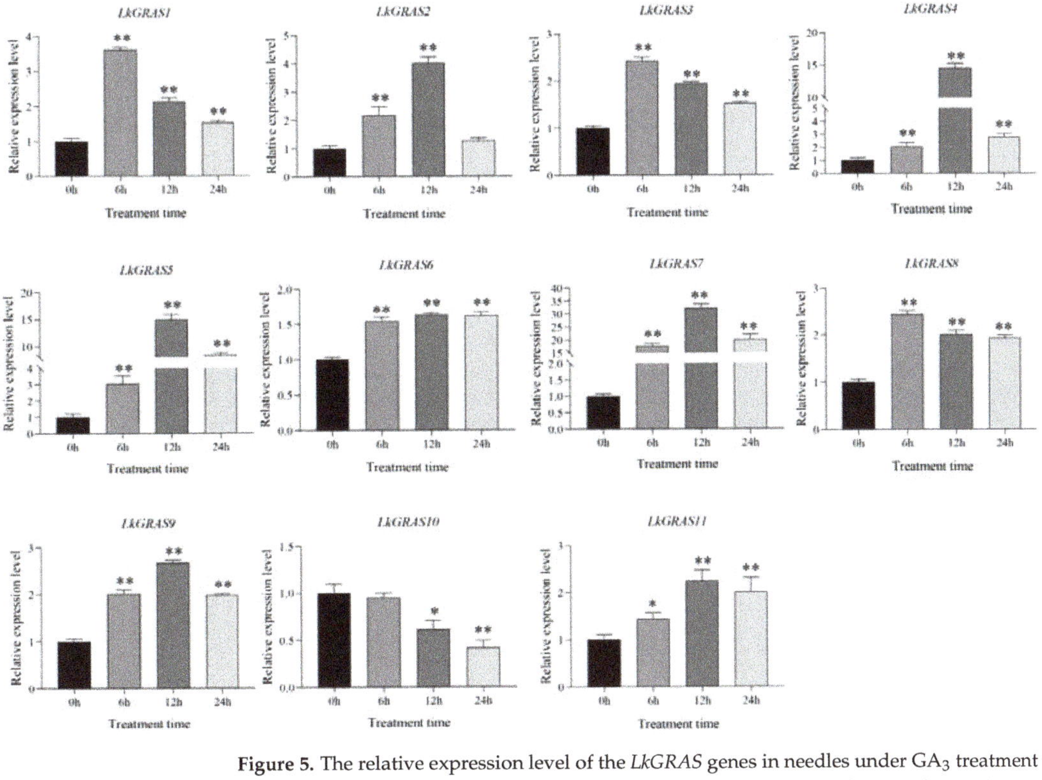

Figure 5. The relative expression level of the *LkGRAS* genes in needles under GA_3 treatment using qRT-PCR. Error bars represent the deviations from three biological replicates. The *x*-axis represents the time points after 100 μM GA3 treatment (* $p < 0.05$, ** $p < 0.01$).

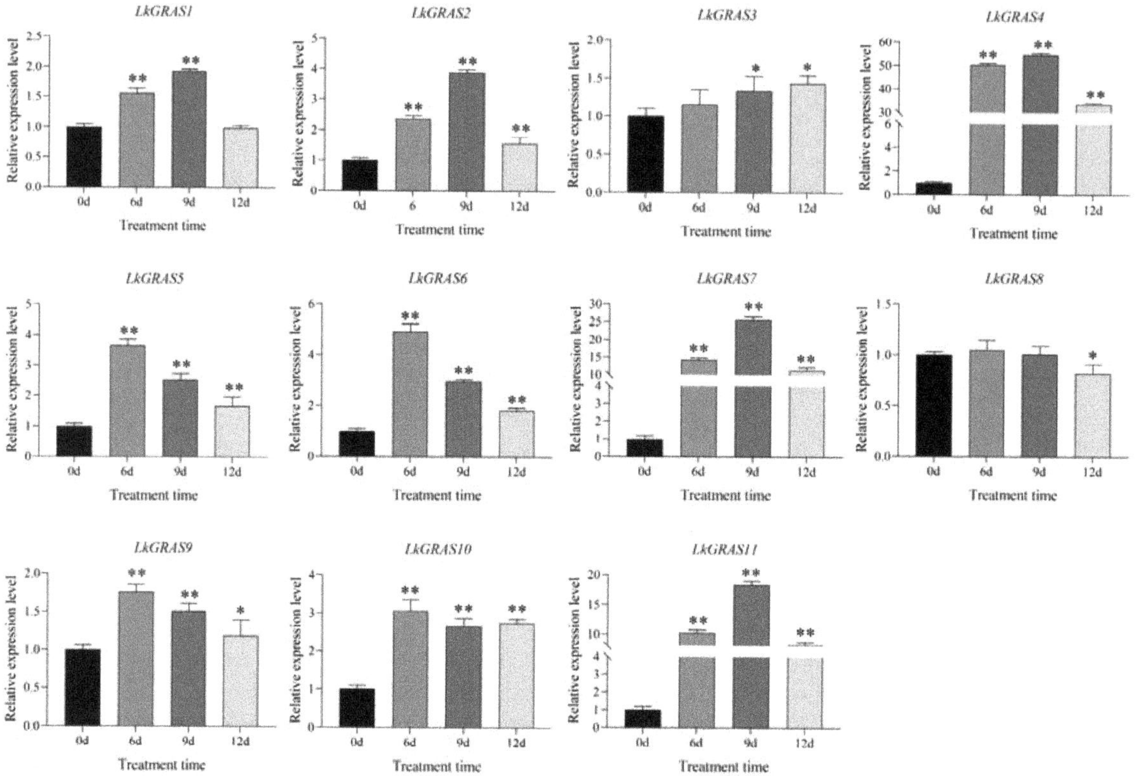

Figure 6. The relative expression levels of the *LkGRAS* genes in needles under drought stress using qRT-PCR. Error bars represent the deviations from three biological replicates. The *x*-axis represents the time points after drought stress (* $p < 0.05$, ** $p < 0.01$).

Collectively, the results showed that these *LkGRAS* genes responded to at least one kind of treatment. For instance, there were nine *LkGRAS* genes upregulated in the GA_3 treatment (*LkGRAS1, 2, 3, 4, 5, 7, 8, 9, 10,* and *11*) in which the *LkGRAS1, 2, 4, 5, 7, 10* and *11* were upregulated in the ABA treatment. Apart from these *LkGRAS* genes, the *LkGRAS3* and *9* also showed opposite expression results. Moreover, among the six drought-inducible genes (*LkGRAS4, 5, 6, 7, 10,* and *11*), five were all upregulated by ABA (*LkGRAS4, 5, 7, 10,* and *11*), and three by GA_3 (*LkGRAS4, 5,* and *7*). Meanwhile, the expression levels of *LkGRAS4, 5, 7, 10,* and *11* in ABA were consistent with those in drought, and *LkGRAS4* and *7* exhibited significantly positive responses to all three kinds of treatments.

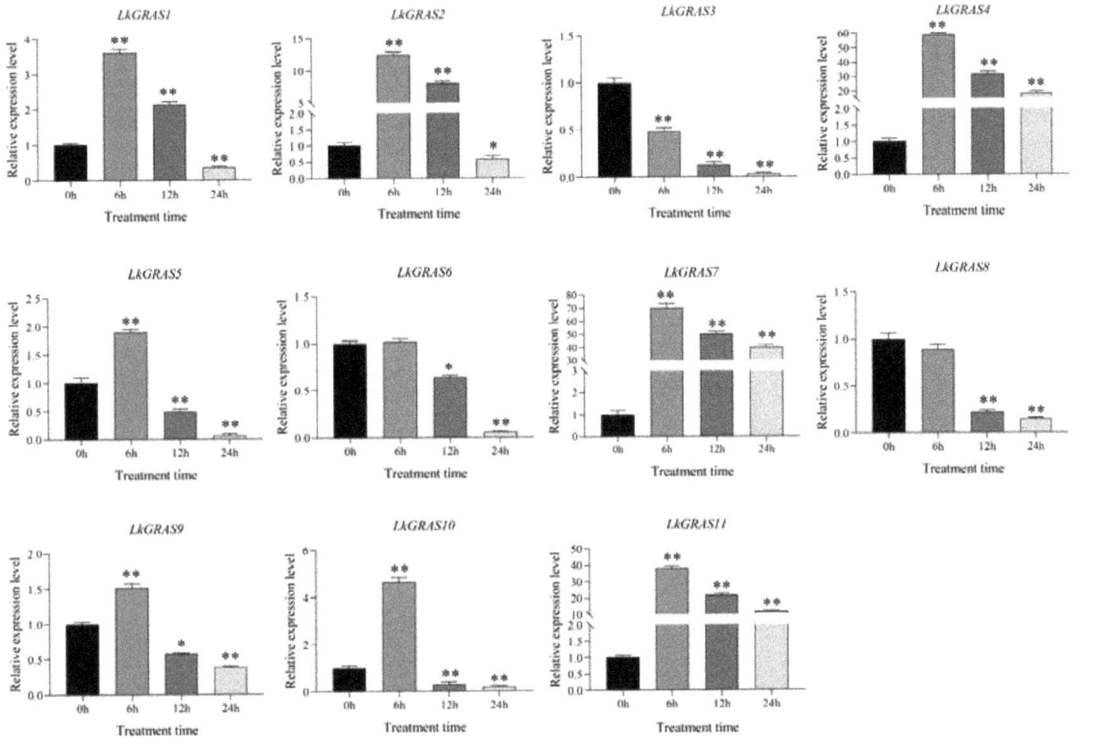

Figure 7. The relative expression levels of the *LkGRAS* genes in needles under ABA treatment using qRT-PCR. Error bars represent the deviations from three biological replicates. The *x*-axis represents the time points after 100 μM ABA treatment (* $p < 0.05$, ** $p < 0.01$).

3.7. Protein Interaction Network of LkGRAS Proteins

Proteins hardly implement their functions independently, but interact with other proteins to regulate cellular biological processes and prediction of the knowledge of protein–protein interactions (PPIs); therefore, they can untangle the cellular behaviors and functionality of the proteins. To identify the relationship of *LkGRAS* proteins with other proteins, we predicted the protein interaction network for *LkGRAS* proteins using STRING. Each *LkGRAS* protein sequence could obtain more than one network, and only the networks with the highest scores are shown in Figure 8. The networks revealed that *LkGRAS* proteins within a subfamily interact with the same proteins. For example, LkGRAS1 and LkGRAS2 of the PAT1 subfamily interact with SCL28, while LkGRAS1 and LkGRAS10 of the same subfamily interact with WAK. LkGRAS6 and LkGRAS7 of the SCL subfamily interact with MYB87, whereas LkGRAS3 and LkGRAS11 of the same subfamily interact with GID1. LkGRAS8 and LkGRAS9 of the HAM subfamily interact with WOX4. It seems that the proteins in a subfamily have highly similar motif alignments and therefore share the same protein targets to interact with each other. In addition, there are several GA, ABA, and drought-stress-related proteins, including SCL28/30, JAZ1, GID1, SLY1, GA3Ox1, PIF3, XBAT35, WDR55, and AT5G67411, among the interacting proteins, implying the interactions between them and *LkGRAS* proteins under GA, ABA, and drought-stress treatment.

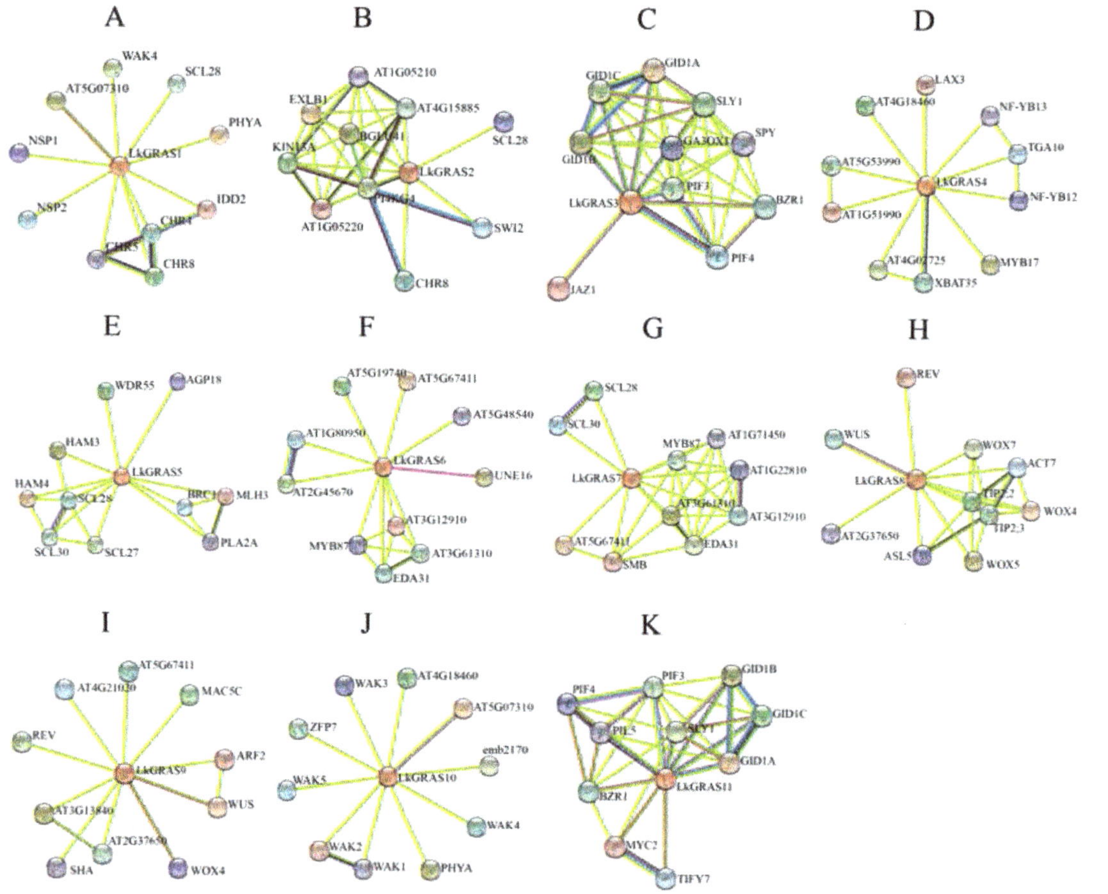

Figure 8. The predicted protein interaction network of *LkGRAS* proteins. (**A**–**K**) The potential protein interaction networks of each protein were predicted by the STRING database. Different colored lines represent different evidence of an interaction.

4. Discussion

The GRAS gene family encodes plant-specific TFs, which play essential roles in various biological processes. To date, the *GRAS* gene family has been extensively reported in various plant species including *A. thaliana* [37], *Brassica campestris* [38], *Brassica juncea* [16], *C. sinensis* [5], *Glycine max* [39], *Gossypium hirsutum* L. [10], *Ipomoea trifida* [40], *Juglans regia* L. [41], *Malus domestica* [42], *Manihot esculenta* [43], *M. truncatula* [7], *Nelumbo nucifera* [44], *O. sativa* [3], *Panax ginseng* [45], *Populus* L. [46], *R. communis* [6], *Solanum lycopersicum* [47], *Triticum aestivum* [48], and *Zea mays* L. [49]. Notably, the GRAS gene family has been largely unexplored in tree species; only reported in cassava [43] and poplar [46]. In our work, we identified the GRAS gene family in *L. kaempferi*, which is an economically and ecologically important tree species in northeastern China, for the first time. Then, we performed comprehensive analyses including phylogenetic analysis, conserved motif, and promoter *cis*-element analyses, tissue-specific and phytohormone and abiotic stress-triggered expression profile analysis, as well as protein interaction network prediction analysis for the *L. kaempferi* GRAS gene family.

Genome-wide identification and phylogenetic analysis revealed that the *LkGRAS* gene family (abbreviation of *LkGRAS* gene family) includes 11 *GRAS* genes which are further classified into four main subfamilies: DELLA, HAM, SCL, and PAT1. Other subfamilies, such as DLT, LAS, LISCL, SCR, and SHR, are not found in the *LkGRAS* gene family; this would probably be due to incompleteness of *L. kaempferi* genome database or unique feature of the *L. kaempferi* species. The structure of *LkGRAS* genes further showed that they have highly conserved motifs at C-terminal regions; conserved motifs were arranged as LHRI-VHIID-LHRII-PFYRE-SAW at C-terminals, while their N-terminal regions showed high variability that may be associated with functional divergence among the *LkGRAS* proteins. All *LkGRAS* proteins except LkGRAS9 and 10 have the SAW motif in the C-terminal region, consistent with previous findings [4] that reported the presence of the SAW motif in the C-terminal region in the *A. thaliana* GRAS family. We also found that the *LkGRAS* proteins in the same subfamily have a similar motif arrangement in the C-terminal region. For example, the *LkGRAS* proteins of the PAT1 subfamily all have a motif5 and a motif7 arranged at the C-terminal region. In addition, the motif2 domain is present in both PAT1 and SCL subfamilies. It postulates that these *LkGRAS* genes might have similar functions in biological processes. In addition, promoter *cis*-element analysis indicated that the promoters of *LkGRAS* genes contain many *cis*-acting elements such as drought-inducibility elements (MBS) and low-temperature responsive elements (LTR), stress- and defense-responsive elements (TC-rich repeats elements), CGTCA/TGACG (MeJA-responsive elements), TCA-element (salicylic-acid-responsive elements), TGA-element (auxin-responsive elements), ABRE elements (abscisic-acid-responsive elements), and TA-rich repeats TC-box (gibberellin-responsive elements), as well as Box4 and G-box (light-responsive elements), suggesting the roles of GRAS TFs in the *L. kaempferi* response to environmental cues (drought, low temperature) and phytohormones (auxin, ABA, gibberellin, MeJA, and salicylic acid).

Due to the presence of putative stress and phytohormone-related *cis*-acting elements in the promoters of *LkGRAS* gene family members, the expression profiles of *LkGRAS* genes were investigated under drought, GA_3, and ABA treatments. Before this, expression of the *LkGRAS* genes was examined in different tissues and it was demonstrated that the *LkGRAS* genes were highly expressed in needles. Then, expression of the *LkGRAS* genes in needles was further investigated under drought, GA_3, and ABA treatments. Upon GA_3 treatment, *LkGRAS4*, *5*, and *7* showed relatively high expression compared to others. *LkGRAS5* belongs to the *DELLA* family gene. The DELLA proteins are known as repressors of gibberellin response in plants [50]; DELLA proteins are essential components in the intracellular GA_3 degradation system, negatively regulating GA_3 signaling in *Arabidopsis*. Many previous studies reported that the GA-DELLA module is conserved and plays a central role in GA signaling in plants [51–53]. Upregulation of *LkGRAS5* (*DELLA* subfamily) upon GA_3 treatment in our work was consistent with the findings in the above previous studies. These findings also verified our result indirectly, that when we apply exogenous GA to *L. kaempferi*, the GA oxidases genes of the *LkGRAS* family will show high expression levels of degraded gibberellin. In addition, under ABA treatment, *LkGRAS* that belong to *PAT1* and *SCL* subfamilies exhibited high expression, indicating that *PAT1* and *SCL* subfamilies are associated with the ABA pathway. *LkGRAS2*, *4*, *5*, *7*, and *11* showed relatively high expression levels compared to the other genes, and among them, *LkGRAS5*, *7*, and *11* were expressed at the highest levels. The presence of ABRE elements in the promoters of *LkGRAS5*, *7*, and *11* would be one of the putative reasons why they showed strong upregulation under ABA treatment. Moreover, the *LkGRAS* genes showed different expression patterns under GA_3 and ABA. This might be due to the antagonistic roles of GA and ABA in plant growth and development [54]. In our work, *LkGRAS2*, *4*, *5*, *7*, *10*, and *11* showed higher expression levels under ABA treatment than under GA_3 treatment (except for *LkGRAS2* and *11*), and *LkGRAS3*, *9*, and *10* showed contrasting patterns of expression under GA_3 and ABA treatments. These results implied that *LkGRAS* genes might be involved in the antagonistic effects of GA and ABA on plant growth and development. In addition to GA and ABA, plant response to drought stress is also known to be related to *GRAS* genes. Previous works revealed that the *AtPAT1* subfamily of the *GRAS* family gene in *Arabidopsis* could increase the tolerance of the plant to abiotic stress, such as cold, drought, and salt [23,55]. The *SCL* subfamily was also demonstrated to participate in drought-stress response [24]. Consistently, our work also manifested that most of the *LkGRAS* genes responded to drought stress; among which *LkGRAS4* belongs to the *PAT1* subfamily and *LkGRAS7* and *11* belong to the *SCL* subfamily. It can be inferred that the *PAT1* and *SCL* subfamilies of *GRAS* genes in *L. kaempferi* are involved in drought-stress response. In addition, drought-stress-related *cis*-acting elements are present in promoters of the differentially expressed *LkGRAS* genes under drought stress. Among the drought-inducible genes, *LkGRAS4, 7, 10,* and *11* showed high expression levels in drought-stress and ABA treatment. In plants, signaling pathways of ABA and drought-stress response are interrelated with each other. It, therefore, appeared that drought and ABA treatments both induced the expression of *LkGRAS4, 7, 10,* and *11*. *LkGRAS4* and *7* also belong to the *PAT1* subfamily, which showed a high expression level under GA and ABA treatments. Overall, expression profiles of the *LkGRAS* genes showed consistency with the prediction from the *cis*-acting elements in promoters of the *LkGRAS* genes. The *LkGRAS* genes with drought-inducibility elements showed high expression levels under drought stress. The highly induced *LkGRAS* genes under GA_3 or ABA treatments also possess GA- or ABA-related *cis*-acting elements in their promoters. Each of the *LkGRAS* genes contains at least two *cis*-elements related to phytohormone or abiotic stress responsiveness.

Moreover, the protein interaction network of *LkGRAS* proteins was predicted using the STRING database, which could provide a supplementary understanding of orthologous proteins' roles in biological processes [56]. We found that the *LkGRAS* proteins within the same subfamily revealed similar protein interaction networks. Among the interacting proteins, we found that several factors, such as SCL28, JAZ1, GID1, SLY1, GA3Ox1, PIF3, XBAT35, WDR55, and AT5G67411, have previously been known to be associated with GA, ABA, and drought-stress responses. SCL28, which is a GRAS type TF in *A. thaliana* [57] and is known to be involved in ABA-mediated stress responses [58], interacts with LkGRTAS1, 2, 5, and 7 proteins. JAZ1, WDR55, and XBAT35 are also the ABA response factors [59–61], which are predicted to interact with LkGRAS3, 4, and 5 proteins, respectively, in our study. In addition, JAZ1 and WDR55, which can regulate drought-stress responses through ABA pathways, interact with LkGRAS3 and 5, respectively. PIF3, which is previously known to enhance resistance to drought stress [62], interacts with LkGRAS3 and 11. In addition to ABA and drought-related factors, there are GA response factors such as GA3Ox1, GID1, and SLY1 among the total interacting factors. GA3Ox1, which is the enzyme for GA biosynthesis, interacts with LkGRAS3 and GID1, which is a gibberellin receptor protein [63,64] and interacts with LkGRAS11. SLY1, which is known to positively regulate GA signaling [65], interacts with both LkGRAS3 and 11.

5. Conclusions

In conclusion, we identified 11 *GRAS* family genes in *L. kaempferi* and analyzed their phylogenetic tree, conserved motifs, and promoter *cis*-elements. The 11 *LkGRAS* genes are classified into four subfamilies, including *DELLA, HAM, SCL,* and *PAT1*. The *LkGRAS* proteins all have conserved LHRI-VHIID-LHRII-PFYRE-SAW motifs at C-terminals and their promoters contain many *cis*-acting elements associated with abiotic stresses and phytohormones. In addition, we evaluated the expression patterns of *LkGRAS* genes in different tissues and under GA_3, ABA, and drought-stress treatments using qRT-PCR. *LkGRAS* genes were mainly expressed in needles and were significantly induced upon exogenous treatment by phytohormones (GA_3 and ABA) and drought stress. We also predicted the protein interaction network of *LkGRAS* proteins. Preliminary results of our work on the *LkGRAS* gene families provided knowledge that would be the basic information for further in-depth functional characterization of *LkGRAS* family genes in *L. kaempferi*.

Supplementary Materials: The following supporting information can be downloaded at: https://www.mdpi.com/article/10.3390/f13091424/s1, Figure S1: Temporal change in the soil water contents under non-watered condition; Figure S2: Conserved GRAS domains (LHRI, VHIID, LHRII,

PFYRE, and SAW domains); Figure S3: The nine putative stress-related and phytohormone-related cis-elements; Table S1: Primers for quantitative qRT-PCR.

Author Contributions: Conceptualization, M.M. and C.L.; experimental design: C.L.; material collection and performing the experiments: M.M., X.W., L.L. and C.Z.; data analysis: M.M.; software, M.M.; writing—original draft: M.M.; writing—review and editing: M.M., S.P. and C.L. All authors approved the final draft. All authors have read and agreed to the published version of the manuscript.

Funding: This work was supported by the Genetically Modified Organisms Breeding Major Projects of China (Grant No. 2018ZX08020003).

Data Availability Statement: The data is available on request from the corresponding author.

Acknowledgments: We thank Jia Yang for proofreading the manuscript.

Conflicts of Interest: The authors declare no conflict of interest.

References

1. Pysh, L.D.; Wysocka-Diller, J.W.; Camilleri, C.; Bouchez, D.; Benfey, P. The GRAS gene family in *Arabidopsis*: Sequence characterization and basic expression analysis of the *SCARECROW-LIKE* genes. *Plant J.* **1999**, *18*, 111–119. [CrossRef] [PubMed]
2. Bolle, C. The role of GRAS proteins in plant signal transduction and development. *Planta* **2004**, *218*, 683–692. [CrossRef] [PubMed]
3. Tian, C.; Wan, P.; Sun, S.; Li, J.; Chen, M. Genome-wide analysis of the GRAS gene family in rice and *Arabidopsis*. *Plant Mol. Biol.* **2004**, *54*, 519–532. [CrossRef] [PubMed]
4. Bolle, C.; Koncz, C.; Chua, N.H. PAT1, a new member of the GRAS family, is involved in phytochrome A signal transduction. *Genes Dev.* **2000**, *14*, 1269–1278. [CrossRef]
5. Zhang, H.; Mi, L.M.; Xu, L.; Yu, C.X.; Li, C. Genome-wide identification, characterization, interaction network and expression profile of GRAS gene family in sweet orange (*Citrus sinensis*). *Sci. Rep.* **2019**, *9*, 2156. [CrossRef]
6. Xu, W.; Chen, Z.; Ahmed, N.; Han, B.; Cui, Q.; Liu, A. Genome-wide identification, evolutionary analysis, and stress responses of the GRAS gene family in *Castor beans*. *Int. J. Mol. Sci.* **2016**, *17*, 1004. [CrossRef]
7. Song, L.; Tao, L.; Cui, H.; Ling, L.; Guo, C. Genome-wide identification and expression analysis of the GRAS family proteins in *Medicago truncatula*. *Acta Physiol. Plant* **2017**, *39*, 93. [CrossRef]
8. Heckmann, A.B.; Lombardo, F.; Miwa, H.; Perry, J.A.; Downie, J.A. *Lotus japonicus* nodulation requires two GRAS domain regulators, one of which is functionally conserved in a non-legume. *Plant Physiol.* **2007**, *142*, 1739–1750. [CrossRef]
9. Hirsch, S.; Oldroyd, G.E. GRAS-domain transcription factors that regulate plant development. *Plant Signal. Behav.* **2009**, *4*, 698–700. [CrossRef]
10. Zhang, B.; Liu, J.; Yang, Z.E.; Chen, E.Y.; Zhang, C.J.; Zhang, X.Y.; Li, F.G. Genome-wide analysis of GRAS transcription factor gene family in *Gossypium hirsutum* L. *BMC Genom.* **2018**, *19*, 348. [CrossRef]
11. Liu, M.; Huang, L.; Ma, Z.; Sun, W.; Chen, H. Genome-wide identification, expression analysis and functional study of the GRAS gene family in *Tartary buckwheat* (*Fagopyrum tataricum*). *BMC Plant Biol.* **2019**, *19*, 342. [CrossRef] [PubMed]
12. Heo, J.O.; Chang, K.S.; Kim, I.A.; Lee, M.H.; Lee, S.A.; Song, S.K.; Lim, J. Funneling of gibberellin signaling by the GRAS transcription regulator SCARECROW-LIKE 3 in the *Arabidopsis* root. *Proc. Natl. Acad. Sci. USA* **2011**, *108*, 2166–2171. [CrossRef] [PubMed]
13. Schulze, S.; Schfer, B.N.; Parizotto, E.A.; Voinnet, O.; Theres, K. *LOST MERISTEMS* genes regulate cell differentiation of central zone descendants in Arabidopsis shoot meristems. *Plant J.* **2010**, *64*, 668–678. [CrossRef] [PubMed]
14. Stuurman, J. Shoot meristem maintenance is controlled by a GRAS-gene mediated signal from differentiating cells. *Genes Dev.* **2002**, *16*, 2213–2218. [CrossRef] [PubMed]
15. Laurenzio, L.D.; Wysocka-Diller, J.; Malamy, J.E.; Pysh, L.; Helariutta, Y.; Freshour, G.; Benfey, P.N. The *SCARECROW* gene regulates an asymmetric cell division that is essential for generating the radial organization of the Arabidopsis root. *Cell* **1996**, *86*, 423–433. [CrossRef]
16. Meng, Y.; Li, B.; Sun, F.J.; Xie, R.G.; Gong, Y.L. Identification of the GRAS gene family in the *Brassica juncea* genome provides insight into its role in tem swelling in stem mustard. *PeerJ* **2019**, *7*, e6682.
17. Feng, G.; Huang, L.; Li, J.; Wang, J.; Xu, L.; Pan, L.; Zhao, X.; Wang, X.; Huang, T.; Zhang, X. Comprehensive transcriptome analysis reveals distinct regulatory programs during vernalization and floral bud development of orchardgrass (*Dactylis glomerata* L.). *BMC Plant Biol.* **2017**, *17*, 216. [CrossRef]
18. Casler, M.D.; Fales, S.L.; Mcelroy, A.R.; Hall, M.H.; Hoffman, L.D.; Leath, K.T. Genetic progress from 40 years of orchardgrass breeding in North America measured under hay management. *Can. J. Plant Sci.* **2001**, *81*, 713–721. [CrossRef]
19. Morohashi, K.; Minami, M.; Takase, H.; Hotta, Y.; Hiratsuka, K. Isolation and characterization of a novel GRAS gene that regulates meiosis-associated gene expression. *J. Biol. Chem.* **2003**, *278*, 20865–20873. [CrossRef]
20. Mayrose, M.; Ekengren, S.K.; Melech-Bonfil, S.; Martin, G.B.; Sessa, G. A novel link between tomato GRAS genes, plant disease resistance and mechanical stress response. *Mol. Plant Pathol.* **2006**, *7*, 593–604. [CrossRef]

21. Benjamin, F.; Tanja, S.; Corinna, T.; Ralf, W. The *Arabidopsis* GRAS protein SCL14 interacts with class II TGA transcription factors and is essential for the activation of stress-inducible promoters. *Plant Cell* **2008**, *20*, 3122–3135.
22. Czikkel, B.E.; Maxwell, D.P. NtGRAS1, a novel stress-induced member of the GRAS family in tobacco, localizes to the nucleus. *J. Plant Physiol.* **2007**, *164*, 1220–1230. [CrossRef] [PubMed]
23. Yuan, Y.; Fang, L.; Karungo, S.K.; Zhang, L.; Gao, Y.; Li, S.; Xin, H. Overexpression of VaPAT1, a GRAS transcription factor from Vitis amurensis, confers abiotic stress tolerance in *Arabidopsis*. *Plant Cell Rep.* **2016**, *35*, 655–666. [CrossRef] [PubMed]
24. Xu, K.; Chen, S.; Li, T.; Ma, X.; Liang, X.; Ding, X.; Liu, H.; Luo, L. *OsGRAS23*, a rice GRAS transcription factor gene, is involved in drought stress response through regulating expression of stress-responsive genes. *BMC Plant Biol.* **2015**, *15*, 141. [CrossRef]
25. Liu, Y.; Huang, W.; Xian, Z.; Hu, N.; Lin, D.; Ren, H.; Chen, J.; Su, D.; Li, Z. Overexpression of *SlGRAS40* in tomato enhances tolerance to abiotic stresses and influences auxin and gibberellin signaling. *Front. Plant Sci.* **2017**, *8*, 1659. [CrossRef]
26. Sun, C.; Xie, Y.H.; Li, Z.; Liu, Y.J.; Sun, X.M.; Li, J.J.; Quan, W.P.; Zeng, Q.Y.; Van, P.Y.; Zhang, S.G. The *Larix kaempferi* genome reveals new insights into wood properties. *J. Integr. Plant Biol.* **2022**, *64*, 1364–1373. [CrossRef]
27. He, L.; Zhao, M.; Wang, Y.; Gai, J.; He, C. Phylogeny, structural evolution and functional diversification of the plant PHOSPHATE1 gene family: A focus on Glycine max. *BMC Evol. Biol.* **2013**, *13*, 103. [CrossRef]
28. Bailey, T.L.; Mikael, B.; Buske, F.A.; Martin, F.; Grant, C.E.; Luca, C.; Ren, J.; Li, W.W.; Noble, W.S. MEME SUITE: Tools for motif discovery and searching. *Nucleic Acids Res.* **2009**, *37*, W202–W208. [CrossRef]
29. Lescot, M.; Déhais, P.; Thijs, G.; Marchal, K.; Moreau, Y.; Peer, Y.; Rouzé, P.; Rombauts, S. PlantCARE, a database of plant cis-acting regulatory elements and a portal to tools for in silico analysis of promoter sequences. *Nucleic Acids Res.* **2002**, *30*, 325–327. [CrossRef]
30. Li, D.; Yu, S.; Zeng, M.; Liu, X.; Yang, J.; Li, C. Selection and validation of Appropriate reference genes for real-time quantitative PCR analysis in needles of *Larix olgensis* under abiotic stresses. *Forests* **2020**, *11*, 193. [CrossRef]
31. Li, W.; Lee, J.; Yu, S.; Wang, F.; Lv, W.; Zhang, X.; Li, C.; Yang, J. Characterization and analysis of the transcriptome response to drought in *Larix kaempferi* using PacBio full-length cDNA sequencing integrated with de novo RNA-seq reads. *Planta* **2021**, *253*, 28. [CrossRef]
32. Pan, C.; Tian, K.; Ban, Q.; Wang, L.; Sun, Q.; He, Y.; Yang, Y.; Pan, Y.; Li, Y.; Jiang, J.; et al. Genome-wide analysis of the biosynthesis and deactivation of gibberellin-dioxygenases gene family in *Camellia sinensis* (L.) O. Kuntze. *Genes.* **2017**, *8*, 235. [CrossRef]
33. Gao, R.; Shi, X.; Lin, W.; Na, H. Drought resistance of one-year-old seedlings of *Larix principis-rupprechtii*. *Sci. Silvae Sin.* **2015**, *51*, 148–156.
34. Jaakola, L.; Pirttil, A.M.; Halonen, M.; Hohtola, A. Isolation of high quality RNA from bilberry (*Vaccinium myrtillus* L.) fruit. *Mol. Biotechnol.* **2001**, *19*, 201–203. [CrossRef]
35. Li, D.; Peng, S.; Chen, S.; Li, Z.; Yang, G. Identification and characterization of 5 walnut MYB genes in response to drought stress involved in ABA signaling. *Physiol. Mol. Biol. Plants* **2021**, *27*, 1323–1335. [CrossRef]
36. Dash, M.; Yordanov, Y.S.; Georgieva, T.; Tschaplinski, T.J.; Busov, V. Poplar PtabZIP1-like enhances lateral root formation and biomass growth under drought stress. *Plant J.* **2017**, *89*, 692–705. [CrossRef]
37. Lee, M.H.; Kim, B.; Song, S.K.; Heo, J.O.; Yu, N.I.; Lee, S.A.; Kim, M.; Kim, D.G.; Sohn, S.O.; Lim, C.E. Large-scale analysis of the GRAS gene family in *Arabidopsis thaliana*. *Plant Mol. Biol.* **2008**, *67*, 659–670. [CrossRef]
38. Song, X.M.; Liu, T.K.; Duan, W.K.; Ma, Q.H.; Ren, J.; Wang, Z.; Li, Y.; Hou, X.L. Genome-wide analysis of the GRAS gene family in Chinese cabbage (*Brassica rapa* ssp. Pekinensis). *Genomics* **2014**, *103*, 135–146. [CrossRef]
39. Wang, L.; Ding, X.; Gao, Y.; Yang, S. Genome-wide identification and characterization of GRAS genes in soybean (*Glycine max*). *BMC Plant Biol.* **2020**, *20*, 415. [CrossRef]
40. Chen, Y.; Zhu, P.; Wu, S.; Lu, Y.; Sun, J.; Cao, Q.; Li, Z.; Xu, T. Identification and expression analysis of GRAS transcription factors in the wild relative of sweet potato *Ipomoea trifida*. *BMC Genom.* **2019**, *20*, 911. [CrossRef]
41. Quan, S.; Niu, J.; Hang, Z.; Qin, Y. Genome-wide identification, classification, expression and duplication analysis of GRAS family genes in *Juglans regia* L. *Sci. Rep.* **2019**, *9*, 11643. [CrossRef]
42. Sheng, F.; Dong, Z.; Cai, G.; Ming, Z.; Wu, H.; Li, Y.; Shen, Y.; Han, M. Identification, classification, and expression analysis of GRAS gene family in *Malus domestica*. *Front. Physiol.* **2017**, *8*, 253.
43. Shan, Z.; Luo, X.; Wu, M.; Wei, L.; Zhu, Y. Genome-wide identification and expression of GRAS gene family members in cassava. *BMC Plant Biol.* **2020**, *20*, 46. [CrossRef]
44. Wang, Y.; Shi, S.; Zhou, Y.; Zhou, Y.; Tang, X. Genome-wide identification and characterization of GRAS transcription factors in sacred lotus (*Nelumbo nucifera*). *PeerJ* **2016**, *4*, e2388. [CrossRef]
45. Wang, N.; Wang, K.; Li, S.; Jiang, Y.; Li, L.; Zhao, M.; Jiang, Y.; Zhu, L.; Wang, Y.; Su, Y.; et al. Transcriptome-wide identification, evolutionary analysis, and GA stress response of the GRAS gene family in *Panax ginseng* C. A. Meyer. *Plants* **2020**, *9*, 190. [CrossRef]
46. Liu, X.; Widmer, A. Genome-wide comparative analysis of the GRAS gene family in *Populus, Arabidopsis* and rice. *Plant Mol. Biol. Rep.* **2014**, *32*, 1129–1145. [CrossRef]
47. Huang, W.; Xian, Z.; Kang, X.; Tang, N.; Li, Z. Genome-wide identification, phylogeny and expression analysis of GRAS gene family in tomato. *BMC Plant Biol.* **2015**, *15*, 209. [CrossRef]
48. Chen, K.; Li, H.; Chen, Y.; Zheng, Q.; Li, B. *TaSCL14*, a novel wheat (*Triticum aestivum* L.) GRAS gene, regulates plant growth, photosynthesis, tolerance to photooxidative stress, and senescence. *J. Genet. Genom.* **2015**, *42*, 21–32. [CrossRef]

49. Guo, Y.; Wu, H.; Li, X.; Li, Q.; Zhao, X.; Duan, X.; An, Y.; Wei, L.; An, H.; Sun, M.X. Identification and expression of GRAS family genes in maize (*Zea mays* L.). *PLoS ONE* **2017**, *12*, e0185418. [CrossRef]
50. Murase, K.; Hlirano, Y.; Sun, T.P.; Hakoshima, T. Gibberellin-induced DELLA recognition by the gibberellin receptor GID1. *Nature* **2008**, *456*, 459–463. [CrossRef]
51. Park, J.; Nguyen, K.T.; Park, E.; Jeon, J.S.; Choi, G. DELLA proteins and their interacting RING finger proteins repress gibberellin responses by binding to the promoters of a subset of gibberellin-responsive genes in *Arabidopsis*. *Plant Cell* **2013**, *25*, 927–943. [CrossRef]
52. Fu, X.; Richards, D.E.; Ait-Ali, T.; Hynes, L.W.; Harberd, N.P. Gibberellin-mediated proteasome-dependent degradation of the barley DELLA protein SLN1 repressor. *Plant Cell* **2002**, *14*, 3191–3200. [CrossRef]
53. Chandler, P.M.; Marion-Poll, A.; Ellis, M.; Gubler, F. Mutants at the Slender1 locus of barley cv Himalaya. Molecular and physiological characterization. *Plant Physiol.* **2002**, *129*, 181–190. [CrossRef]
54. Gubler, F.; Chandler, P.M.; White, R.G.; Llewellyn, D.J.; Jacobsen, J.V. Gibberellin signaling in barley aleurone cells. Control of SLN1 and GAMYB expression. *Plant Physiol.* **2002**, *129*, 191–200. [CrossRef]
55. Torres-Galea, P.; Huang, L.F.; Chua, N.H.; Bolle, C. The GRAS protein SCL13 is a positive regulator of phytochrome-dependent red light signaling but can also modulate phytochrome A responses. *Mol. Genet. Genom.* **2006**, *276*, 13–30. [CrossRef]
56. Hao, T.; Peng, W.; Wang, Q.; Wang, B.; Sun, J. Reconstruction and application of protein-protein interaction network. *Int. J. Mol. Sci.* **2016**, *17*, 907. [CrossRef]
57. Goldy, C.; Pedroza-Garcia, J.A.; Breakfield, N.; Cools, T.; Rodriguez, R.E. The *Arabidopsis* GRAS-type SCL28 transcription factor controls the mitotic cell cycle and division plane orientation. *Proc. Natl. Acad. Sci. USA* **2021**, *118*, e2005256118. [CrossRef]
58. Cruz, T.; Carvalho, R.F.; Richardson, D.N.; Duque, P. Abscisic acid (ABA) regulation of *Arabidopsis* SR protein gene expression. *Int. J. Mol. Sci.* **2014**, *15*, 17541–17564. [CrossRef]
59. Yu, F.; Cao, X.; Liu, G.; Wang, Q.; Xia, R.; Zhang, X.; Xie, Q. ESCRT-I component VPS23A is targeted by E3 ubiquitin ligase XBAT35 for proteasome-mediated degradation in modulating ABA signaling. *Mol. Plant* **2020**, *13*, 1556–1569. [CrossRef]
60. Jie, F.; Hua, W.; Ma, S.; Xiang, D.; Liu, R.; Xiong, L. OSJAZ1 attenuates drought resistance by regulating JA and ABA signaling in rice. *Front. Plant Sci.* **2017**, *8*, 2108.
61. Park, S.R.; Hwang, J.; Kim, M. The Arabidopsis WDR55 is positively involved in ABA-mediated drought tolerance response. *Plant Biotechnol. Rep.* **2020**, *14*, 407–418. [CrossRef]
62. Gao, Y.; Jiang, W.; Dai, Y.; Xiao, N.; Zhang, C.; Li, H.; Lu, Y.; Wu, M.; Tao, X.; Deng, D. A maize phytochrome-interacting factor 3 improves drought and salt stress tolerance in rice. *Plant Mol. Biol.* **2015**, *87*, 413–428. [CrossRef]
63. Xu, P.; Chen, H.; Li, T.; Xu, F.; Mao, Z.; Cao, X.; Miao, L.; Du, S.; Hua, J.; Zhao, J. Blue light-dependent interactions of CRY1 with GID1 and DELLA proteins regulate gibberellin signaling and photomorphogenesis in *Arabidopsis*. *Plant Cell* **2021**, *33*, 2375–2394. [CrossRef]
64. Wang, Y.; Deng, D. Molecular basis and evolutionary pattern of GA–GID1–DELLA regulatory module. *Mol. Genet. Genom.* **2014**, *289*, 1–9. [CrossRef]
65. Ariizumi, T.; Lawrence, P.K.; Steber, C.M. The role of two f-box proteins, SLEEPY1 and SNEEZY, in *Arabidopsis* gibberellin signaling. *Plant Physiol.* **2011**, *155*, 765–775. [CrossRef]

forests MDPI

Article

Transcriptome Analysis of Response to Aluminum Stress in *Pinus massoniana*

Ting Wang [1,†], Ying Hu [1,2,†], Hu Chen [1,3,4], Jianhui Tan [1], Huilan Xu [1,3], Peng Li [1,3], Dongshan Wu [1,4], Jie Jia [1,4] and Zhangqi Yang [1,3,4,*]

1 Guangxi Key Laboratory of Superior Timber Trees Resource Cultivation, Guangxi Forestry Research Institute, Nanning 530002, China; tingwang_dev@163.com (T.W.); huying19863008@163.com (Y.H.); chenhubeijing-2008@163.com (H.C.); tanjianhui94@126.com (J.T.); xuhuilan1220@163.com (H.X.); lipeng121@sina.com (P.L.); wudshan@126.com (D.W.); jiajia04@163.com (J.J.)
2 Masson Pine Engineering Research Center of the State Forestry Administration, Nanning 530002, China
3 Masson Pine Engineering Research Center of Guangxi, Nanning 530002, China
4 College of Forestry, Guangxi University, Nanning 530004, China
* Correspondence: yangzhangqi@163.com
† These authors contributed equally to this work.

Abstract: *Pinus massoniana* is a vital kind of coniferous species rich in rosin. Aluminum stress is a severe problem for *P. massoniana* growth in acidic soil causing root poisoning. However, the molecular mechanisms of aluminum-responsive are still unclear. We performed a transcriptome analysis of the *P. massoniana* root in response to aluminum stress. Through WGCNA analysis, we identified 338 early and 743 late response genes to aluminum stress. Gene Ontology analysis found many critical functional pathways, such as carbohydrate binding, cellulase activity, and phenylalanine ammonia-lyase activity. In addition, KEGG analysis revealed a significant enrichment of phenylpropanoid biosynthesis pathways. Further analysis showed that the expression of lignin synthesis genes *4CL*, *CAD*, and *COMT* were up-regulated, indicating that they may play a crucial role in the process of aluminum tolerance in *P. massoniana* roots. These results provide method support for studying the regulation mechanism of *P. massoniana* aluminum stress.

Keywords: *Pinus massoniana*; aluminum stress; transcriptomic; WGCNA analysis; phenylpropanoid biosynthesis

Citation: Wang, T.; Hu, Y.; Chen, H.; Tan, J.; Xu, H.; Li, P.; Wu, D.; Jia, J.; Yang, Z. Transcriptome Analysis of Response to Aluminum Stress in *Pinus massoniana*. *Forests* **2022**, *13*, 837. https://doi.org/10.3390/f13060837

Academic Editor: Yuepeng Song

Received: 8 April 2022
Accepted: 23 May 2022
Published: 27 May 2022

1. Introduction

Aluminum stress is a severe problem for plant growth in acidic soil [1,2]. In neutral or weakly acidic conditions, most of the aluminum in the soil exists in the fixed form of silicates and oxides, and it is generally not harmful to plants. However, under acidic conditions (pH < 5), fixed aluminum is easily activated to produce soluble aluminum, mainly in the form of trivalent aluminum (Al^{3+}), which has strong toxicity to plants and inhibits plant growth [3,4]. Over the past decades, the research on plant root growth and the physiology of aluminum stress has made great progress [5–7]. The root tip is the first tissue to contact and feel the aluminum toxicity [8]. A large number of experiments have shown that aluminum toxicity can inhibit the growth and division of root cells, thereby affecting the absorption of water and nutrients [9], such as rice [10], maize [11], and tea plant [12]. Meanwhile, aluminum toxicity will break the original balance of physiological reactions in plants, promotes excessive accumulation of active oxygen, and causes oxidative stress in plants [13,14].

Recently, with the rapid development of sequencing technology, many researches have described the internal response mechanism of plants to aluminum ions. The effect of Al on maize grown in acidic soil was studied, and it was discovered that there were several specifically expressed genes, and the expression of genes encoding citric acid cycle enzymes was up-regulated at the same time [15]. The MATE gene family encodes citrate transporter and plays a vital role in aluminum stress response [16]. Studies on alfalfa showed that Al treatment induced the expression of MsMATE transporters, indicating that these genes might be related to the aluminum tolerance to alfalfa [17]. In addition, ASR5 and STOP1 are key transcription factors that play an important role in the expression of al-responsive genes in rice [18] and *Arabidopsis* [19].

P. massoniana, one of the most important coniferous species in southern China, is widely cultivated with strong adaptability, rapid growth, and good stress tolerance [20,21]. *P. massoniana* has an important position in forestry production. It provides an essential source of timber and rosin, and it can be used as building materials and as an important raw material for chemical production [22]. In recent years, due to the increase of acidic sediments caused by industrialization, acid aluminum stress has seriously poisoned the roots of *P. massoniana*, causing the decline and even death [23,24]. Previous research of *P. massoniana* found that under aluminum stress the accumulation of Al ions primarily was in roots and small amounts of Al were transported aboveground [23]. However, there is little research on molecular mechanisms of A1 toxicity in *P. massoniana*. This suggested it will be meaningful to study the response mechanism of *P. massoniana* under aluminum stress. In this study, *P. massoniana* seedlings were under aluminum treatments, and the ones under water treatments served as controls. We analyzed the root transcriptome changes at different times of treatment stages. The aim was to identify the Al responsive genes and gain new insights into molecular mechanisms of aluminum stress and tolerance.

2. Materials and Methods

2.1. Plant Materials and Experimental Treatments

The seeds of *P. massoniana* were from Guangxi Forestry Research Institute (Nanning city, Guangxi Province, China). Seeds were sterilized with sodium hypochlorite for 30 min, washed with sterile water 3–5 times, then with pregermination in an artificial climate box at 28 °C for 24 h. After that, every 60 seeds were sowed evenly in 15 plastic pots containing yellow soil and coconut bran (3:1) and cultured under constant temperature conditions (25 ± 2 °C) and photoperiod (14/10 h light/dark cycle) for 30 days.

Then healthy seedlings of similar size were selected for treatment. We selected seedlings with consistent growth status for aluminum treatment. In the formalized treatment, seedlings were cultivated with 50 μM $AlCl_3$ (50 mL of every pot and 1 time every 2 days for three times). The control group was treated with water instead of $AlCl_3$ Samples of roots were collected. 0 d (sample of control group denoted as C0, as well as A0), 3 d (C1 and A1), and 6 d (C2 and A2) after treatment.

Samples for transcriptome sequencing and real-time quantitative PCR were root tips with a length of 2 cm and collected 0.5 g. All samples had 3 biological replicates and were stored at minus 80 °C for further use. Meanwhile, Samples for microstructure observation were 0.5 cm tissue away from the root tip fixed in FAA (70% alcohol 90 mL, glacial acetic acid 5 mL, and formaldehyde 5 mL).

2.2. Preparation of Paraffin Section

Paraffin sections were used to observe the cell morphology, according to previous research methods [25]. After embedding, slicing, staining, and making slices, permanent slices were made. The longitudinal sections were cut using the RM2235 rotary microtome (Leica, Wetzlar, Germany), and the slice thickness was 8 μm. Photomicrographs were taken using the Nikon Eclipse E100 microscope (Nikon, Tokyo, Japan) and edited using the Nikon DS-U3 imaging system. We observed the structure of root tips, measured cell wall thickness, and so on. The data were tested for differences by Student's Test ($p < 0.05$).

2.3. RNA Extraction, Transcriptome Sequencing and Genome Mapping

Total RNA from each sample was extracted using the RNeasy Plant Mini Kit (Qiagen, Venlo, The Netherlands) according to the manufacturer's instructions. The RNA integrity was confirmed using the 2100 Bioanalyzer (Agilent, Santa Clara, CA, USA), and the samples with RNA integrity >8 were used for library preparation. The prepared library was sequenced using the Illumina HiSeq 4000 platform (Illumina, San Diego, CA, USA). To obtain high-quality read data for sequence analysis, the raw reads containing adapter sequences and low-quality sequences were removed. After that, the clean reads were assembled into unigenes as the reference sequences using the Trinity (v2.4.0) [26] and were mapped back to reference sequences using the Bowtie2 (v2.2.5) [27]. Then, the read numbers of each gene were calculated by the RSEM (v1.2.8) [28]. Transcriptome sequencing and analysis were performed by BGI (Shenzhen, China).

2.4. Identification of Significant DEGs

The expression levels of each gene were estimated by fragments per kilobase per transcript per million mapped reads (FPKM). Differentially expressed genes (DEGs) analysis was performed using the R package DESeq2 (v1.32.0) [29]. The significant DEGs were filtered with $|\log_2 FC| \geq 2$ and FDR ≤ 0.001 in each pairwise comparison, and the number of DEGs in different comparisons was visualized using the R package UpSetR (v1.4.0) [30]. To infer the putative functions of DEGs, the Gene Ontology (GO) term and Kyoto Encyclopedia of Genes and Genomes (KEGG) pathway analysis was performed using the OmicShare tools, an online platform for data analysis (https://www.omicshare.com/tools (accessed on 15 February 2022)).

2.5. Construction and Visualization of Co-Expression Network

The WGCNA is a systems biology method aimed at finding modules of highly correlated genes and describing the co-expression network [31]. Modules are clusters of highly interconnected genes, and genes within the same cluster have high correlation coefficients among them. To reveal patterns of gene expression, the DEGs were used to perform the weighted gene co-expression network analysis, and co-expression networks were constructed using the R package WGCNA (v1.70.3) [31]. In this study, the genes with a low expression (FPKM < 0.05) and low variation of median absolute deviation (MAD < 0.01) were discarded. Based on the criterion of an approximate scale-free network, the adjacency matrix was calculated with a soft threshold ($\beta = 14$). The co-expression network was visualized using the Cytoscape (v3.8.2) [32]. We used Mapman (https://mapman.gabipd.org/ (accessed on 15 February 2022)) for drawing metabolic pathway maps and used Tbtools [33] and GraphPad Prism 7.0 (GraphPad Software, San Diego, CA, USA) for mapping gene cluster heat maps.

2.6. Real-Time Quantitative PCR (qRT-PCR) Analysis

To verify the results of RNA-Sequence, 10 DEGs were randomly selected to perform qRT-PCR, and *ACT1* was selected as the housekeeping gene according to a previous report [34]. The specific primers for the qRT-PCR were designed using the primer primier5.0 (Table S1). Total RNA from each sample was extracted using the RNeasy Plant Mini Kit (Qiagen, Venlo, The Netherlands) according to the manufacturer's instructions. Total RNA was reverse transcribed to cDNA using PrimeScript RT reagent Kit (Takara, Kusatsu, Japan). The qRT-PCR was carried out using a Light-Cycler 96 Real-Time PCR System (Roche, Basel, Switzerland), and the relative expression levels of each gene were calculated using the $2^{-\Delta\Delta Ct}$ method [35].

3. Results

3.1. Root Tip Microstructure under Aluminum Stress

Roots play a vital role in plant growth, and root growth inhibition is a typical symptom of plants poisoned by aluminum. In order to reveal the root changes under aluminum stress, we observed the longitudinal section structures of the root tip by microscopy (Figure 1A). In the absence of aluminum stress, the cell structure of the root tip meristem is clear, arranged tightly and orderly, and the elongation zone cells are full and regular in shape.

Compared with the control, aluminum stress changed the shape of root tip cells, the cells in the root tip meristem were loosely arranged, the cortical cells in the elongation zone became flattened, the intercellular space became smaller, the cell wall became wrinkled and uneven, and the meristem under severe aluminum stress cells are ruptured and the structure is blurred. Further analysis of the root tip cell structure showed that the root tip width and cell wall thickness under aluminum stress were significantly lower than those of the control (Figure 1B). Cell wall thickness was changed from 2.93 μm in C0 to 2.01 μm in A1 and 1.91 in A2. Root tip width was changed from 434.95 μm in C0 to 387.90 μm in A1 and 324.20 μm in A2. These structural transformations indicate that the root tip structure significantly changed when exposed to aluminum stress, and aluminum toxicity destroys the typical morphological structure of root tip cells.

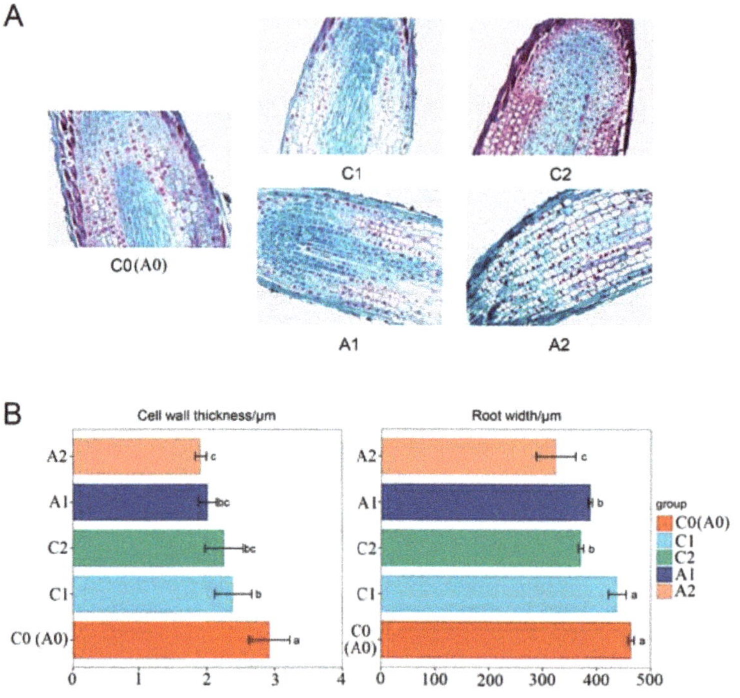

Figure 1. Effects of aluminum stress on root. (**A**) Longitudinal anatomy of root at different aluminum stress stages (samples of C0, C1, C2, A1, and A2). (**B**) Statistics of cell wall thickness and root width. The values are represented as Mean ± SE, and different lowercase between samples indicate significance at $p < 0.05$. C0 and A0 were same samples and C0 was used to represent the two samples in this paper.

3.2. The Global Transcriptome Analysis under Aluminum Stress

RNA sequencing generated 63.24 to 68.51 million (M) of 150 bp pair-end reads from 15 different root samples at different aluminum stress stages (C0, C1, C2, A1, and A2). After trimming the adapter and low-quality reads, the counts of clean reads ranged from 54.89 to 58.86 M, the percentages of Q30 were greater than 80%, and the percentages of mapped reads were more significant than 70% (Table S2). The three components of PCA could divide the time series groups well. Principal component analysis revealed that the samples could be clearly assigned into five groups (C0, C1, C2, A1, and A2). Furthermore, a high correlation between three biological replicates was observed (Figure 2A,B), indicating the consistency between samples was good. Figure 2E showed a heat map of a total of

41,107 different expression genes among samples and classification situations. The heat map showed that three replicates of the same sample were clustered together, and genes responded differently at different times of treatment.

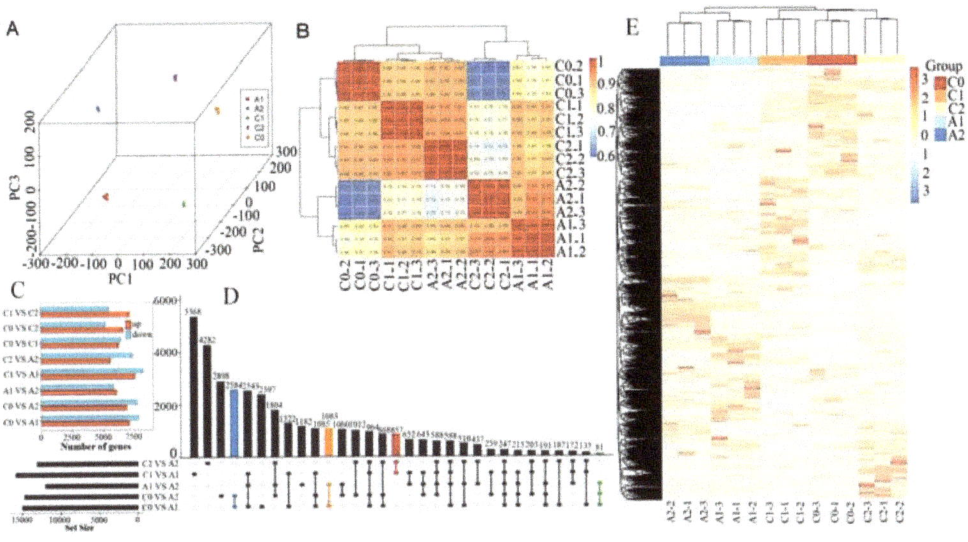

Figure 2. Global analysis of transcriptome data. (**A**) Principal Component Analysis (PCA) of the RNA-Seq data. (**B**) Sample correlation coefficient clustering heat–map. (**C**) Comparison of the differential expressed genes of each pair. The red or blue bar indicates the up-regulated or down−regulated genes, respectively. (**D**) The Venn analysis for the number of DEGs between different-sample vs sample-groups. (**E**) Heat map indicating the different expression pattern of DEGs for classification.

3.3. *Identification of Differentially Expressed Genes (DEGs) under Aluminum Stress*

For a better understanding of gene expression dynamics under aluminum stress, the differentially expressed genes (DEGs) were identified between these samples of C0, C1, C2, A1, and A2. The differential expression analysis revealed a large number of candidate DEGs (Figure 2C,D). In total, 15,000 DEGs were identified in the comparison C0_vs_A1, with 7129 DEGs up-regulated and 7871 DEGs down-regulated. 14702 DEGs were identified in the comparison C0_vs_A2, with 6929 DEGs up-regulated and 7773 DEGs down-regulated. 11996 DEGs were identified in the comparison A1_vs_A2, with 6104 DEGs up-regulated and 5892 DEGs down-regulated. Among all these comparisons above, the total number of down-regulated genes was higher than up-regulated genes. Meanwhile, 857 DEGs were identified simultaneously between C1_vs_A1 and C2_vs_A2. 2584 DEGs were identified simultaneously between the comparisons of C0_vs_A1 and C0_vs_A2. 1083 DEGs were identified simultaneously between the comparisons of C0_vs_A1 and A2_vs_A2. Only 81 DEGs were identified simultaneously in the three comparisons of C0_vs_A1, C0_vs_A1, and A2_vs_A2. The results indicated that aluminum stress induces transcripts changes in *P. massoniana* shoots.

3.4. *The Co-Expression Network Analysis of DEGs under Aluminum Stress*

To identify genes related to aluminum stress, Weighted gene co-expression network analysis (WGCNA) be performed using non-redundant DEGs. After filtering out the genes with a low expression (FPKM < 0.05), 6641DEGs were retained for the WGCNA. The analysis identified 17 modules (labeled with different colors) shown in the cluster dendrogram, in which major tree branches define the modules, and each leaf in the branch is one gene (Figure 3A). The module eigengene is defined as the first principal component of a given module and represents the module's gene expression profile. The correlation coefficient is calculated between module eigengene and sample traits for the module-trait relationships (Figure 3B). Notably, 2 out of 17 co-expression modules were highly expressed under aluminum stress (Figure 3C,D). The black module identified 338 genes and was highly correlated with the A1 sample ($r = 0.94$, $p = 1 \times 10^{-7}$); the expression levels of the genes belonging to the module were considered early aluminum-responsive genes. Through co-expression networks, we found genes, including Rab GDP dissociation inhibitor (*RabGDI1*), 5-methyltetrahydropteroyltriglutamate—homocysteine methyltransferase (*METE*), and heat shock proteins (*HSP90*), that play key early regulatory roles (Figure S2A). The brown module identified 743 genes and was highly correlated with the A2 sample ($r = 0.99$, $p = 4 \times 10^{-13}$). The expression levels of the genes belonging to the module were considered late aluminum-responsive genes. These genes mainly included beta-galactosidase (*GLB1*), xylose isomerase (*xylA*), interleukin-1 receptor-associated kinase 4 (IRAK4), zinc finger FYVE domain-containing protein 26 (*ZFYVE26*), and so on (Figure S2B).

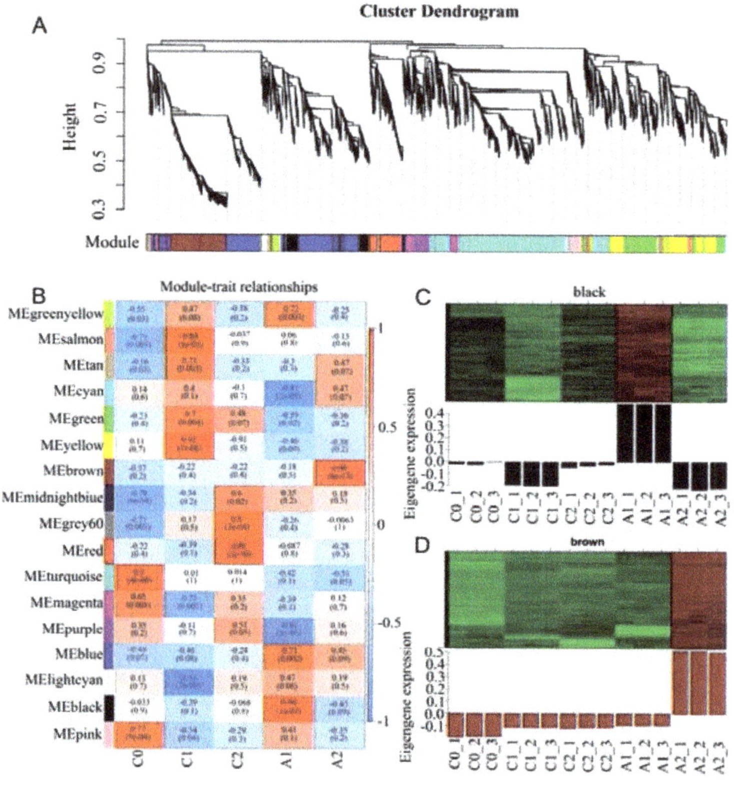

Figure 3. Weighted gene co-expression network analysis (WGCNA) of DEGs identified. (**A**) Gene cluster dendrogram tree showing 17 modules of co-expressed genes. Each of the 6641 DEGs is represented by a tree leaf and each of the modules by a major tree branch. The color row underneath the dendrogram shows the module assignment determined by the Dynamic Tree Cut. (**B**) Module−trait relationships. Each row corresponds to a module, and each column corresponds to a time point. The number of each cell at the row−column intersection indicates the module−trait correlations and corresponding p−values (in parentheses) between the module and the time points. The left panel shows the 17 modules. The right panel shows the color scale of module−trait correlations from blue to red (−1 to 1). (**C**) Gene expression of the black module. (**D**) Gene expression of the brown module.

3.5. Functional Annotations of Aluminum-Responsive Genes

To further assess the biological functions of aluminum-responsive genes, we performed the Gene Ontology (GO) and Kyoto Encyclopedia of Genes and Genomes (KEGG) enrichment analyses GO term enrichment analysis of the early aluminum-responsive genes was mainly enriched in carbohydrate binding, cellulase activity, and phenylalanine ammonia-lyase activity (Figure 4A). However, the GO term enrichment analysis of the early aluminum-responsive genes was mainly enriched in oxidoreductase activity, hydrogen peroxide metabolic process, and catalytic activity (Figure 4B). It could be seen that most of the genes involved in aluminum stress were classified as molecular functions, followed by those classified as biological processes. The KEGG pathway enrichment analysis of the early aluminum-responsive genes was mainly enriched in Metabolic pathways, Biosynthesis of secondary metabolites, and Phenylpropanoid biosynthesis (Figure 4C). However, the KEGG pathway enrichment analysis of the late aluminum-responsive genes was mainly enriched in Phenylpropanoid biosynthesis, Metabolic pathways, and MAPK signaling pathway-plant (Figure 4D). The results indicated that aluminum stress has a significant impact on the root of *P. massoniana*, and the significant enrichment of biological functions is related to the response of *P. massoniana* to aluminum stress.

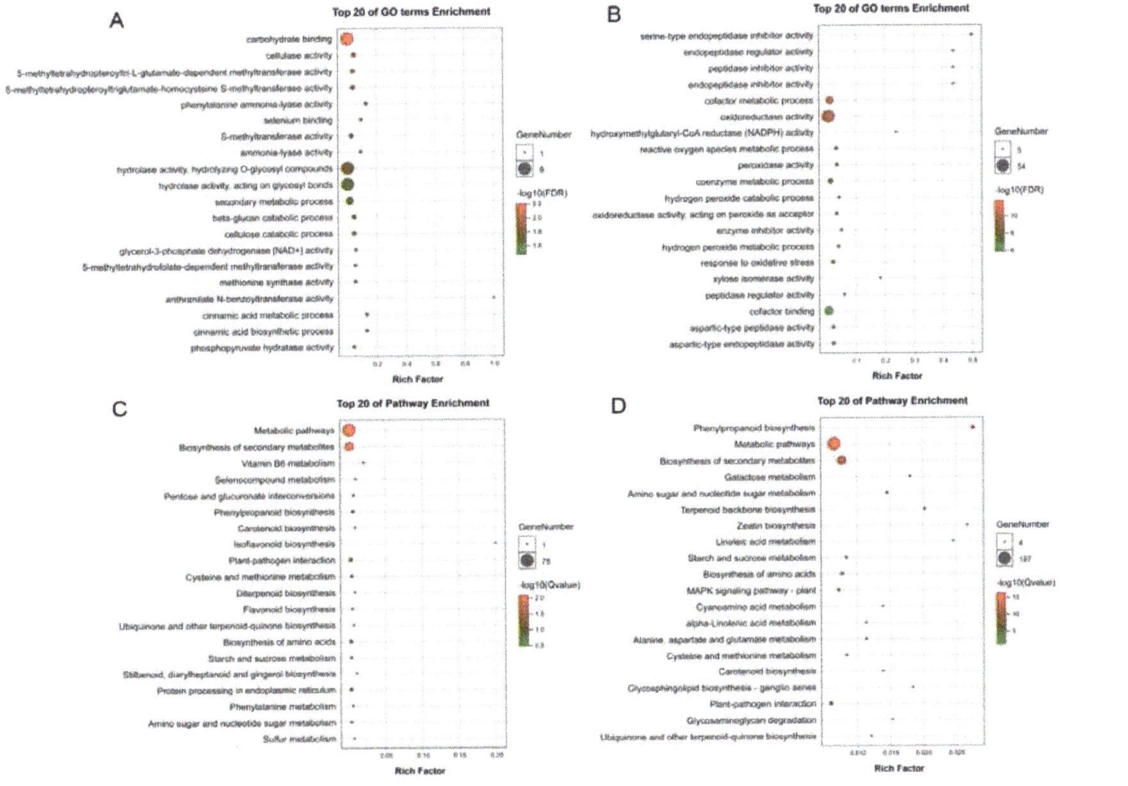

Figure 4. Gene enriched on different GO terms and KEGG pathway. (**A**) GO term enrichment analysis of the early aluminum-responsive genes. (**B**) GO term enrichment analysis of the late aluminum−responsive genes. (**C**) KEGG pathway enrichment analysis of the early aluminum−responsive genes. (**D**) KEGG pathway enrichment analysis of the late aluminum−responsive genes.

3.6. Expression of Genes Related to Phenylpropanoid Biosynthesis

Phenylpropanoid metabolism is one of the essential metabolisms in plants, associated with plant development and plant-environment interplay. Due to most of the early and late aluminum-responsive genes were enriched in phenylpropanoid biosynthesis, a closer pathway analysis of these genes was conducted (Figure 5). The key enzymes that catalyzed the reactions of the phenylpropanoid pathway exhibited different expression levels in the five samples. The phenylalanine ammonia-lyase (PAL) is the gateway enzyme of the general phenylpropanoid pathway, which guides metabolic flux from the shikimate pathway to the numerous branches of phenylpropanoid metabolism. The *PAL* gene (isoform_9515, isoform_19788) was up-regulated in A1 compared with those in C0, C1, C2, and A2. The 4-coumarate-CoA ligase (4CL) enzyme plays a crucial role in generating CoA esters. The *4CL* gene (isoform_17392, isoform_243903, and isoform_248945) were significantly up-regulated in A2 compared with those in C0, C1, C2, and A1. The enzyme of caffeic acid 3-O-methyltransferase (COMT) catalyzes the methylation of caffeic acid. The *COMT* gene (isoform_26178, isoform_277382, isoform_289634, isoform_77630, and isoform_81932) was markedly higher in A2 compared with those in C0, C1, C2, and A1. The cinnamyl alcohol dehydrogenase (CAD) enzyme catalyzes the formation of p-coumaryl alcohol and coniferyl alcohol. The *CAD* gene was significantly higher in A2 compared with those in C0, C1, C2, and A1. Overall, the data indicate that differences in the expression levels of genes associated with the phenylpropanoid pathway may be a key factor in response to aluminum stress.

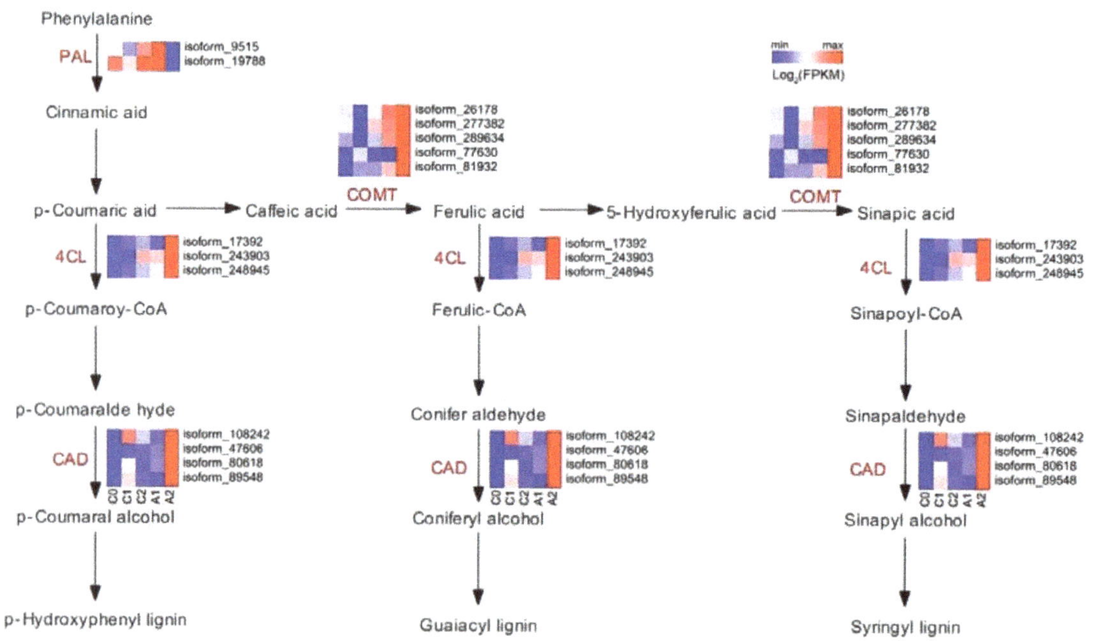

Figure 5. Expression of genes associated with phenylpropanoid biosynthesis. The colors scale ranging from green to red (row min to row max) indicates the expression of genes, and the expressions were presented as FPKM and normalized by log2.

3.7. The qRT-PCR Validation of DEGs under Aluminum Stress

To verify the results of RNA-Seq, 10 DEGs were selected to perform qRT-PCR. The expression trends of 10 genes were similar to the transcriptome data, and the correlation between the two data sets is strong (Figure S1). These results verified the reliability of RNA-Seq data.

4. Discussion

The root tip is the main site of aluminum toxicity, which is a common material used to study the response of plants to aluminum stress [36]. Aluminum causes extensive root damage, leading to poor absorption of ions and water [37]. Due to the changes in root tissue structure and biochemical processes caused by aluminum stress, it has been observed that root tip growth is inhibited by aluminum stress in many plants [38]. Based on microscopic observations, we observed that aluminum stress has a tremendous negative effect on the roots of the *P. massoniana*. Compared with the control, the root anatomical characteristics of the roots under aluminum stress are pretty different. The cortical parenchyma cells are large, and the cell wall is wavy. That is similar to the observation results of the morphological anatomy of the corn root under aluminum toxicity [39]. Studies have shown that aluminum stress can reduce the size of plant root tips, among which root cap cells, meristems, and elongation cells are the most severely affected parts [40]. Counted the cell wall thickness and root tip width, we found that it was significantly lower than the control level, further confirming the loss of root tip caused by aluminum toxicity. Aluminum stress destroys the integrity of the cell structure, and affects the composition of the cell wall.

We found that *RabGDI1*, *METE*, and *HSP90* might play important roles in the initial stage of response to aluminum stress. *RabGDI1* has been confirmed that played an important role in improving rough dwarf disease resistance of maize [41]. *RabGDI1* regulated intracellular vesicular trafficking to enhance salt tolerance of Chilense, and is highly expressed in roots [42]. *HSP* has been proven to play important roles in a variety of abiotic stresses [43]. Under heavy metal stress, plants improved responses by enhancing the expression of *METE*, or by intensifying the cytoskeleton and stimulation of ethylene metabolism [44]. After plants received a signal of abiotic stress, the content of beta-galactosidase increased, thus alleviating plant injury [45]. The higher expression level of xylose Isomerase indicated that the plant had started glucose metabolism in response to stress. This discovery of Xylose Isomerase provided a new idea for improving the aluminum tolerance of the plant through exogenous addition treatment in the future.

To adapt to the environment, plants have evolved complex response mechanisms [46]. The external morphology and physiological response in a stress environment are related to the gene expression in the body [47,48]. In this study, 220,204 genes were identified from the root tips of *P. massoniana*, and most DEGs were down-regulated under aluminum stress. Weighted correlation network analysis (WGCNA) can aggregate modules of highly correlated genes and associate the modules with sample traits to identify sample-specific modules and candidate central genes [31]. In this study, WGCAN analysis was used to identify the early and late modules of the aluminum stress response. GO analysis of the genes in these modules showed that most of the genes were enriched in carbohydrate binding, cellulase activity, and phenylalanine ammonia-lyase activity. KEGG enrichment analysis showed that Phenylpropanoid biosynthesis is a strong response metabolic pathway. The biosynthetic pathway of phenylalanine starts with phenylalanine. After a series of enzymatic reactions, phenylalanine can be converted into aromatic compounds, including benzene, flavonoids, and lignin. These compounds are usually involved in plant development and plant-environmental interactions [49].

As the primary binding site of aluminum, the cell wall of the root tip is the first position to contact and feel Al^{3+}, so it is the first barrier for cells to resist aluminum poisoning [50]. Al^{3+} inhibits root cell elongation and cell division by interacting with plant cell walls and plasma membranes. The toughness of the cell wall regulates the elongation of the cell. Lignin is an essential part of the cell wall and determines the toughness of the cell wall [51]. Further analysis of Phenylpropanoid biosynthesis metabolism revealed that most of the catalytic genes known to be involved in lignin biosynthesis were lower expressed at the first time, and then more highly expressed under aluminum stress after treatment of 6 d, such as PAL, 4CL, COMT, and CAD. Previous studies have shown that when subjected to external stresses, such as mechanical damage, pathogen invasion, and heavy metals, the lignin content of plant root tip cell walls will all show varying degrees of increase [52]. Lignin provides mechanical strength and toughness to the cell wall and promotes the formation of xylem vessels. Moreover, lignin accumulation helps plants respond to various abiotic stresses [44]. It was found in cotton that the *4CL* up-regulated gene leads to the accumulation of lignin, which promotes the thickening of cell walls and reduces the permeability of cells to water, which may help plants fight drought stress. The transcription of lignin synthesis genes *4CL* and *CAD* are up-regulated, leading to lignin deposition, thickening of secondary cell walls, and enhancing the salt-resistance and permeability of birch and apple [53,54]. The expression of lignin synthesis gene *PAL* and *4CL* in loquat is induced, and lignin accumulation may benefit plants adapting to the cold environment [55]. In this study, the up-regulated expression of lignin synthesis genes *4CL*, *CAD*, and *COMT* genes might help *P. massoniana* survive under aluminum stress and involve cell rebuild in the late stage of stress.

The experimental results showed that there were many differentially expressed genes at different control sampling time points of C0, C1, and C2, which was related to the experimental materials used in this study. We used about 30-day of *P. massoniana* seedlings, which were in the rapid growth period of seedlings, and there might be a large number of differentially expressed genes regulating growth and development. In the future, we will further verify whether the removal of different genes at different time points will be more conducive to the discovery of key regulatory genes in *P. massoniana* responding to aluminum stress. *P. massoniana* is a typical ectomycorrhizal tree. Studies have shown that Mycorrhizal *P. massoniana* can improve plant drought resistance [56] and low phosphorus tolerance [57], regulate root heavy metal migration [58], and aluminum tolerance [24]. Mycorrhizal promotes the expression of the protective enzyme system gene and MATE gene in *P. massoniana* under aluminum stress, thus enhancing the activity of plant protective enzyme and nutrient metabolism, so as to enhance the absorption and utilization of nutrient elements and water in plants [24]. Under stress conditions, root secretion is involved in improving the utilization efficiency of plant resources, promoting the "dialogue" between plants and soil microorganisms to improve stress, and carboxylate plays an important role in aluminum decomposition [59]. Therefore, the interaction between mycorrhizal *P. massoniana* and root exudates, root microorganisms, and N and P elements will be the focus of future research.

In conclusion, we used WGCNA analysis to aggregate genes related to the Al stress response. Further functional analysis shows that the stress response is related to multiple pathways. The phenylpropanoid biosynthesis pathway is significantly enriched. Gene expression catalyzed lignin synthesis helps *P. massoniana* adapt to aluminum stress.

Supplementary Materials: The following supporting information can be downloaded at: https://www.mdpi.com/article/10.3390/f13060837/s1, Figure S1. The correlation between qPCR and RNA sequencing. Figure S2. Co-representation network. Table S1. Gene and primers for qRT-PCR. Table S2. Statistics on the quality and output of the RNA-Seq libraries.

Author Contributions: T.W., Y.H. and H.C. designed and conducted the experiments and wrote the manuscript; J.T. and H.X. contributed to manuscript writing and editing; P.L. executed the bioinformatics tools; D.W. and J.J. performed the physiology and biochemistry experiments and analyzed the data; and Z.Y. contributed to the experimental design and editing. All authors have read and agreed to the published version of the manuscript.

Funding: This research was supported by Guangxi Key Laboratory of Superior Timber Trees Resource Cultivation (17-B-01-01), the Guangxi Natural Science Foundation (2019GXNSFDA245033, 2019GXNSFBA245064), the Special Fund for Bagui Scholar and Bagui Young Scholar, the Natural Science Foundation of China (32060348, 31660219), and the Guangxi Science and Technology and Talents Special Project (AD19254004).

Institutional Review Board Statement: Not applicable.

Informed Consent Statement: Not applicable.

Data Availability Statement: Data is contained within the article or Supplementary Materials. It is also available from the correspondence author (yangzhangqi@163.com).

Conflicts of Interest: The authors declare no conflict of interest.

References

1. Godbold, D.L.; Fritz, E.; Huttermann, A. Aluminum toxicity and forest decline. *Proc. Natl. Acad. Sci. USA* **1988**, *85*, 3888–3892. [CrossRef] [PubMed]
2. Kochian, L.V.; Piñeros, M.A.; Liu, J.; Magalhaes, J.V. Plant adaptation to acid soils: The molecular basis for crop aluminum resistance. *Annu. Rev. Plant Biol.* **2015**, *66*, 571–598. [CrossRef] [PubMed]
3. Liu, J.; Piñeros, M.A.; Kochian, L.V. The role of aluminum sensing and signaling in plant aluminum resistance. *J. Integr. Plant Biol.* **2014**, *56*, 221–230. [CrossRef] [PubMed]
4. Sade, H.; Meriga, B.; Surapu, V.; Gadi, J.; Sunita, M.S.L.; Suravajhala, P.; Kavi Kishor, P.B. Toxicity and tolerance of aluminum in plants: Tailoring plants to suit to acid soils. *Biometals* **2016**, *29*, 187–210. [CrossRef]
5. Zhu, X.; Wang, P.; Bai, Z.M.; Herde, M.; Ma, Y.Q.; Li, N.; Liu, S.; Huang, C.F.; Cui, R.X.; Ma, H.Y.; et al. Calmodulin-like protein CML24 interacts with CAMTA2 and WRKY46 to regulate ALMT1-dependent Al resistance in *Arabidopsis thaliana*. *New Phytol.* **2022**, *233*, 2471–2487. [CrossRef]
6. Li, C.; Liu, G.; Geng, X.; He, C.; Quan, T.; Hayashi, K.I.; De Smet, I.; Robert, H.S.; Ding, Z.; Yang, Z.B. Local regulation of auxin transport in root-apex transition zone mediates aluminium-induced Arabidopsis root-growth inhibition. *Plant J.* **2021**, *108*, 55–66. [CrossRef]
7. Sawaki, Y.; Iuchi, S.; Kobayashi, Y.; Ikka, T.; Sakurai, N.; Fujita, M.; Shinozaki, K.; Shibata, D.; Kobayashi, M.; Koyama, H. STOP1 regulates multiple genes that protect arabidopsis from proton and aluminum toxicities. *Plant Physiol.* **2009**, *150*, 281–294. [CrossRef]
8. Kochian, L.V. Cellular Mechanisms of Aluminum Toxicity and Resistance in Plants. *Annu. Rev. Plant Biol.* **1995**, *46*, 237–260.
9. Poschenrieder, C.; Gunsé, B.; Corrales, I.; Barceló, J. A glance into aluminum toxicity and resistance in plants. *Sci. Total Environ.* **2008**, *400*, 356–368. [CrossRef]
10. Prakash, A.J.; Bedabrata, S.; Preetom, R.; Smita, S.; Umakanta, C.; Amit, P.; Anupam, R.; Kumar, P.S.; Manoj, P. Morpho-physiological analysis of tolerance to aluminum toxicity in rice varieties of North East India. *PLoS ONE* **2017**, *12*, e0176357.
11. Luma, C.; Deise, C.; Liliane, C.; Thays, C.C.; Ana, B. Nitrogen compounds, proteins and amino acids in corn subjected to doses of aluminum. *Afr. J. Agric. Res.* **2016**, *11*, 1519–1524. [CrossRef]
12. Ghanati, F.; Morita, A.; Yokota, H. Effects of Aluminum on the Growth of Tea Plant and Activation of Antioxidant System. *Plant Soil* **2005**, *276*, 133–141. [CrossRef]
13. Huang, W.; Yang, X.; Yao, S.; LwinOo, T.; He, H. Reactive oxygen species burst induced by aluminum stress triggers mitochondria-dependent programmed cell death in peanut root tip cells. *Plant Physiol. Biochem.* **2014**, *82*, 76–84. [CrossRef] [PubMed]
14. Li, N.N.; Meng, H.J.; Xing, H.T.; Liang, L.; Zhao, X.; Luo, K.M. Genome-wide analysis of MATE transporters and molecular characterization of aluminum resistance in *Populus*. *J. Exp. Bot.* **2017**, *68*, 5669–5683. [CrossRef]
15. Mattiello, L.; Begcy, K.; da Silva, F.R.; Jorge, R.A.; Menossi, M. Transcriptome analysis highlights changes in the leaves of maize plants cultivated in acidic soil containing toxic levels of Al^{3+}. *Mol. Biol. Rep.* **2014**, *41*, 8107–8116. [CrossRef]
16. Magalhaes, J.V.; Liu, J.; Guimarães, C.T.; Lana, U.G.P.; Kochian, L.V. A gene in the multidrug and toxic compound extrusion (MATE) family confers aluminum tolerance in sorghum. *Nat. Genet.* **2007**, *39*, 1156–1161. [CrossRef]
17. Min, X.; Jin, X.; Liu, W.; Wei, X.; Zhang, Z.; Ndayambaza, B.; Wang, Y. Transcriptome-wide characterization and functional analysis of MATE transporters in response to aluminum toxicity in *Medicago sativa* L. *PeerJ* **2019**, *7*, e6302. [CrossRef]
18. Arenhart, R.A.; Bai, Y.; Oliveira, L.F.V.D.; Neto, L.B.; Schunemann, M.; Maraschin, F.D.S.; Mariath, J. New insights into aluminum tolerance in rice: The ASR5 protein binds the STAR1 promoter and other aluminum-responsive genes. *Mol. Plant* **2014**, *7*, 709–721. [CrossRef]
19. Iuchi, S.; Koyama, H.; Iuchi, A.; Kobayashi, Y.; Kitabayashi, S.; Kobayashi, Y.; Ikka, T.; Hirayama, T.; Shinozaki, K.; Kobayashi, M. Zinc finger protein STOP1 is critical for proton tolerance in Arabidopsis and coregulates a key gene in aluminum tolerance. *Proc. Natl. Acad. Sci. USA* **2007**, *104*, 9900–9905. [CrossRef]
20. Kuang, Y.W.; Sun, F.F.; Wen, D.Z.; Zhou, G.Y.; Zhao, P. Tree-ring growth patterns of Masson pine (*Pinus massoniana* L.) during the recent decades in the acidification Pearl River Delta of China. *For. Ecol. Manag.* **2008**, *255*, 3534–3540. [CrossRef]
21. Yang, Z.Q.; Chen, H.; Tan, J.H.; Xu, H.L.; Jia, J.; Feng, Y.H. Cloning of three genes involved in the flavonoid metabolic pathway and their expression during insect resistance in *Pinus massoniana* Lamb. *Genet. Mol. Res.* **2016**, *15*, gmr15049332. [CrossRef] [PubMed]
22. Zeng, W.S.; Tang, S.Z. Modeling compatible single-tree aboveground biomass equations for masson pine (*Pinus massoniana*) in southern China. *J. For. Res.* **2012**, *23*, 593–598. [CrossRef]
23. Zhang, H.H.; Jiang, Z.; Qin, R.; Zhang, H.; Zou, J.; Jiang, W.; Liu, D. Accumulation and cellular toxicity of aluminum in seedling of *Pinus massoniana*. *BMC Plant Biol.* **2014**, *14*, 264. [CrossRef] [PubMed]

24. Liu, H.Y.; Chen, H.Y.; Ding, G.J.; Li, K.F.; Ren, Q.F. Identification of candidate genes conferring tolerance to aluminum stress in *Pinus massoniana* inoculated with ectomycorrhizal fungus. *BMC Plant Biol.* **2020**, *20*, 521. [CrossRef]
25. Pi, N.; Tam, N.F.Y.; Wu, Y.; Wong, M.H. Root anatomy and spatial pattern of radial oxygen loss of eight true mangrove species. *Aquat. Bot.* **2009**, *90*, 222–230. [CrossRef]
26. Grabherr, M.G.; Haas, B.J.; Yassour, M.; Levin, J.Z.; Thompson, D.A.; Amit, I.A.; Xian, F.L.; Raychowdhury, R.; Zeng, Q.D. Full-length transcriptome assembly from RNA-Seq data without a reference genome. *Nat. Biotechnol.* **2011**, *29*, 644–652. [CrossRef]
27. Langmead, B.; Salzberg, S.L. Fast gapped-read alignment with Bowtie 2. *Nat. Methods* **2012**, *9*, 357–359. [CrossRef]
28. Dewey, C.N.; Li, B. RSEM: Accurate transcript quantification from RNA-Seq data with or without a reference genome. *BMC Bioinform.* **2011**, *12*, 323.
29. Wang, L.; Feng, Z.; Wang, X.; Wang, X.W. DEGseq: An R package for identifying differentially expressed genes from RNA-seq data. *Bioinformatics* **2010**, *26*, 136–138. [CrossRef]
30. Conway, J.R.; Lex, A.; Gehlenborg, N. UpSetR: An R package for the visualization of intersecting sets and their properties. *Bioinformatics* **2017**, *33*, 2938–2940. [CrossRef]
31. Langfelder, P.; Horvath, S. WGCNA: An R package for weighted correlation network analysis. *BMC Bioinform.* **2008**, *9*, 559. [CrossRef] [PubMed]
32. Shannon, P.; Markiel, A.; Ozier, O.; Baliga, N.S.; Wang, J.T.; Ramage, D.; Amin, N.; Schwikowski, B.; Ideker, T. Cytoscape: A software environment for integrated models of biomolecular interaction networks. *Genome Res.* **2003**, *13*, 2498–2504. [CrossRef] [PubMed]
33. Chen, C.; Chen, H.; Zhang, Y.; Thomas, H.R.; Frank, M.H.; He, Y.H.; Xia, R. TBtools: An tntegrative toolkit developed for interactive analyses of big biological data. *Mol. Plant* **2020**, *13*, 1194–1202. [CrossRef] [PubMed]
34. Chen, H.; Yang, Z.Q.; Hu, Y.; Tan, J.H.; Jia, J.; Xu, H.L.; Chen, X.H. Reference genes selection for quantitative gene expression studies in *Pinus massoniana* L. *Trees-Struct. Funct.* **2016**, *30*, 685–696. [CrossRef]
35. Livak, K.J.; Schmittgen, T.D. Analysis of relative gene expression data using real-time quantitative PCR and the $2^{-\Delta\Delta CT}$ method. *Methods* **2002**, *25*, 402–408. [CrossRef]
36. Samac, D.A.; Tesfaye, M. Plant improvement for tolerance to aluminum in acid soils—A review. *Plant Cell Tissue Organ Cult.* **2003**, *75*, 189–207. [CrossRef]
37. Sasaki, M.; Yamamoto, Y.; Ma, J.F.; Matsumoto, H. Early events induced by aluminum stress in elongating cells of wheat root. *Soil Sci. Plant Nutr.* **1997**, *41*, 1009–1014. [CrossRef]
38. Yang, J.L.; Zheng, S.J.; He, Y.F.; Tang, C.X.; Zhou, G.D. Genotypic differences among plant species in response to aluminum stress. *J. Plant Nutr.* **2005**, *28*, 949–961. [CrossRef]
39. Batista, M.F.; Moscheta, I.S.; Bonato, C.M.; MA, de.; Almeida, O.J.G.; Inoue, T.T. Aluminum in corn plants: Influence on growth and morpho-anatomy of root and leaf. *Rev. Bras. Ciênc. Solo* **2013**, *37*, 177–187. [CrossRef]
40. Ozyigit, I.I.; Vardar, F.; Yasar, U.; Sener, A. Long-term effects of aluminum and cadmium on growth, leaf anatomy, and photosynthetic pigments of cotton. *Commun. Soil Sci. Plant Anal.* **2013**, *44*, 3076–3091. [CrossRef]
41. Liu, Q.C.; Deng, S.N.; Liu, B.S.; Tao, Y.F.; Ai, H.Y.; Liu, J.J.; Zhang, Y.Z.; Zhao, Y.; Xu, M.L. A helitron-induced RabGDIα variant causes quantitative recessive resistance to maize rough dwarf disease. *Nat. Commun.* **2020**, *11*, 495. [CrossRef] [PubMed]
42. Martín-Davison, A.S.; Pérez-Díaz, R.; Soto, F.; Madrid-Espinoza, J.; González-Villanueva, E.; Pizarro, L.; Norambuena, L.; Tapia, J.; Tajima, H.; Blumwald, E. Involvement of SchRabGDI1 from Solanum chilense in endocytic trafficking and tolerance to salt stress. *Plant Sci. Int. J. Exp. Plant Biol.* **2017**, *263*, 1–11. [CrossRef] [PubMed]
43. Waters, E.R.; Vierling, E. Plant small heat shock proteins—Evolutionary and functional diversity. *New Phytol.* **2020**, *227*, 24–37. [CrossRef] [PubMed]
44. Hego, E.; Bes, C.M.; Bedon, F.; Palagi, P.M.; Chaumeil, P.; Barré, A.; Claverol, S.; Dupuy, J.W.; Bonneu, M.; Lalanne, C.; et al. Differential accumulation of soluble proteins in roots of metallicolous and nonmetallicolous populations of *Agrostis capillaris* L. exposed to Cu. *Proteomics* **2014**, *14*, 1746–1758. [CrossRef] [PubMed]
45. Ramalingam, J.; Pathan, M.S.; Feril, O.; Ross, M.K.; Ma, X.F.; Mahmoud, A.A.; Layton, J.; Rodriguez-Milla, M.A.; Chikmawati, T.; Valliyodan, B.; et al. Structural and functional analyses of the wheat genomes based on expressed sequence tags (ESTs) related to abiotic stresses. *Genome* **2006**, *49*, 1324–1340. [CrossRef] [PubMed]
46. Osakabe, Y.; Osakabe, K.; Shinozaki, K.; Tran, L.S.P. Response of plants to water stress. *Front. Plant Sci.* **2014**, *5*, 86. [CrossRef]
47. Hirayama, T.; Shinozaki, K. Research on plant abiotic stress responses in the post-genome era: Past, present and future. *Plant J.* **2010**, *61*, 1041–1052. [CrossRef]
48. Belal, R.; Tang, R.j.; Li, Y.P.; Mabrouk, Y.; Badr, E.; Luan, S. An ABC transporter complex encoded by Aluminum Sensitive 3 and NAP3 is required for phosphate deficiency responses in Arabidopsis. *Biochem. Biophys. Res. Commun.* **2015**, *463*, 18–23. [CrossRef]
49. Dong, N.Q.; Lin, H.X. Contribution of phenylpropanoid metabolism to plant development and plant-environment interactions. *J. Integr. Plant Biol.* **2021**, *63*, 180–209. [CrossRef]
50. Kobayashi, Y.; Kobayashi, Y.; Watanabe, T.; Shaff, J.E.; Ohta, H.; Kochian, L.V.; Wagatsuma, T.; Kinraide, T.B.; Koyama, H. Molecular and physiological analysis of Al^{3+} and H^+ rhizotoxicities at moderately acidic conditions. *Plant Physiol.* **2013**, *163*, 180–192. [CrossRef]

51. Sun, P.; Tian, Q.Y.; Chen, J.; Zhang, W.H. Aluminium-induced inhibition of root elongation in Arabidopsis is mediated by ethylene and auxin. *J. Exp. Bot.* **2010**, *61*, 347–356. [CrossRef] [PubMed]
52. Meychik, N.R.; Yermakov, I.P. Swelling of root cell walls as an indicator of their functional state. *Biochemistry* **2001**, *66*, 178–187. [PubMed]
53. Hu, P.; Zhang, K.; Yang, C. BpNAC012 Positively Regulates Abiotic Stress Responses and Secondary Wall Biosynthesis. *Plant Physiol.* **2019**, *179*, 700–717. [CrossRef] [PubMed]
54. Chen, K.Q.; Song, M.R.; Guo, Y.N.; Liu, L.F.; Xue, H.; Dai, H.Y.; Zhang, Z.H. MdMYB46 could enhance salt and osmotic stress tolerance in apple by directly activating stress-responsive signals. *Plant Biotechnol. J.* **2019**, *17*, 2341–2355. [CrossRef] [PubMed]
55. Zhang, J.; Yin, X.R.; Li, H.; Xu, M.; Zhang, M.X.; Li, S.J.; Liu, X.F.; Shi, Y.N.; Donald, G.; Chen, K.S. ETHYLENE RESPONSE FACTOR39–MYB8 complex regulates low-temperature-induced lignification of loquat fruit. *J. Exp. Bot.* **2020**, *71*, 3172–3184. [CrossRef] [PubMed]
56. Wang, Y.; Ding, G. Physiological responses of mycorrhizal *Pinus massoniana* seedlings to drought stress and drought resistance evaluation. *Chin. J. Appl. Ecol.* **2013**, *24*, 639–645.
57. Zhang, T.; Wen, X.; Ding, G. Ectomycorrhizal symbiosis enhances tolerance to low phosphorous through expression of phosphate transporter genes in masson pine (*Pinus massoniana*). *Acta Physiol. Plant.* **2017**, *39*, 101. [CrossRef]
58. Yu, P.; Sun, Y.; Huang, Z.; Zhua, F.; Sun, Y.; Jiang, L. The effects of ectomycorrhizal fungi on heavy metals' transport in *Pinus massoniana* and bacteria community in rhizosphere soil in mine tailing area. *J. Hazard. Mater.* **2020**, *381*, 121203. [CrossRef]
59. Chai, Y.N.; Schachtman, D.P. Root exudates impact plant performance under abiotic stress. *Trends Plant Sci.* **2022**, *27*, 80–91. [CrossRef]

forests

MDPI

Article

Coexpression Network Analysis Based Characterisation of the R2R3-MYB Family Genes in Tolerant Poplar Infected with *Melampsora larici-populina*

Qiaoli Chen [1,2], Feng Wang [1,2] and Danlei Li [1,2,*]

1 Key Laboratory of Alien Forest Pests Monitoring and Control-Heilongjiang Province, School of Forestry, Northeast Forestry University, Harbin 150040, China
2 Key Laboratory of Sustainable Forest Ecosystem Management-Ministry of Education, Northeast Forestry University, Harbin 150040, China
* Correspondence: danleili@nefu.edu.cn; Tel.: +86-0451-82190384

Abstract: R2R3-MYB protein is the most abundant class of MYB transcription factor family in plants. The transcript profiles of two E4 races of *Melampsora larici-populina*-tolerant poplars and an intolerant poplar were investigated to characterise the role of the R2R3-MYB family genes in the poplar–E4 interaction. In this study, 217 R2R3-MYBs were identified, and 83 *R2R3-MYB* genes were assigned to 22 different coexpression modules by weighted gene coexpression network analysis. Most *R2R3-MYB* genes were unchanged in the early period of E4 infection in both tolerant and intolerant poplars. However, there were obvious increases in differentially expressed *R2R3-MYB* genes in tolerant poplars at 2 and 4 dpi when defence responses occurred, suggesting that differently expressed *R2R3-MYB* genes at these time points may play an important role in poplar resistance to E4 infection. In total, 34 *R2R3-MYB* genes showed differential expression at 2 and 4 dpi between tolerant and intolerant poplars. Among them, 16 differentially expressed *R2R3-MYB* genes were related to 43 defence-related genes that had significant differences between tolerant and intolerant poplars. There might be coregulatory relationships between R2R3-MYBs and other TFs during poplar–E4 interaction. Some differentially expressed *R2R3-MYB* genes were related to genes involved in flavonoid biosynthesis and IAA or free SA signal transduction and might help activate defence response during poplar–E4 interaction. MYB194 could be an important node in the convergence of IAA and SA signalling.

Keywords: R2R3-MYB; *Populus*; rust; *Melampsora larici-populina*; transcription factor

Citation: Chen, Q.; Wang, F.; Li, D. Coexpression Network Analysis Based Characterisation of the R2R3-MYB Family Genes in Tolerant Poplar Infected with *Melampsora larici-populina*. *Forests* **2022**, *13*, 1255. https://doi.org/10.3390/f13081255

Academic Editor: Andrea Coppi

Received: 31 May 2022
Accepted: 28 July 2022
Published: 9 August 2022

1. Introduction

Melampsora larici-populina causes serious foliar rust disease of poplar worldwide [1,2]. Hybrids between *Populus deltoides*, *P. nigra*, or *P. trichocarpa* were selected for their immunity to rust in the mid-20th century in Europe. However, outbreaks of rust on clones of these hybrids were caused by E4, a new rust race that occurred in the late 20th century. Nevertheless, some hybrid poplars were found to be tolerant to infection by E4 and developed hypersensitive response (HR) cell death at the infection site [3]. Further study to identify the genetic basis of host susceptibility or tolerance will provide new insight into how rust overcomes poplar resistance.

To ensure the optimal intensity and duration of immune responses, plant innate immunity is regulated at different levels. In plant immune signalling pathways, mitogen-activated protein kinases (MAPKs) [4–6] and calcium-dependent protein kinases (CDPKs) [7] can regulate the expression of plant immune-related genes by phosphorylating downstream transcription factors (TFs). TFs are proteins that control target gene expression levels and modulate rates of transcription. Many TFs can regulate plant immunity by regulating the expression of downstream defence-related genes [8] and participate in regulating the crosstalk between different defensive signalling pathways [9]. Several small messenger molecules, such as salicylic acid (SA), jasmonic acid (JA), and ethylene (ET), are involved in translating pathogen-induced early signalling events into the activation of effective defence responses [10,11]. Some TFs are important nodes for the convergence of phytohormone signalling and play an important role in the regulation of phytohormone-responsive genes [12]. Therefore, many TFs are key participants in plant immune responses and are considered key targets for genetic engineering to enhance adaptation to not only abiotic but also biotic stresses in valuable plants.

The transcript levels of TFs, such as MYBs, AP2/ERFs, and WRKYs, were previously found to be altered in poplars infected with E4 [3,13,14]. The MYB family is one of the largest groups of TFs in plants. The MYB domain, which is a highly conserved DNA-binding domain, is characteristic of MYB proteins. Structurally, the MYB domain consists of 1–4 MYB repeat units, and the MYB family can be divided into four types according to the number of replicates, namely 4R-MYB (containing 4 MYB replicates), 3R-MYB (containing 3 MYB replicates), R2R3-MYB (containing 2 MYB replicates), and MYB-related (containing 1 MYB). In plants, the R2R3-MYB protein is the most abundant class of MYB family, containing two MYB domains at the N-terminus. The plant MYB family has selectively expanded, particularly through the large R2R3-MYB family, and many (if not all) R2R3-MYBs play central roles in plant-specific processes [15,16]. As the MYB family is expected to play an important role in plant defence against biotic and abiotic stresses [17,18], placing each member in an organised nomenclature system and providing maps of MYB family gene expression should contribute to unravelling the complexity of the transcriptional regulation of defence-related genes in poplar–rust interactions.

Our previous study indicated that the hybrid poplar *P. nigra* × *P. deltoides* ('Intolerant') is susceptible to the virulent E4 race of *M. larici-populina*, whereas *P. deltoides* × *P. trichocarpa* ('Tolerant 2') and *P. trichocarpa* × *P. deltoides* ('Tolerant 1') are tolerant to E4 and timely activation or inhibition of the SA or JA pathways is the key difference between tolerant and intolerant poplars [3]. However, the molecular characteristics of TFs, especially defence-related TFs such as R2R3-MYBs, need to be further analysed and explored [19,20]. Therefore, in this study, weighted gene coexpression network analysis (WGCNA) [21] was further performed to obtain a better understanding of the expression patterns of *R2R3-MYB* genes and their related genes in E4-infected poplars to provide a better understanding of the interaction between poplar and rust on the basis of the previous study [3]. The changes in transcriptome profiling and contents of indole-3-acetic acid (IAA) [22] and free SA [20] after E4 infection were also investigated to study whether and how R2R3-MYBs might interfere with phytohormone signalling pathways.

2. Materials and Methods

2.1. E4 Isolates, Plant Materials, and Inoculation Procedure

E4-infected poplar leaves were collected from *P. trichocarpa* cv. Trichobel at Markington (northern England) [2], and E4 rust isolates were obtained from single uredinial pustules as previously reported [2,3,13,14]. One-year-old hybrid poplars, including an intolerant poplar, *P. nigra* × *P. deltoides* ('Intolerant'), and two tolerant poplars, *P. deltoides* × *P. trichocarpa* ('Tolerant 2') and *P. trichocarpa* × *P. deltoides* ('Tolerant 1'), were used as plant tissue sources. E4 completed its vegetative cycle in 7 days on 'Intolerant'. The growth of E4 was inhibited in 'Tolerant 2' and 'Tolerant 1'. By 7 days post-inoculation (dpi), only a few new or barely mature urediniospores were found, and visible scattered lesions and confluent necrosis appeared on 'Tolerant 2' and 'Tolerant 1', respectively [3].

These hybrid poplars were grown as described previously [3,13,14,23]. Leaves of hybrid poplars were inoculated with E4 as described by Pei et al. [2] and Chen et al. [3,13,14]. In brief, fully expanded leaves from leaf plastochrony index 5–9 were detached from plants and spray-inoculated on their abaxial surface with a rust spore suspension in deionised water containing 0.004% Tween 20 adjusted to 100,000 spores ml^{-1}, or with deionised water containing 0.004% Tween 20 as a control. After inoculation, the E4-inoculated leaves were incubated in a phytotron with 16 h day^{-1} illumination (80 $\mu E\ m^{-2}\ s^{-1}$) for different periods as described previously (2, 6, and 12 h and 1, 2, 4, and 7 d) [2,3,13,14,23]. The control groups contained E4-free leaves (leaves were treated with deionised water only) incubated under the same conditions. Each leaf sample was frozen and ground using liquid nitrogen for RNA extraction.

2.2. Transcriptome Library Preparation and Sequencing

The CTAB (cetyltrimethylammonium bromide) method was used to extract total RNA. The RNA Nano 6000 Assay Kit of the Agilent Bioanalyzer 2100 system (Agilent Technologies, Santa Clara, CA, USA) was used to assess RNA quality. A NanoDrop™ spectrophotometer (Thermo Scientific, Wilmington, DE, USA) was used to determine the RNA purity. Genomic DNA was removed using DNase I, Amplification Grade (Invitrogen, Foster, CA, USA, cat. no. 18068-015). The construction of the libraries and sequencing of the three poplars were performed on a BGISEQ-500 RNA-seq platform (BGI, Shenzhen, China). The average insert size for the paired-end libraries was 300 bp (±50 bp).

Low-quality (more than 20% of bases in the total read have quality score lower than 15), adaptor-polluted and high content (5%) of unknown base (N) reads were trimmed to acquire clean reads using SOAPnuke (v1.5.2, https://github.com/BGI-flexlab/SOAPnuke, accessed on 24 November 2016). After clean reads were obtained, HISAT2 (Hierarchical Indexing for Spliced Alignment of Transcripts, v2.0.4, http://www.ccb.jhu.edu/software/hisat, accessed on 18 May 2016) was used to align clean reads to the genome sequence of *P. trichocarpa* (version 3.0, http://www.phytozome.net/poplar.php, accessed on 26 November 2018) [24,25]. The uniformity of the mapping result for each sample suggested that the samples were comparable. After comparison with the reference genome, StringTie (v1.0.4, http://ccb.jhu.edu/software/stringtie, accessed on 19 May 2016) [26] was used to reconstruct the transcriptome of each sample.

2.3. Identification of R2R3-MYBs

Extensive BLAST (Basic Local Alignment Search Tool) searches (https://www.ncbi.nlm.nih.gov/, accessed on 15 May 2020) were conducted to select R2R3-MYB family members based on transcriptome sequencing results. The sequences of 192 R2R3-MYBs identified by Wilkins et al. [27] were used as the query sequences to perform BLASTP local search with the E value set to 1×10^{-10}. The selected candidate sequences were then assessed by the hidden Markov model of Pfam (PF00249, http://xfam.org/, accessed on 15 May 2020) with the E value set to 1×10^{-3}. After that, the sequences were submitted to the website InterPro (http://www.ebi.ac.uk/interpro/, accessed on 15 May 2020) and SMART (http://smart.embl-heidelberg.de/, accessed on 15 May 2020) for DNA-binding domain test, and proteins containing two repeated sequences (R2 and R3) in the DNA-binding domain were taken as R2R3-MYB family members in the three hybrid poplars.

2.4. WGCNA

WGCNA (https://horvath.genetics.ucla.edu/html/CoexpressionNetwork/Rpackages/WGCNA/, accessed on 19 May 2020) [21] was performed to obtain a better understanding of the expression patterns of *R2R3-MYB* genes and their related genes in E4-infected poplars to provide a better understanding of the interaction between poplar and rust on the basis of our previous study [3]. WGCNA was implemented in the R software package (http://www.r-project.org/, accessed on 19 May 2020). The input data and parameter settings were consistent with our previous study [3]. After filtering out genes with median FPKM levels that did not exceed 1 [3], the expression of 97 of the *R2R3-MYB* genes were included in the construction of the coexpression module (grey module was reserved for unassigned genes) with WGCNA package tools.

According to the tutorials of the WGCNA package, the module eigengene (ME) is defined as the first principal component of a given module, which can be considered representative of the gene expression profiles in a module. Intramodular connectivity (IC) was defined only for the genes inside a given module and was calculated for each gene by summing the connection strengths with those of other module genes and dividing this number by the maximum intramodular connectivity. IC measures how connected, or coexpressed, a given gene is with respect to the genes of a particular module. Correlation analyses between MEs and external traits (SA and IAA levels) were performed to look for the most significant associations. For each expression profile, gene significance (GS) was calculated as the absolute value of the Pearson correlation between the expression profile and each trait. Module membership (MM) was defined based on the Pearson correlation of the expression profile and each ME. Genes with higher MM were defined as the more important (central) elements of the modules.

The eigengene dendrogram and heatmap were used to identify groups of correlated eigengenes. Gene coexpression network depictions were constructed using Cytoscape software [28]. In gene coexpression networks and *R2R3-MYB*-gene-focused gene networks, genes with weight values were connected by lines and the colour of the line represented the weight value between two genes. Kyoto Encyclopaedia of Genes and Genomes (KEGG) enrichment analysis was performed on selected genes [29] using the hypergeometric test to determine which pathway was significantly enriched in selected genes compared with the whole genome background and p value ≤ 0.05 after the correction was used as the threshold. PossionDis algorithms were used to detect the DEG (differentially expression genes). In this study, DEGs were defined by default as those with a false discovery rate (FDR) ≤ 0.05 and differences of more than twofold ($\log_2$ value of E4-inoculated expression to E4-free expression ≥ 1 or ≤ -1). The depictions of heatmaps of gene expression pattern were constructed using the OmicShare tools, a free online platform for data analysis (http://www.omicshare.com/tools, accessed on 21 April 2022).

2.5. Quantification of IAA and Free SA Levels

SA and IAA were quantified by high-performance liquid chromatography (HPLC) mass spectrometry from crude plant extracts based on the method of Pan et al. [30] and Chen et al. [3]. According to Pan et al., poplar leaves were frozen in liquid nitrogen and then ground followed by adding working solution and extraction solvent. Then, SA and IAA were extracted and quantified by an ultra-HPLC-Q-Exactive™ system (Thermo Scientific, San José, CA, USA) using an ODS column (µ-Bondasphere C_{18}, 5 µm, 3.9 × 150 mm; Waters, Milford, MA, USA). Authentic SA and IAA (Sigma Aldrich, Burlington, MA, USA, cat. no. S5922 and I3750) were used as external standards. The amounts of SA and IAA were calculated by comparing with the corresponding internal standard. Three separate biological replicates of each treatment were performed, and each replicate was assessed three times.

2.6. RT-qPCR

Quantitative real-time reverse transcription PCR (RT–qPCR) was performed with the GoTaq 2-Step RT–qPCR System Kit (Promega, Madison, WI, USA, cat. no. A6010) and the Stratagene Mx3000P qPCR system (Agilent Technologies, Santa Clara, CA, USA) to validate the transcript levels of selected genes at 2 hpi, 6 hpi, 12 hpi, 1 dpi, 2 dpi, 4 dpi, and 7 dpi. The 18S ribosomal RNA was used as the internal control. All primers used in this study are listed in Table S1. The PCR program was 95 °C for 10 s and 40 cycles of 95 °C for 5 s, 58 °C for 30 s, and 72 °C for 30 s. Each treatment was performed on three separate biological replicates, and each replicate was measured three times. Quantification of the RT–qPCR results was conducted as previously described [3]. The normalisation of the data followed the instructions of the GoTaq 2-Step RT–qPCR System Kit and the $2^{-\Delta\Delta C_T}$ method [31]. Significance was determined by Student's t-test.

3. Results

3.1. Identification of R2R3-MYBs

The transcriptomes of the leaves of the three poplars were measured, producing an average of 6.54 GB of data per sample (Table S2). The average genome mapping rate is 79.58% and the average gene mapping rate is 77.77%. A total of 34,279 genes were detected, including 33,896 known genes and 406 predicted new genes. A total of 12,994 new transcripts were detected, of which 11,604 were new alternative splicing subtypes of known protein-coding genes, 406 were transcripts of new protein-coding genes, and the remaining 984 were long noncoding RNAs. Based on the results of Wilkins et al. [27], BLASTP was used to search for translated nucleotides of R2R3-MYBs in two tolerant poplars, *P. trichocarpa* × *P. deltoides* ('Tolerant 1') and *P. deltoides* × *P. trichocarpa* ('Tolerant 2'), and an intolerant poplar, *P. nigra* × *P. deltoides* ('Intolerant'). Our extensive search for R2R3 MYB DNA binding-domain-containing proteins identified 217 putative distinct R2R3-MYBs in the three poplars and named them MYB001 to MYB217 (Table S3).

3.2. Expression Characteristics of *R2R3-MYB* Genes Based on WGCNA

Of the 97 *R2R3-MYB* genes analysed by WGCNA, 83 were assigned to 22 different coexpression modules (Table S4), and 14 were assigned to the grey module. As the grey module was reserved for unassigned genes, the modules that contained more than 5 of the *R2R3-MYB* genes were blue, brown, black, green, green–yellow, yellow, and turquoise. Correlation of MEs was analysed and groups of correlated MEs of modules that these *R2R3-MYB* genes assigned to were identified, indicating that the correlation of expressions of most *R2R3-MYB* genes was low (Figure S1A,B). This also suggested that different *R2R3-MYB* genes might play different roles in response of poplar to rust infestation. To further analyse the gene expression differences in *R2R3-MYB* genes in the three poplars at different time points, the expression patterns of those 83 *R2R3-MYB* genes in the 22 modules were compared (Figure 1A). The results indicated that obvious differences in the expression levels between tolerant and intolerant poplars were observable mainly at 4 dpi.

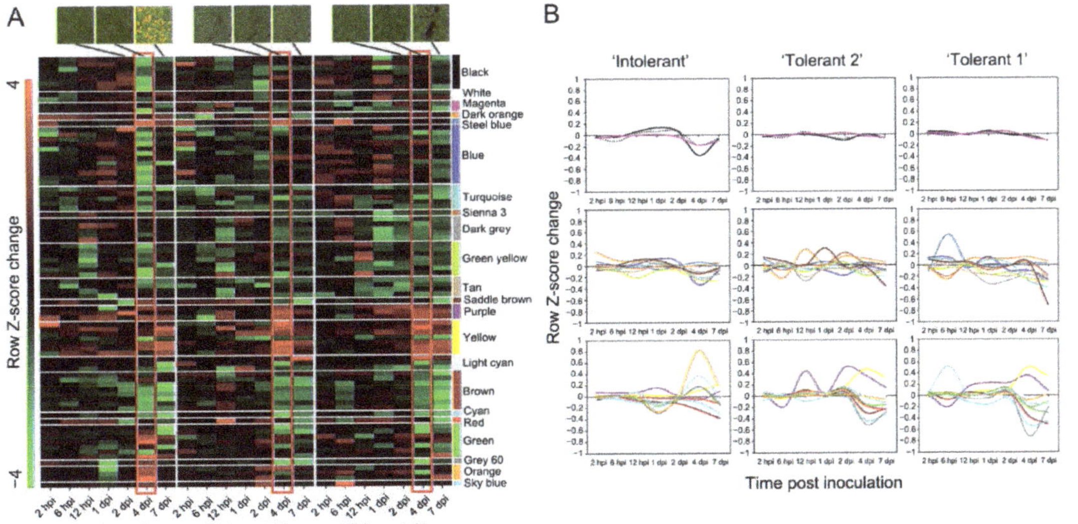

Figure 1. WGCNA revealed the expression characteristics of *R2R3-MYB* genes. (**A**) Heatmap of *R2R3-MYB* genes (rows, the Z-score change was calculated for the expression of each gene based on FPKM, and the results are presented in E4-inoculated samples minus E4-free samples) across the samples (columns). The red rectangle highlights the difference in expression. (**B**) Comparison of eigengene expressions (*y*-axis, row Z-score change in eigengene expression, E4-inoculated sample minus E4-free sample for each time point) in different modules across the samples (*x*-axis). Coexpression modules were shown in different colours.

The MEs of the 22 modules to which those 83 *R2R3-MYB* genes were assigned were compared across the samples (Figure 1B). The most obvious difference among the three poplars was that changes in MEs for different modules were more similar in 'Tolerant 1' and 'Tolerant 2'. In many modules, the MEs did not change obviously in 'Intolerant' at many time points, but the MEs in 'Tolerant 1' and 'Tolerant 2' at the same time points showed obvious increasing or decreasing trends. In contrast, when the MEs did not change obviously in 'Tolerant 1' and 'Tolerant 2', the MEs in 'Intolerant' showed obvious increasing or decreasing trends. These situations occurred at 4 dpi. The results showed that there were significant differences in the MEs between tolerant and intolerant poplars mainly at 4 dpi, which agreed with the expression characteristics of *R2R3-MYB* genes, indicating that the expression of these *R2R3-MYB* genes was consistent with the basic expression characteristics of the genes in the specific modules.

The MM and IC for each *R2R3-MYB* gene in the 22 modules to which the 83 *R2R3-MYB* genes were assigned were calculated (Table S5). Among them, 4 *R2R3-MYB* genes had relatively high MMs (>0.90), and 21 *R2R3-MYB* genes had high MMs (>0.80), indicating that these *R2R3-MYB* genes were central elements in their respective modules. The results also indicated that highly connected intramodular *R2R3-MYB* genes tend to have high MMs in the respective modules. Therefore, *R2R3-MYB* genes with high MMs and ICs of a particular module should occupy important regulatory positions within the module. This result was also found in other TF genes associated with plant–pathogen interactions, such as *AP2*, *ERF*, *NAC*, and *WRKY*.

To identify the functions and pathways of *R2R3-MYB* gene-related genes, KEGG enrichment analyses were performed on genes in the 22 modules with which those 83 *R2R3-MYB* genes had weight values (Table S6). Genes with weight values of *R2R3-MYB* genes were enriched in plant hormone signal transduction (ko04075) in 15 modules, plant–pathogen interaction (ko04626) in 13 modules, and MAPK signalling-pathway–plant (ko04016) in 13 modules. Genes enriched in these three pathways accounted for a large proportion of all genes in approximately half of all these modules, especially the yellow and grey 60 modules. These results suggested that R2R3-MYBs played an important role in the interaction between poplar and E4.

3.3. Differential Expression of *R2R3-MYB* Genes

To further explore the effect of the differential expression of *R2R3-MYB* genes on the interaction between poplar and E4, the expression of *R2R3-MYB* genes at different time points in different poplars after E4 infection was analysed (Figure 2). We focused on genes with different expression patterns between the tolerant and intolerant poplars. It was found that most *R2R3-MYB* genes were unchanged during the interaction between poplar and E4, but there were many more unchanged *R2R3-MYB* genes at 2 and 4 dpi in 'Intolerant' than in both 'Tolerant 2' and 'Tolerant 1'. Additionally, many *R2R3-MYB* genes were downregulated at 4 and 7 dpi, which were the late stages of E4 infection, in both tolerant and intolerant poplars. To find the main difference between tolerant and intolerant poplars, the *R2R3-MYB* genes with different expression patterns between tolerant and intolerant poplars at 2 and 4 dpi were screened. In total, 15 of the 22 modules had 34 *R2R3-MYB* genes with differential expression at 2 and 4 dpi between tolerant and intolerant poplars (Table S7).

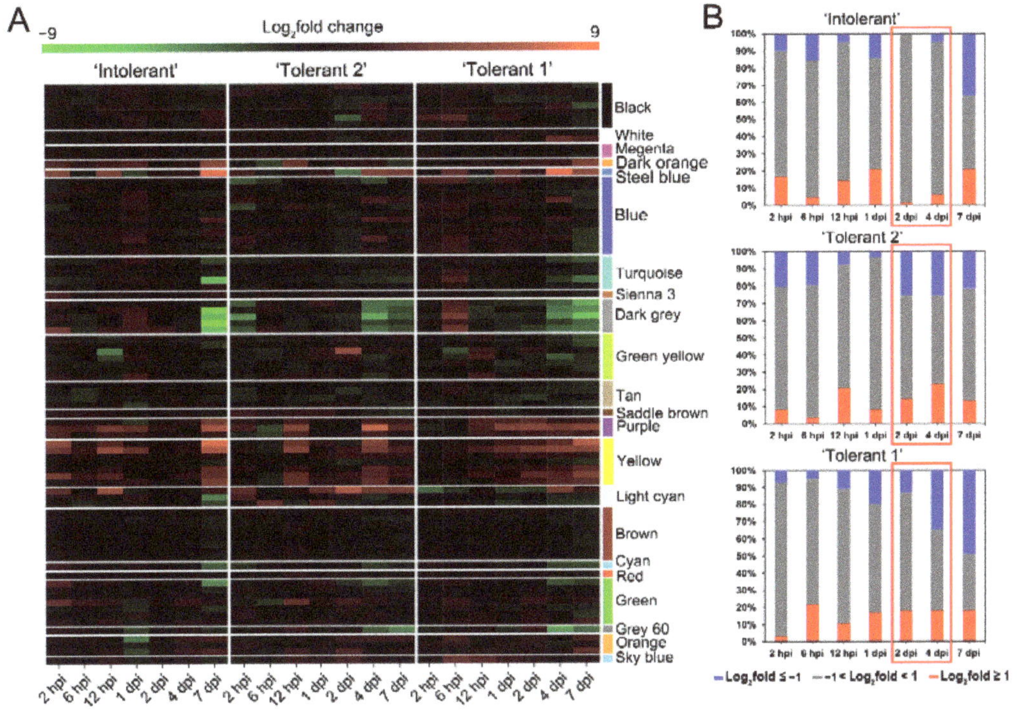

Figure 2. Expression analysis of *R2R3-MYB* genes at different time points in different poplars after E4 infection. (**A**) Heatmap of *R2R3-MYB* genes. (**B**) Proportion of differentially expressed *R2R3-MYB* genes. Coexpression modules were shown in different colours. The red rectangle highlights the difference in proportion.

The expression of 16 *R2R3-MYB* genes in 'Intolerant' was unchanged or significantly lower than that in 'Tolerant 2' and 'Tolerant 1' at 2 or 4 dpi (Figure S2A–C and Table S7), and 6 of these 16 *R2R3-MYB* genes were assigned to the yellow module (Figure S3A–C and Table S7). These 6 *R2R3-MYB* genes from the yellow module showed a gradual increase in expression with increasing inoculation time in 'Tolerant 1'. However, in 'Tolerant 2' and 'Intolerant', the expression of these genes fluctuated greatly, with no obvious change patterns. On the other hand, the expression levels of 18 *R2R3-MYB* genes in 'Intolerant' were unchanged or significantly higher than those in 'Tolerant 2' and 'Tolerant 1' at 2 or 4 dpi (Figure S2D–F and Table S7), and 5 of these *R2R3-MYB* genes were assigned to the dark grey module (Figure S3D–F and Table S7). These 5 *R2R3-MYB* genes from the dark grey module showed a gradual decrease in expression after 6 hpi with the increase in inoculation time in 'Tolerant 1'. However, in 'Tolerant 2' and 'Intolerant', there was no noticeable change in the expression of such genes.

To further explore the relationship between the expression of the *R2R3-MYB* genes and their related genes in the interactions between different poplars and E4, KEGG enrichment results on genes with weighted values of all these differently expressed *R2R3-MYB* genes were further analysed. The modules with more genes enriched in plant–pathogen interaction (ko04626) and signalling-related pathways (ko04016 and ko04075) were in the black, purple, yellow, light cyan, and grey 60 modules (Figure 3 and Table S8). Additionally, many genes were enriched in pathways related to the biosynthesis of other secondary metabolites that were important in plant stress resistance and development, such as phenylpropanoid biosynthesis (ko00940), flavonoid biosynthesis (ko00941), isoflavonoid biosynthesis (ko00943), anthocyanin biosynthesis (ko00942), and flavone and flavonol biosynthesis (ko00944). The modules with more genes enriched in these biosynthesis of other secondary metabolite pathways were black, dark grey, yellow, and light cyan (Figure 3 and Table S8). Thus, *R2R3-MYB* genes and their related genes in the six modules, black, dark grey, purple, yellow, light cyan, and grey 60 modules, were more likely to be involved in the interaction between poplars and E4.

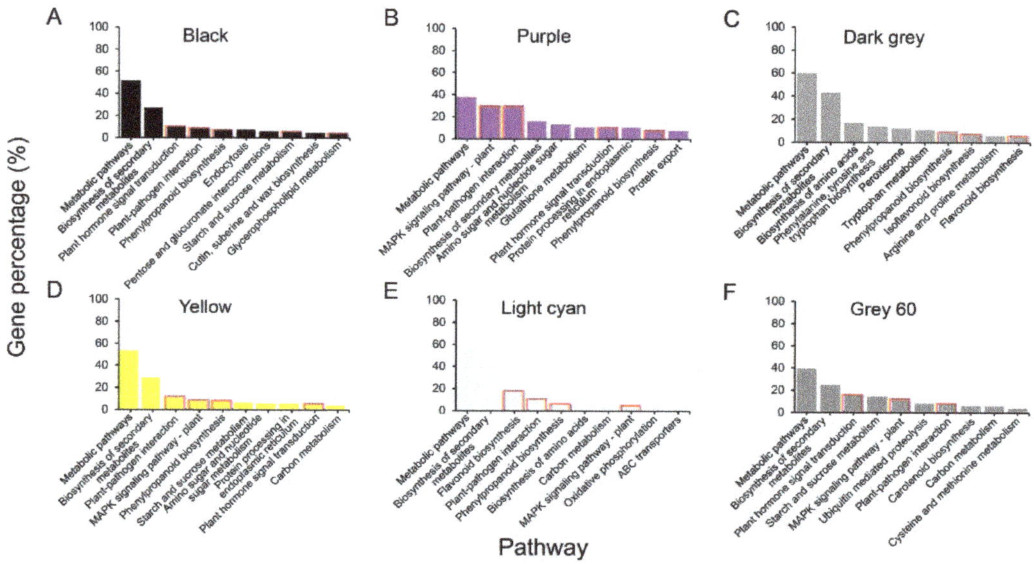

Figure 3. KEGG enrichment analysis for genes that differentially expressed *R2R3-MYB* genes had weight values within selected modules (top 10 pathways, detailed in Table S8). (**A**) The black module. (**B**) The purple module. (**C**) The dark grey module. (**D**) The yellow module. (**E**) The light cyan module. (**F**) The grey 60 module. The red rectangle highlights the pathways involved in the interaction between poplars and E4.

To further explore the regulatory mechanisms of *R2R3-MYB* genes, the functions of the proteins encoded by genes from the six modules, black, dark grey, purple, yellow, light cyan, and grey 60, were annotated with BLASTP based on the NR (nonredundant proteins) database. We focused on the functions of the genes that were enriched in plant–pathogen interaction pathways (ko04626), plant hormone signal transduction (ko04075), MAPK signalling pathway—plant (ko04016), phenylpropanoid biosynthesis (ko00940), flavonoid biosynthesis (ko00941), isoflavonoid biosynthesis (ko00943), anthocyanin biosynthesis (ko00942), flavone and flavonol biosynthesis (ko00944), and cutin, suberine and wax biosynthesis (ko00073), which were all related to the interaction between poplars and E4.

Among the six modules, the yellow module had the largest number of genes enriched in these pathways, at 147. Sixty-nine of these genes were annotated with specific functions (Table S9). Many of the genes encoded pathogenesis-related family proteins, peroxidase family proteins, WRKY transcription factors, and disease-resistance proteins (Figure 4A). The light cyan module had the second highest number of genes enriched in these pathways at 49. Twenty-three of these genes were annotated as having specific functions (Table S9). Many of the genes encoded naregenin-chalcone synthase family proteins, receptor-like proteins, and leucoanthocyanidin dioxygenase family proteins (Figure 4B).

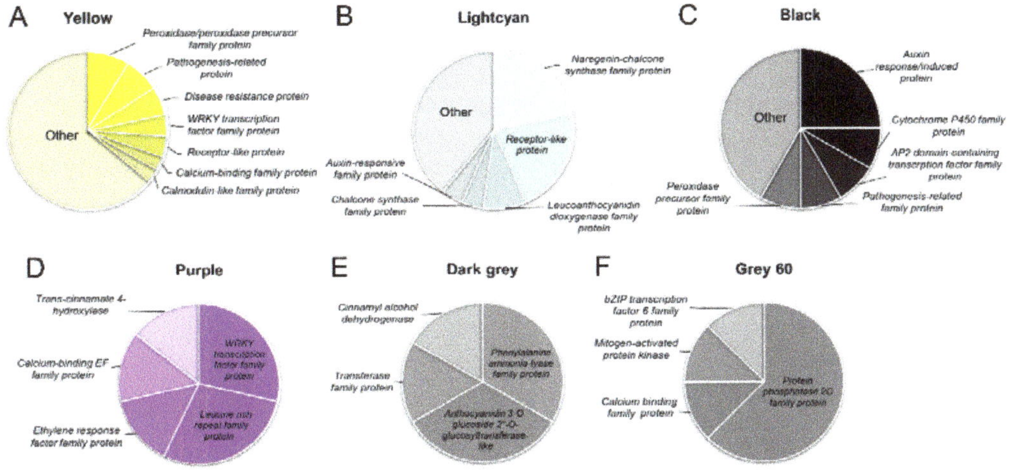

Figure 4. Proportion of differently expressed *R2R3-MYB* gene-related genes involved in poplar–E4 interactions in the selected modules. (**A**) The yellow module. (**B**) The light cyan module. (**C**) The black module. (**D**) The purple module. (**E**) The dark grey module. (**F**) The grey 60 module.

The numbers of genes enriched in these pathways in the black, dark grey, purple, and grey 60 modules were 24, 19, 16, and 13, respectively. Of these, 12, 6, 7, and 8 genes were annotated with specific functions, respectively (Table S9). Many of the genes encoded auxin responsive/induced proteins in the black module; WRKY transcription factors, ethylene response factors, leucine-rich repeat family proteins, and calcium-binding EF hand family proteins in the purple module; phenylalanine ammonia-lyase family proteins and anthocyanidin 3-O-glucoside 2″-O-glucosyltransferase-like proteins in the dark grey module; and protein phosphatase 2C family proteins in the grey 60 module (Figure 4C–F). Thus, different R2R3-MYBs and their related genes should regulate different genes in the pathways.

3.4. Gene Networks of Differently Expressed R2R3-MYB Genes

To further comprehensively analyse the regulatory interactions between differentially expressed *R2R3-MYB* genes and their related genes (genes have weight value with *R2R3-MYB* genes) enriched in the pathways involved in the interaction between poplars and E4 in the black, dark grey, purple, yellow, light cyan, and grey 60 modules, gene networks were predicted based on the weight values between genes for each selected module (Figure 5). In the yellow module, the weight value between *linamarase family protein* (No. 261), *beta-glucosidase 12-like* genes (No. 182 and 205), and *peroxidase* genes (No. 393 and 408) was higher (top 1%, weight value > 0.4, Figure S4). Therefore, there were high correlations among these genes. *K^+ rectifying channel family protein* (No. 32) had the highest IC in the network, indicating that this gene occupied a central position among these genes and that the regulation between *MYB169* (No. 204), *MYB194* (No. 371), *MYB024* (No. 653), *MYB129* (No. 1040), *MYB046* (No. 1588), and *MYB011* (No. 1654) and the other genes was affected by *K^+ rectifying channel family protein* (No. 32). Additionally, many genes were related to both MYBs (Nos. 204, 371, 653, 1040, 1588, and 1654) and WRKYs (Nos. 200, 305, and 509).

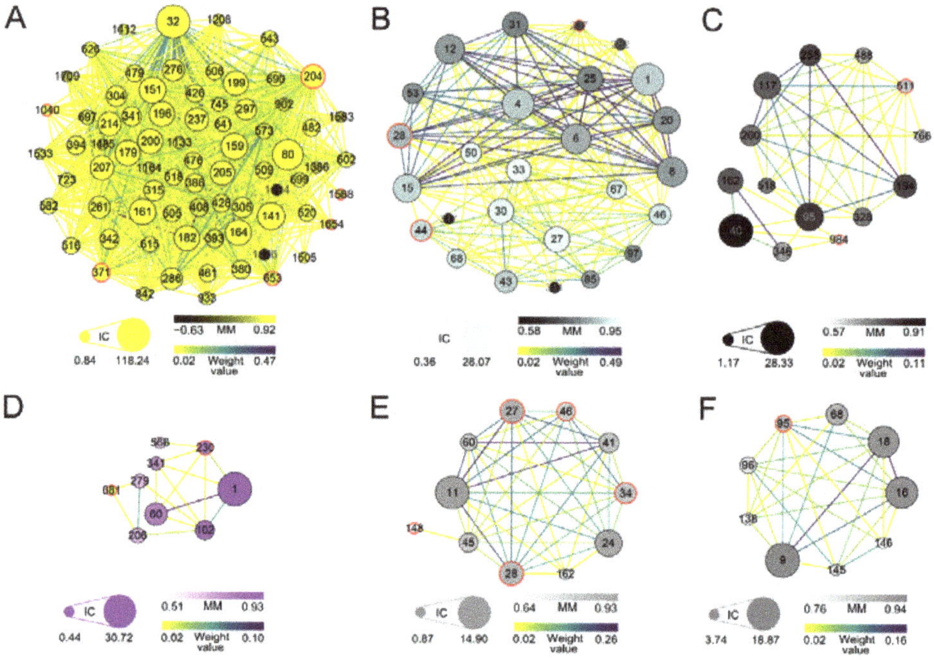

Figure 5. WGCNA revealed gene networks for differentially expressed *R2R3-MYB* genes and their related genes (detailed in Table S10) involved in poplar–E4 interactions in the selected modules. (**A**) The yellow module. (**B**) The light cyan module. (**C**) The black module. (**D**) The purple module. (**E**) The dark grey module. (**F**) The grey 60 module. The size of the dots represents IC. The colour of the dot represents MM. The colour of the line represents the weight value between two genes. The label of the dot is listed based on IC of the specific gene in the specific module. The *R2R3-MYB* genes are highlighted with red borders. IC, intramodular connectivity; MM, module membership.

In the light cyan module, based on weight values between genes, the correlations between *MYB162* (No. 28) and the other genes were greater than the correlations between *MYB086* (No. 44) or *MYB202* (No. 237) and the other genes (Figure 5B). *MYB162* (No. 28) was highly related to *dihydroflavonol reductase family protein* (No. 1), which was the gene with the highest IC in the light cyan module. Additionally, *dihydroflavonol reductase family protein* (No. 1), *chalcone synthase family protein* (No. 4), *naregenin-chalcone synthase family protein* genes (No. 6, 8, 12, 15, 20, and 53), *leucoanthocyanidin dioxygenase family protein* (No. 25), *MYB162* (No. 28), and *leucoanthocyanidin reductase family protein* (No. 31) had high weight values with each other, suggesting that they were highly related (Figure S5).

In the black module, *MYB146* (No. 511) and *MYB101* (No. 984) were involved in regulating the genes of different groups. *Auxin-responsive family protein* (No. 40) was the gene with the highest IC in the network, and this gene was only related to *MYB101* (No. 984). However, *glucose-methanol-choline oxidoreductase family protein* (No. 95) played a role in the connection between *MYB146* (No. 511) and *MYB101* (No. 984). Except for *glucose-methanol-choline oxidoreductase family protein* (No. 95), *Fe(III)-Zn(II) purple acid phosphatase family protein* (No. 117), *AP2 domain-containing transcription factor family protein* (No. 194) *cytochrome P450 family protein* (No. 258), and *beta-ketoacyl-CoA synthase family protein* (No. 260), which had higher weight values with each other, *quinone oxidoreductase family protein* (No. 162) and *transferase family protein* (No. 346) also had high weight values with each other, indicating that they were highly related (Figure 5C). Additionally, *auxin response factor 2 family protein* (No. 488) was related to both *MYB146* (No. 511) and *AP2 domain-containing transcription factor family protein* (No. 194), indicating that it was regulated by both MYB and AP2.

In the purple module, *MYB137* (No. 230) was predicted to be related to *leucine-rich repeat family protein* (No. 1), which was the gene with the highest IC in the module. *The calcium-binding EF hand family protein* (No. 279) played a role in the connection between *MYB137* (No. 230) and *MYB104* (No. 681). *WRKY transcription factor 51 family protein* (No. 60) and *leucine-rich repeat family protein* (No. 1) had a high weight value, indicating they were highly related (Figure 5D). Additionally, *leucine-rich repeat family protein* genes (Nos. 1 and 102) were related to both *MYB137* (No. 230) and *WRKY transcription factor 51 family protein* (No. 60), and *calcium-binding EF hand family protein* (No. 279) was related to both *MYB104* (No. 681) and *WRKY transcription factor 42 family protein* (No. 206), indicating that these genes were regulated by both MYBs and WRKYs.

In the dark grey module, *cinnamyl alcohol dehydrogenase 6* (No. 11) was the gene with the highest IC in the network and was related to *MYB078* (No. 27), *MYB214* (No. 28), *MYB195* (No. 34), and *MYB156* (No. 46). *Phenylalanine ammonia-lyase family protein* (No. 45) played a role in the connection between *MYB082* (No. 148) and the other four *R2R3-MYB* genes. Additionally, *cinnamyl alcohol dehydrogenase 6* (No. 11) and *anthocyanidin 3-O-glucoside 2″-O-glucosyltransferase-like* genes (No. 41 and 60) had high weight values with each other, indicating that they were highly related (Figure 5E).

In the grey 60 module, *MYB049* (No. 95) had a high weight value with *protein phosphatase 2C family protein* genes (No. 16 and 18). *MYB049* (No. 95) also had a high weight value with *bZIP transcription factor 6 family protein* (No. 9), which was the gene with the highest IC in the network and had a high weight value with *protein phosphatase 2C family protein* genes (No. 16 and No. 18, Figure 5F). Additionally, *protein phosphatase 2C family protein* genes (Nos. 16, 18, 96, and 145), *mitogen-activated protein kinase homologues* (No. 68), and *calcium binding family protein* (No. 138) were related to both *MYB049* (No. 95) and *bZIP transcription factor 6 family protein* (No. 9), indicating that they were regulated by both MYB and bZIP.

3.5. Analysis of Interactions between R2R3-MYB Genes, IAA, and Free SA

Many genes in these modules involved in the interaction between poplars and E4 were related to auxin. In the yellow module, *aux/IAA family protein* (No. 1494) and *auxin-responsive GH3 family protein* (No. 1583) were related to 31 genes (Figure S6A). In the light cyan module, *auxin-responsive family protein* (No. 33) was related to 24 genes (Figure S6B). In the black module, *auxin-responsive family protein* genes (Nos. 40 and 488) and *auxin-induced protein IAA4* were related to 10 genes (Figure S6C). This result suggested that R2R3-MYBs could be involved in the interaction between poplars and E4 by regulating the expression of genes associated with IAA.

In the light cyan module, compared to other genes, two *receptor-like protein 12 isoform X9* genes (Nos. 85 and 97) had higher GSs with IAA (>0.20). In the black module, GSs for all auxin-related genes with IAA were negative or very low. However, in the yellow module, six genes (Nos. 304, 371, 505, 543, 602, and 1583) had higher GSs with IAA, and there was a certain correlation between them (Figure 6A,B). A *MYB194* (No. 371) was included here. There was a high correlation between genes in the yellow module and free SA (Figure 6B). These results suggested that *MYB194* (No. 371) was involved in regulating IAA and free SA changes in the poplar interaction with E4 and affecting the defence response of poplar.

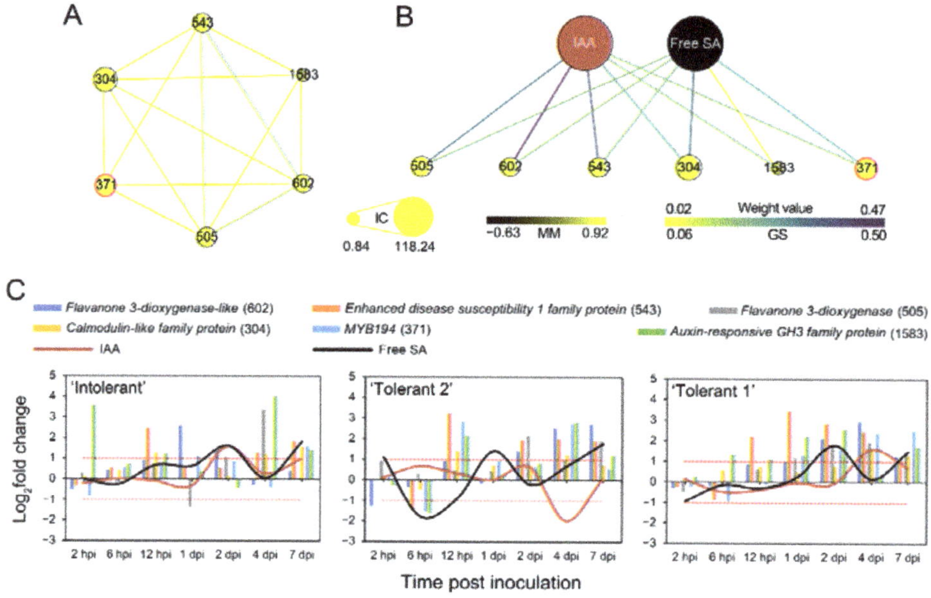

Figure 6. Coexpression characteristics of auxin-related genes. (**A**) Coexpression relationship between auxin-related genes. (**B**) Correlation between auxin-related genes and auxin (indole-3-acetic acid, IAA) and free salicylic acid (free SA). The size of the dots represents IC. The colour of the dot represents MM. The colour of the line represents the weight value between two genes in (**A**) and the GS for IAA or free SA in (**B**). The label of the dot is listed based on IC for the specific gene in the specific module. The R2R3-MYB genes are highlighted with a red border. GS, gene significance; IC, intramodular connectivity. (**C**) Comparison of the changes in the expression ($\log_2$E4-inoculated/E4-free) of auxin-related genes and levels of IAA and free SA.

Comparison of the expression changes in these auxin-related genes and levels of IAA and free SA indicated that changes in gene expression should play a role in affecting the levels of IAA and free SA (Figure 6B,C). In 'Intolerant', the changes in the levels of IAA and free SA showed the same trend, while in 'Tolerant 2' and 'Tolerant 1', the changes in the levels of IAA and free SA showed the opposite trend, which indicated completely different strategies between tolerant and intolerant poplars. On the other hand, 'Tolerant 2' and 'Tolerant 1' also likely adopt different strategies in terms of IAA and free SA-related tolerance as they changed differently at different time points. These results suggested that interactions between IAA and free SA were significantly different in the tolerant and intolerant poplars and that IAA and free SA acted in different ways in poplars with different tolerances.

3.6. Expression Characteristics of R2R3-MYB Gene-Related Genes Involved in the Poplar–E4 Interaction

To further explore the differences between tolerant and intolerant poplars in defence-related genes after E4 infection, the expression of *R2R3-MYB* gene-related genes involved in poplar–E4 interactions was investigated. RT-qPCR was performed to validate the changes in transcript levels of the genes. In the yellow module, we found that the expression of many genes was downregulated in the middle and late stages of E4 infection in 'Intolerant'. For example, *enhanced disease susceptibility 1 family protein* (*EDS1*; No. 543) was continuously upregulated after 2 dpi in 'Tolerant 2' and 'Tolerant 1'. However, although the expression of *EDS1* was upregulated at 4 and 7 dpi, there was no significant change in the expression of the gene at 2 dpi in 'Intolerant'. Similar situations occurred for the *pathogenesis-related protein*

(No. 179), *calcium-binding family protein* (No. 196), and *NBS-LRR type disease resistance protein* (No. 516). Some *pathogenesis-related family protein* genes (*PRs*; Nos. 164, 141, 237, and 626) were continuously upregulated after 12 hpi in 'Tolerant 1' and continuously upregulated after 2 dpi in 'Tolerant 2'. However, there were no significant changes in the expression of these *PRs* at 4 dpi in 'Intolerant'. The upregulations of *WRKY transcription factor 72* (No. 305), *calcium-dependent protein kinase* (No. 341), *peroxidase family protein* (No. 408), calmodulin-like protein (No. 699), and *NBS-LRR type disease resistance protein* (No. 1208) did not happen at 4 dpi in 'Intolerant'. The upregulations of *Leucine-rich repeat receptor-like protein kinase* genes (Nos. 380 and 386) and *disease resistance protein* genes (Nos. 428 and 506) did not happen at 2 and 4 dpi in 'Intolerant'. In addition, the upregulations of *Cytochrome P450 family protein* (No. 207), *flavonol synthase/flavanone 3-hydroxylase-like* (No. 276), *peroxidase family protein* (No. 315), *WRKY transcription factor 47 family protein* (No. 509), and *flavanone 3-dioxygenase-like* (No. 602) did not happen at 4 and 7 dpi in 'Intolerant'. However, the upregulation of *auxin-responsive GH3 family protein* (No. 1583) did not happen at 4 dpi, and the downregulation of *aux/IAA family protein* (No. 1494) happened at 4 and 7 dpi in 'Tolerant 1' (Figures S7 and S13 and Table S10).

In the light cyan module, *LRR receptor-like serine/threonine-protein kinase* genes (Nos. 27 and 30), *receptor-like protein* genes (Nos. 43, 85, and 97), and *chitinase* (No. 317) were only downregulated at 2 dpi and/or 4 dpi in 'Intolerant' (Figures S8 and S14 and Table S10). In the black module, *cytochrome P450 family protein* (No. 258) and *PR* (No. 766) were only downregulated at 4 dpi in 'Intolerant' (Figures S9 and S15 and Table S10). In the purple module, *leucine-rich repeat family protein* (No. 1) and *WRKY transcription factor 42 family protein* (No. 206) were only not upregulated at 2 and 4 dpi in 'Intolerant' (Figures S10 and S16 and Table S10). In the dark grey module, *phenylalanine ammonia-lyase family protein* (No. 24) and *cinnamyl alcohol dehydrogenase 6* (No. 11) were only upregulated at 12 hpi and 1 dpi in 'Intolerant' (Figures S11 and S17 and Table S10). In the grey 60 module, most genes were only upregulated at 4 dpi in 'Intolerant', including *protein phosphatase* genes (Nos. 16, 18, 96, 145, and 146), *bZIP transcription factor 6 family protein* (No. 9), and *calcium binding family protein* (No. 138, Figures S12 and S18 and Table S10).

These results suggested that the continuous expression of some defence-related genes in the middle and late stages of E4 infection could play an important role in the timely development of the defence response in tolerant poplars. Additionally, genes involved in flavonoid biosynthesis, phenylpropanoid biosynthesis, and anthocyanin biosynthesis were likely to play an important role in poplar defence, but their responses might be delayed or suppressed in 'Intolerant' in the late stages of E4 infection (Figure 7, Figures S13–S18 and Table S10).

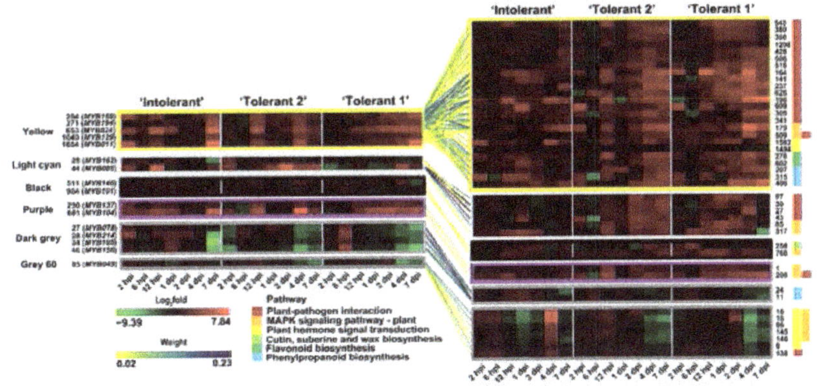

Figure 7. Expression characteristics of *R2R3-MYB* gene-related genes (detailed in Table S10) involved in poplar–E4 interactions.

4. Discussion

In this study, 217 R2R3-MYB-encoding genes were identified by transcriptome sequencing based on the genome sequence of *P. trichocarpa* [24,27]. The transcript profiles of two tolerant poplars ('Tolerant 1' and 'Tolerant 2') and an intolerant poplar ('Intolerant') at different time points after E4 inoculation were investigated to study the roles of *R2R3-MYB* genes and their related genes in the interaction between poplar and E4. Our previous study indicated that the expression of most TF genes of 'Intolerant' did not react to E4 infection, particularly many defence-related TF genes, including *MYB* genes [3,13,14]. The susceptibility of 'Intolerant' may be related to the lack of response of most TFs at the infection phase of E4 [13]. Here, we found that the expression of *R2R3-MYB* genes and their related genes varied greatly at different time points in different poplars, revealing the special role of R2R3-MYBs in poplar defence against E4 infection.

By weighted gene coexpression network analysis, we found that 83 *R2R3-MYB* genes were assigned to 22 different coexpression modules, which had different expression patterns. Among them, 25 *R2R3-MYB* genes were central elements in their respective modules. However, it was found that the expression of most *R2R3-MYB* genes did not change significantly in both tolerant and intolerant poplars in the early period of E4 infection (2 hpi to 1 dpi), suggesting that the defence response was not actively mobilised or that only a few R2R3-MYBs were needed in the early interaction. Then, at 2 and 4 dpi, there were almost no differentially expressed *R2R3-MYB* genes in 'Intolerant', but there was an obvious increase in differentially expressed genes in 'Tolerant 1' and 'Tolerant 2', indicating that the response of some *R2R3-MYB* genes at this period played an important role in the defence response in tolerant poplars.

The major stages of *M. larici-populina* infection include germination and penetration, early colonisation of plant tissue, colonisation of plant mesophyll, and uredinia formation [14,32,33]. In the compatible interaction, stomatal penetration normally occurs from 2 to 6 hpi after the germination of urediniospores, followed by substomatal vesicle formation from 6 to 12 hpi and the development of infectious hyphae from 12 hpi to 1 dpi. Then, the formation of haustoria occurs from 1 to 2 dpi, and dense infection hyphae and haustorial networks grow from 2 to 4 dpi. By 4 dpi, the whole plant mesophyll is colonised by infection structures, and differentiation of the first sporogenous hyphae is observed at this time point [14,32,34]. Therefore, 2–4 dpi can be considered the important biotrophic growth period of E4 during its infection in 'Intolerant'. As an obligate biotrophic fungus, *M. larici-populina* must produce haustoria to derive nutrients from the host to achieve spore production and sporulation and to suppress host defences, enabling its proliferation between 1 and 4 dpi [34,35]. 'Intolerant' should have responded to rust invasion, as several *R2R3-MYB* genes, some other TF genes, and defence-related genes were differentially expressed during the infection phase [13,14,36]. Therefore, E4 must overcome plant surveillance systems in 'Intolerant' [37], and the manipulation of plant defence mechanisms in poplar is most likely to begin at 2 dpi. Differences in gene expression between poplars at 2–4 dpi should be related to differences in their susceptibility.

Many *R2R3-MYB* gene-related genes are involved in plant–pathogen interaction- and signalling-related pathways. We concentrated on genes with different expression patterns between tolerant and intolerant poplars at 2 and 4 dpi. In total, 15 of the 22 modules had 34 *R2R3-MYB* genes with differential expression at 2 and 4 dpi between tolerant and intolerant poplars. Of the 34 differentially expressed *R2R3-MYB* genes, 11 had distinct expression characteristics in 'Tolerant 1', which had confluent necrosis at 7 dpi [3]. Six of these *R2R3-MYB* genes were assigned to the yellow module and showed a gradual increase in expression with increasing inoculation time in 'Tolerant 1'. The other 5 *R2R3-MYB* genes were assigned to the dark grey module and showed a gradual decrease in expression after 6 hpi with the increase in inoculation time in 'Tolerant 1'. The largest difference between 'Tolerant 1' and the other two poplars was that 'Tolerant 1' gradually showed programmed cell death (PCD) in the late stage of infection [3]. Therefore, these results suggested that these *R2R3-MYB* genes might be associated with the regulation of PCD in 'Tolerant 1' after E4 infection [38–42]. However, this correlation needs to be further verified.

KEGG enrichment analysis of genes related to differentially expressed *R2R3-MYB* genes showed that different *R2R3-MYB* genes were involved in regulating different levels of defence-related responses (Figure 8). We focused on modules with more genes enriched in plant–pathogen interaction- and signalling-related pathways, as well as pathways related to the biosynthesis of secondary metabolites that are important in plant stress resistance. The groups of genes associated with defence in different modules were quite different, indicating that different R2R3-MYBs participated in different defence-related processes. It is generally accepted that eukaryotic genes are regulated by more than one TF and that their target genes are also dependent on several TFs [43]. The regulation of plant tolerance to disease is complex, with a number of TF families playing important roles [44]. Here, we found coregulatory relationships between R2R3-MYBs. Additionally, other TFs, such as the AP2 domain-containing transcription factors bZIP and WRKY, were coregulated with R2R3-MYBs.

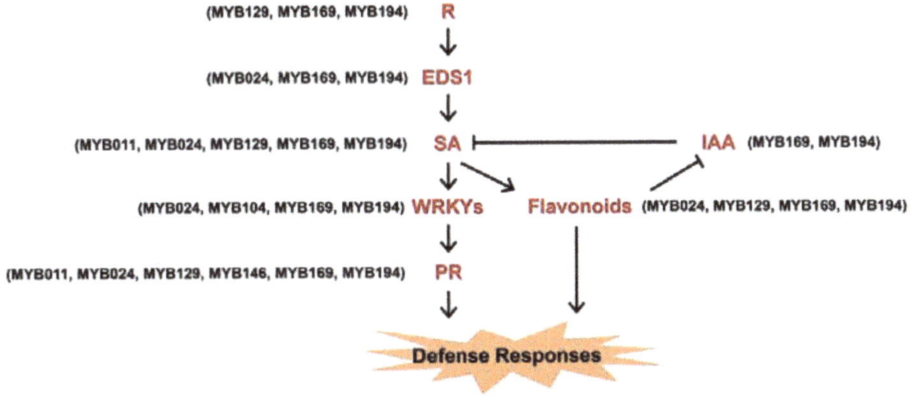

Figure 8. Model of how R2R3-MYBs are involved in regulating defence-related responses. R, disease resistance protein; EDS1, enhanced disease susceptibility 1 family protein; SA, salicylic acid; IAA, indole-3-acetic acid; WRKY, WRKY transcription factor; PR, pathogenesis-related family protein.

We analysed the expression patterns of all differentially expressed *R2R3-MYB* gene-related genes and identified 43 defence-related genes with significant differences between tolerant and intolerant poplars. These genes related to 16 different *R2R3-MYB* genes, most of which were assigned to the yellow module, which was related to free SA [3], followed by the dark grey module, suggesting that many differentially expressed *R2R3-MYB* genes and *R2R3-MYB* gene-related genes might be associated with free SA. Among them, *MYB169* had the highest MM (0.92) and IC (72.81), indicating that it had regulatory relationships with more genes within the module and therefore played an essential regulatory role. Most of the differentially expressed genes were enriched in plant–pathogen interactions and were expressed at low levels or downregulated in 'Intolerant' at 2 and/or 4 dpi. These genes included *enhanced disease susceptibility 1 family protein*, *leucine-rich repeat receptor-like protein kinase*, *disease resistance protein*, *pathogenesis-related family protein*, and *calcium-binding family protein*. These results suggested that most of the differentially expressed *R2R3-MYB* genes and their associated genes played a positive role in the interaction between poplars and E4 to improve defence.

However, genes related to *MYB049* were mainly enriched in plant hormone signal transduction and the MAPK signalling pathway, and their significant downregulation at 1 dpi and upregulation at 4 dpi in 'Intolerant' might have been manipulated by E4, as these two time points were exactly the time points when E4 began and ended bioaccumulation [34]. This suggested that E4 might no longer inhibit the transmission of defensive signals after colonisation, possibly to maintain the physiological state of the leaf and enable it to successfully complete its life cycle [3]. On the other hand, the downregulated expression of these genes in tolerant poplars might weaken signal transmission and reduce the involvement of the defence response in the late stage of E4 infection [3].

Some *flavanone 3-hydroxylase* (also named *flavanone 3-dioxygenase*) genes were downregulated in 'Intolerant' at 2 and/or 4 dpi. Flavanone 3-hydroxylase is involved in the flavonoid biosynthesis pathway, which is part of secondary metabolite biosynthesis. Flavonoids are widely distributed in plants and play important roles. Plants promote the accumulation of flavonoids under stress conditions, resulting in the production of compounds that have, for example, antimicrobial activity (phytoalexins), thereby protecting themselves [45–47]. Overexpression of the *flavanone 3-hydroxylase* gene confers tolerance to both biotic and abiotic stresses, and flavanone 3-hydroxylase induction significantly promoted SA and inhibited JA accumulation [48,49]. The R2R3-MYB family has been demonstrated to act as the main flavonoid biosynthesis regulator in many plant species [50–55]. The ME of the yellow module to which these *flavanone 3-hydroxylase* genes were assigned was associated with free SA [3], so their related R2R3-MYBs (MYB011, MYB024, MYB129, MYB169, and MYB194) might play an important regulatory role in flavonoid biosynthesis and free SA signal transduction, and the accumulation of flavonoids and free SA should be highly correlated with the defence response of poplar to E4 infection.

However, *phenylalanine ammonia-lyase family protein* and *cinnamyl alcohol dehydrogenase 6* were only upregulated at 12 hpi and 1 dpi in 'Intolerant'. A pathogen contacting the plant cell wall is the first signal that triggers the phenylpropanoid pathway for plant defence [56]. Recently, some key players, such as phenylalanine ammonia-lyases and cinnamyl alcohol dehydrogenase from the phenylpropanoid pathway, have been proposed to have broad-spectrum disease resistance [56–59]. Their gene expression is regulated by R2R3-MYBs [60–62]. The upregulation of *phenylalanine ammonia-lyase family protein* and *cinnamyl alcohol dehydrogenase 6* at the infective hyphae development period suggested that in the early developmental stages of the E4 hyphae, some defence-related metabolites were produced in the intolerant poplar, but these substances might not be key defensive substances and were not enough to inhibit further E4 infection.

Although the genes associated with *R2R3-MYB* genes varied greatly from one module to another, we found that some of the genes associated with *R2R3-MYB* genes in the different modules were associated with IAA, namely, *auxin-responsive Gretchen Hagen3 (GH3) family protein*, *aux/IAA family protein*, *auxin-responsive family protein*, and *auxin-induced protein IAA4*. Auxin has long been recognised as a regulator of plant defence [63]. Auxin biosynthesis, transport, and signalling antagonise SA biosynthesis and signalling that is required for resistance to biotrophic pathogens [63]. The changes in free SA and IAA in 'Intolerant' showed the same trend all the time, suggesting that IAA might have been inhibiting the accumulation of free SA. In 'Tolerant 1' and 'Tolerant 2', the free SA and IAA contents showed a more complementary pattern, suggesting that IAA would decrease at some time point to promote SA-dominated defence responses. This difference might eventually lead to different susceptibilities of poplar to E4 infection.

Plants can quickly sense and respond to changes in auxin levels, and these responses involve several major classes of auxin-responsive genes, including the *auxin/indole-3-acetic acid* family, the *auxin response factor* family, *small auxin upregulated RNA*, and the *auxin-responsive GH3 family* [64]. This suggested that there might be multiple R2R3-MYBs involved in IAA-related pathways, and the differential expression of IAA-related genes might be associated with the occurrence of defence responses in poplar–E4 interactions. Among them, the *R2R3-MYB* gene with the highest correlation with IAA levels was *MYB194*, which was assigned to the yellow module. Genes that were related to *MYB194* in the yellow module and had high relationships with IAA levels included *flavanone 3-dioxygenase*, *enhanced disease susceptibility 1 family protein*, *calmodulin-like family protein*, and *auxin-responsive GH3 family protein*. As one of the three major auxin-responsive families, the auxin-responsive GH3 family maintains hormonal homeostasis by conjugating excess IAA, SA, and JA to amino acids during hormone and stress-related signalling [65–68]. Our previous study found that these genes were also related to free SA [3], which is essential in poplar–E4 interactions and defence responses against E4, indicating that MYB194 might be an important node for the convergence of IAA and SA signalling. Flavonoids act as endogenous negative regulators of auxin transport [69] and are possibly involved in SA-related stress signalling [70]. Therefore, R2R3-MYBs were involved in multiple processes from IAA or free SA signal transduction to flavonoid biosynthesis during poplar–E4 interactions (Figure 8).

Considerable interest exists in identifying and utilising key TFs in plant defence to engineer increased resistance to plant pathogens. A comprehensive analysis of the physiological functions and biological roles of the R2R3-MYB family and their related genes in tolerant and intolerant poplars when under attack by E4 is required to fully describe the R2R3-MYB family, and such an analysis will provide rich resources and opportunities to understand rust tolerance in poplar and to screen effective *R2R3-MYB* genes for utilisation of transgenic technology to improve poplar resistance to *M. larici-populina*.

5. Conclusions

In this study, 217 R2R3-MYBs were identified and 83 *R2R3-MYB* genes were assigned to 22 different coexpression modules. Most *R2R3-MYB* genes were unchanged in the early period of E4 infection (2 hpi to 1 dpi) in both tolerant and intolerant poplars. However, there were obvious increases in differentially expressed *R2R3-MYB* genes in 'Tolerant 1' and 'Tolerant 2' later at 2 and 4 dpi, which was an important biotrophic growth period of E4 during its infection of 'Intolerant' and the period when 'Tolerant 1' and 'Tolerant 2' developed hypersensitive cell death responses at the infection sites. These results suggested that the expression of *R2R3-MYB* genes is associated with the occurrence of defence responses, and differently expressed *R2R3-MYB* genes at 2 and 4 dpi between tolerant and intolerant poplars may play an important role in poplar resistance to E4 infection. In total, 34 *R2R3-MYB* genes showed differential expression at 2 and 4 dpi between tolerant and intolerant poplars, and they may participate in different defence-related processes. Among them, 16 differentially expressed *R2R3-MYB* genes were related to 43 defence-related genes that had significant differences between tolerant and intolerant poplars. There might be coregulatory relationships between R2R3-MYBs and other TFs during poplar–E4 interaction. Some differentially expressed *R2R3-MYB* genes were related to genes involved in flavonoid biosynthesis and IAA or free SA signal transduction and might help activate defence response during poplar–E4 interaction. MYB194 could be an important node in the convergence of IAA and SA signalling.

Supplementary Materials: The following supporting information can be downloaded at: https://www.mdpi.com/article/10.3390/f13081255/s1, Figure S1: Correlation of module eigengenes of the modules to which the *R2R3-MYB* genes were assigned; Figure S2: Comparison (log_2E4-inoculated/E4-free) of the differently expressed *R2R3-MYB* genes at different time points in different poplars after E4 infection (colours of the lines represent different genes); Figure S3: Comparison (log_2E4-inoculated/E4-free) of the differentially expressed *R2R3-MYB* genes at different time points in different poplars after E4 infection (colours of the lines represent different modules); Figure S4: High weight value genes in yellow module (detailed in Table S9); Figure S5: High weight value genes in light cyan module (detailed in Table S9); Figure S6: Gene networks for auxin-related genes (detailed in Table S9); Figure S7: Expression characteristics of *R2R3-MYB* gene-related genes involved in poplar–E4 interaction in yellow module; Figure S8: Expression characteristics of *R2R3-MYB* gene-related genes involved in poplar–E4 interaction in light cyan module; Figure S9: Expression characteristics of *R2R3-MYB* gene-related genes involved in poplar–E4 interaction in black module; Figure S10: Expression characteristics of *R2R3-MYB* gene-related genes involved in poplar–E4 interaction in purple module; Figure S11: Expression characteristics of *R2R3-MYB* gene-related genes involved in poplar–E4 interaction in dark grey module; Figure S12: Expression characteristics of *R2R3-MYB* gene-related genes involved in poplar–E4 interaction in grey 60 module; Figure S13: Expression validation of genes in yellow module by RT–qPCR; Figure S14: Expression validation of genes in light cyan module by RT–qPCR; Figure S15: Expression validation of genes in black module by

RT–qPCR; Figure S16: Expression validation of genes in purple module by RT–qPCR; Figure S17: Expression validation of genes in dark grey module by RT–qPCR; Figure S18: Expression validation of genes in grey 60 module by RT–qPCR; Table S1: Primers used in this study; Table S2: Quality statistics of filtered reads; Table S3: Basic information of *R2R3-MYB* genes; Table S4: Number of *R2R3-MYB* genes in each module; Table S5. MMs and ICs of *R2R3-MYB* genes in each module; Table S6: Top 10 pathways for genes that *R2R3-MYB* genes had weight values within each module; Table S7: Differentially expressed *R2R3-MYB* genes; Table S8: Top 10 pathways for genes that differently expressed *R2R3-MYB* genes had weight values with in each module; Table S9: Annotation of differently expressed *R2R3-MYB* genes-related genes that involved in poplar–E4 interaction in the selected modules; Table S10: Differently expressed *R2R3-MYB* genes and their related defence-related genes.

Author Contributions: Conceptualisation, Q.C. and F.W.; methodology, Q.C., F.W. and D.L.; software, Q.C.; validation, Q.C., F.W. and D.L.; formal analysis, Q.C.; investigation, Q.C.; resources, D.L.; data curation, Q.C., F.W. and D.L.; writing—original draft preparation, Q.C.; writing—review and editing, F.W. and D.L.; visualisation, Q.C.; supervision, D.L.; project administration, D.L.; funding acquisition, D.L. All authors have read and agreed to the published version of the manuscript.

Funding: This research was funded by National Natural Science Foundation of China, grant number 31870632.

Data Availability Statement: The data presented in this study are available in the Sequence Read Archive (SRA, accession No. SRR4302070) and Supplementary Materials.

Conflicts of Interest: The authors declare no conflict of interest.

References

1. Pinon, J.; Van Dam, B.C.; Genetet, I.; De Kam, M. Two pathogenic races of *Melampsora larici-populina* in Northwestern Europe. *Eur. J. Forest Pathol.* **1987**, *17*, 47–53. [CrossRef]
2. Pei, M.; Ruiz, C.; Bayon, C.; Hunter, T.; Lonsdale, D. Pathogenic variation in poplar rust *Melampsora larici-populina* from England. *Eur. J. Plant Pathol.* **2005**, *111*, 147–155. [CrossRef]
3. Chen, Q.; Zhang, R.; Li, D.; Wang, F. Integrating transcriptome and coexpression network analyses to characterize salicylic acid- and jasmonic acid-related genes in tolerant poplars infected with rust. *Int. J. Mol. Sci.* **2021**, *22*, 5001. [CrossRef]
4. Zhang, S.; Klessig, D.F. MAPK cascades in plant defense signaling. *Trends Plant Sci.* **2001**, *6*, 520–527. [CrossRef]
5. Meng, X.; Zhang, S. MAPK cascades in plant disease resistance signaling. *Annu. Rev. Phytopathol.* **2013**, *51*, 245–266. [CrossRef]
6. Thulasi Devendrakumar, K.; Li, X.; Zhang, Y. MAP kinase signalling: Interplays between plant PAMP- and effector-triggered immunity. *Cell. Mol. Life Sci.* **2018**, *75*, 2981–2989. [CrossRef]
7. Boudsocq, M.; Sheen, J. CDPKs in immune and stress signaling. *Trends Plant Sci.* **2013**, *18*, 30–40. [CrossRef]
8. Ng, D.W.; Abeysinghe, J.K.; Kamali, M. Regulating the regulators: The control of transcription factors in plant defense signaling. *Int. J. Mol. Sci.* **2018**, *19*, 3737. [CrossRef]
9. Chen, W.; Provart, N.J.; Glazebrook, J.; Katagiri, F.; Chang, H.S.; Eulgem, T.; Mauch, F.; Luan, S.; Zou, G.; Whitham, S.A.; et al. Expression profile matrix of Arabidopsis transcription factor genes suggests their putative functions in response to environmental stresses. *Plant Cell* **2002**, *14*, 559–574. [CrossRef]
10. Nimchuk, Z.; Eulgem, T.; Holt III, B.F.; Dangl, J.L. Recognition and response in the plant immune system. *Annu. Rev. Genet.* **2003**, *37*, 579–609. [CrossRef]
11. Glazebrook, J. Contrasting mechanisms of defense against biotrophic and necrotrophic pathogens. *Annu. Rev. Phytopathol.* **2005**, *43*, 205–227. [CrossRef] [PubMed]
12. Pieterse, C.M.J.; Leon-Reyes, A.; Van der Ent, S.; Van Wees, S.C.M. Networking by small-molecule hormones in plant immunity. *Nat. Chem. Biol.* **2009**, *5*, 308–316. [CrossRef]
13. Chen, Q.; Wang, J.; Li, D.; Wang, Z.; Wang, F.; Zhang, R. Molecular characterization of the transcription factors in susceptible poplar infected with virulent *Melampsora larici-populina*. *Int. J. Mol. Sci.* **2019**, *20*, 4806. [CrossRef] [PubMed]
14. Chen, Q.; Wang, Z.; Li, D.; Wang, F.; Zhang, R.; Wang, J. Molecular characterization of the ERF family in susceptible poplar infected by virulent *Melampsora larici-populina*. *Physiol. Mol. Plant. Pathol.* **2019**, *108*, 101437. [CrossRef]
15. Martin, C.; Paz-Ares, J. MYB transcription factors in plants. *Trends Genet.* **1997**, *13*, 67–73. [CrossRef]
16. Dubos, C.; Stracke, R.; Grotewold, E.; Weisshaar, B.; Martin, C.; Lepiniec, L. MYB transcription factors in *Arabidopsis*. *Trends Plant Sci.* **2010**, *15*, 573–581. [CrossRef]
17. Liu, J.; Osbourn, A.; Ma, P. MYB transcription factors as regulators of phenylpropanoid metabolism in plants. *Mol. Plant* **2015**, *8*, 689–708. [CrossRef]
18. Chezem, W.R.; Memon, A.; Li, F.S.; Weng, J.K.; Clay, N.K. SG2-type R2R3-MYB transcription factor MYB15 controls defense-induced lignification and basal immunity in *Arabidopsis*. *Plant Cell* **2017**, *29*, 1907–1926. [CrossRef]

19. Lee, M.W.; Qi, M.; Yang, Y. A novel jasmonic acid-inducible rice myb gene associates with fungal infection and host cell death. *Mol. Plant Microbe Interact.* **2001**, *14*, 527–535. [CrossRef]
20. Zhang, Z.; Liu, X.; Wang, X.; Zhou, M.; Zhou, X.; Ye, X.; Wei, X. An R2R3 MYB transcription factor in wheat, TaPIMP1, mediates host resistance to *Bipolaris sorokiniana* and drought stresses through regulation of defense- and stress-related genes. *New Phytol.* **2012**, *196*, 1155–1170. [CrossRef]
21. Langfelder, P.; Horvath, S. WGCNA: An R package for weighted correlation network analysis. *BMC Bioinform.* **2008**, *9*, 559. [CrossRef] [PubMed]
22. Wang, R.K.; Cao, Z.H.; Hao, Y.J. Overexpression of a R2R3 MYB gene *MdSIMYB1* increases tolerance to multiple stresses in transgenic tobacco and apples. *Physiol. Plant* **2014**, *150*, 76–87. [CrossRef] [PubMed]
23. Li, D.; Wang, F.; Wang, C.; Zou, L.; Wang, Z.; Chen, Q.; Niu, C.; Zhang, R.; Ling, Y.; Wang, B. MicroRNA-mediated susceptible poplar gene expression regulation associated with the infection of virulent *Melampsora larici-populina*. *BMC Genom.* **2016**, *17*, 59. [CrossRef] [PubMed]
24. Tuskan, G.A.; Difazio, S.; Jansson, S.; Bohlmann, J.; Grigoriev, I.; Hellsten, U.; Putnam, N.; Ralph, S.; Rombauts, S.; Salamov, A.; et al. The genome of black cottonwood, *Populus trichocarpa* (Torr. & Gray). *Science* **2006**, *313*, 1596–1604. [CrossRef]
25. Kim, D.; Langmead, B.; Salzberg, S.L. HISAT: A fast spliced aligner with low memory requirements. *Nat. Methods* **2015**, *12*, 357–360. [CrossRef]
26. Pertea, M.; Pertea, G.M.; Antonescu, C.M.; Chang, T.C.; Mendell, J.T.; Salzberg, S.L. StringTie enables improved reconstruction of a transcriptome from RNA-seq reads. *Nat. Biotechnol.* **2015**, *33*, 290–295. [CrossRef]
27. Wilkins, O.; Nahal, H.; Foong, J.; Provart, N.J.; Campbell, M.M. Expansion and diversification of the Populus R2R3-MYB family of transcription factors. *Plant Physiol.* **2009**, *149*, 981–993. [CrossRef]
28. Shannon, P.; Markiel, A.; Ozier, O.; Baliga, N.S.; Wang, J.T.; Ramage, D.; Amin, N.; Schwikowski, B.; Ideker, T. Cytoscape: A software environment for integrated models of biomolecular interaction networks. *Genome Res.* **2003**, *13*, 2498–2504. [CrossRef]
29. Kanehisa, M.; Goto, S. KEGG: Kyoto encyclopedia of genes and genomes. *Nucleic Acids Res.* **2000**, *28*, 27–30. [CrossRef]
30. Pan, X.; Welti, R.; Wang, X. Quantitative analysis of major plant hormones in crude plant extracts by high-performance liquid chromatography-mass spectrometry. *Nat. Protoc.* **2010**, *5*, 986–992. [CrossRef]
31. Livak, K.J.; Schmittgen, T.D. Analysis of relative gene expression data using real-time quantitative PCR and the $2^{-\Delta\Delta CT}$ method. *Methods* **2001**, *25*, 402–408. [CrossRef] [PubMed]
32. Duplessis, S.; Hacquard, S.; Delaruelle, C.; Tisserant, E.; Frey, P.; Martin, F.; Kohler, A. *Melampsora larici-populina* transcript profiling during germination and timecourse infection of poplar leaves reveals dynamic expression patterns associated with virulence and biotrophy. *Mol. Plant Microbe Interact.* **2011**, *24*, 808–818. [CrossRef] [PubMed]
33. Hacquard, S.; Petre, B.; Frey, P.; Hecker, A.; Rouhier, N.; Duplessis, S. The poplar-poplar rust interaction: Insights from genomics and transcriptomics. *J. Pathog.* **2011**, *2011*, 716041. [CrossRef] [PubMed]
34. Hacquard, S.; Delaruelle, C.; Legué, V.; Tisserant, E.; Kohler, A.; Frey, P.; Martin, F.; Duplessis, S. Laser capture microdissection of uredinia formed by *Melampsora larici-populina* revealed a transcriptional switch between biotrophy and sporulation. *Mol. Plant Microbe Interact.* **2010**, *23*, 1275–1286. [CrossRef]
35. Rinaldi, C.; Kohler, A.; Frey, P.; Duchaussoy, F.; Ningre, N.; Couloux, A.; Wincker, P.; Thiec, D.L.; Fluch, S.; Martin, F.; et al. Transcript profiling of poplar leaves upon infection with compatible and incompatible strains of the foliar rust *Melampsora larici-populina*. *Plant Physiol.* **2007**, *144*, 347–366. [CrossRef] [PubMed]
36. Hamel, L.P.; Benchabane, M.; Major, I.T.; Nicole, M.C.; Sheen, J.; Séguin, A. The genomics of poplar-rust interactions to improve tree resistance against fungal disease. *BMC Proc.* **2011**, *5*, 12. [CrossRef]
37. Dodds, P.N.; Rathjen, J.P. Plant immunity: Towards an integrated view of plant–pathogen interactions. *Nat. Rev. Genet.* **2010**, *11*, 539–548. [CrossRef]
38. Vailleau, F.; Daniel, X.; Tronchet, M.; Montillet, J.L.; Triantaphylidès, C.; Roby, D. A R2R3-MYB gene, AtMYB30, acts as a positive regulator of the hypersensitive cell death program in plants in response to pathogen attack. *Proc. Natl. Acad. Sci. USA* **2002**, *99*, 10179–10184. [CrossRef]
39. Peng, S.Q.; Wu, K.X.; Huang, G.X.; Chen, S.C. HbMyb1, a Myb transcription factor from *Hevea brasiliensis*, suppresses stress induced cell death in transgenic tobacco. *Plant Physiol. Biochem.* **2011**, *49*, 1429–1435. [CrossRef]
40. Kaurilind, E.; Xu, E.; Brosché, M. A genetic framework for H_2O_2 induced cell death in *Arabidopsis thaliana*. *BMC Genom.* **2015**, *16*, 837. [CrossRef]
41. Chen, B.; Niu, F.; Liu, W.Z.; Yang, B.; Zhang, J.; Ma, J.; Cheng, H.; Han, F.; Jiang, Y.Q. Identification, cloning and characterization of R2R3-MYB gene family in canola (*Brassica napus* L.) identify a novel member modulating ROS accumulation and hypersensitive-like cell death. *DNA Res.* **2016**, *23*, 101–114. [CrossRef] [PubMed]
42. Yao, L.; Yang, B.; Xian, B.; Chen, B.; Yan, J.; Chen, Q.; Gao, S.; Zhao, P.; Han, F.; Xu, J.; et al. The R2R3-MYB transcription factor BnaMYB111L from rapeseed modulates reactive oxygen species accumulation and hypersensitive-like cell death. *Plant Physiol. Biochem.* **2020**, *147*, 280–288. [CrossRef] [PubMed]
43. Sorrells, T.R.; Johnson, A.D. Making sense of transcription networks. *Cell* **2015**, *161*, 714–723. [CrossRef] [PubMed]

44. Singh, K.; Foley, R.C.; Oñate-Sánchez, L. Transcription factors in plant defense and stress responses. *Curr. Opin. Plant Biol.* **2002**, *5*, 430–436. [CrossRef]
45. Treutter, D. Significance of flavonoids in plant resistance and enhancement of their biosynthesis. *Plant Biol.* **2005**, *7*, 581–591. [CrossRef]
46. Pourcel, L.; Routaboul, J.M.; Cheynier, V.; Lepiniec, L.; Debeaujon, I. Flavonoid oxidation in plants: From biochemical properties to physiological functions. *Trends Plant Sci.* **2007**, *12*, 29–36. [CrossRef]
47. Hernández, I.; Alegre, L.; Van Breusegem, F.; Munné-Bosch, S. How relevant are flavonoids as antioxidants in plants? *Trends Plant Sci.* **2009**, *14*, 125–132. [CrossRef]
48. Mahajan, M.; Yadav, S.K. Overexpression of a tea flavanone 3-hydroxylase gene confers tolerance to salt stress and *Alternaria solani* in transgenic tobacco. *Plant Mol. Biol.* **2014**, *85*, 551–573. [CrossRef]
49. Jan, R.; Aaqil Khan, M.; Asaf, S.; Lubna; Park, J.R.; Lee, I.J.; Kim, K.M. Flavonone 3-hydroxylase relieves bacterial leaf blight stress in rice via overaccumulation of antioxidant flavonoids and induction of defense genes and hormones. *Int. J. Mol. Sci.* **2021**, *22*, 6152. [CrossRef]
50. Espley, R.V.; Hellens, R.P.; Putterill, J.; Stevenson, D.E.; Kutty-Amma, S.; Allan, A.C. Red colouration in apple fruit is due to the activity of the MYB transcription factor, MdMYB10. *Plant J.* **2007**, *49*, 414–427. [CrossRef]
51. Stracke, R.; Ishihara, H.; Huep, G.; Barsch, A.; Mehrtens, F.; Niehaus, K.; Weisshaar, B. Differential regulation of closely related R2R3-MYB transcription factors controls flavonol accumulation in different parts of the *Arabidopsis thaliana* seedling. *Plant J.* **2007**, *50*, 660–677. [CrossRef] [PubMed]
52. Zhou, H.; Lin-Wang, K.; Wang, H.; Gu, C.; Dare, A.P.; Espley, R.V.; He, H.; Allan, A.C.; Han, Y. Molecular genetics of blood-fleshed peach reveals activation of anthocyanin biosynthesis by NAC transcription factors. *Plant J.* **2015**, *82*, 105–121. [CrossRef] [PubMed]
53. Liu, C.; Long, J.; Zhu, K.; Liu, L.; Yang, W.; Zhang, H.; Li, L.; Xu, Q.; Deng, X. Characterization of a citrus R2R3-MYB transcription factor that regulates the flavonol and hydroxycinnamic acid biosynthesis. *Sci. Rep.* **2016**, *6*, 25352. [CrossRef] [PubMed]
54. Wang, N.; Liu, W.; Zhang, T.; Jiang, S.; Xu, H.; Wang, Y.; Zhang, Z.; Wang, C.; Chen, X. Transcriptomic analysis of red-fleshed apples reveals the novel role of MdWRKY11 in flavonoid and anthocyanin biosynthesis. *J. Agric. Food Chem.* **2018**, *66*, 7076–7086. [CrossRef]
55. Zhai, R.; Zhao, Y.; Wu, M.; Yang, J.; Li, X.; Liu, H.; Wu, T.; Liang, F.; Yang, C.; Wang, Z.; et al. The MYB transcription factor PbMYB12b positively regulates flavonol biosynthesis in pear fruit. *BMC Plant Biol.* **2019**, *19*, 85. [CrossRef]
56. Yadav, V.; Wang, Z.; Wei, C.; Amo, A.; Ahmed, B.; Yang, X.; Zhang, X. Phenylpropanoid pathway engineering: An emerging approach towards plant defense. *Pathogens* **2020**, *9*, 312. [CrossRef]
57. Tronchet, M.; Balagué, C.; Kroj, T.; Jouanin, L.; Roby, D. Cinnamyl alcohol dehydrogenases-C and D, key enzymes in lignin biosynthesis, play an essential role in disease resistance in *Arabidopsis*. *Mol. Plant Pathol.* **2010**, *11*, 83–92. [CrossRef]
58. Rong, W.; Luo, M.; Shan, T.; Wei, X.; Du, L.; Xu, H.; Zhang, Z. A wheat cinnamyl alcohol dehydrogenase TaCAD12 contributes to host resistance to the sharp eyespot disease. *Front. Plant Sci.* **2016**, *7*, 1723. [CrossRef]
59. Liu, X.; Cui, X.; Ji, D.; Zhang, Z.; Li, B.; Xu, Y.; Chen, T.; Tian, S. Luteolin-induced activation of the phenylpropanoid metabolic pathway contributes to quality maintenance and disease resistance of sweet cherry. *Food Chem.* **2021**, *342*, 128309. [CrossRef]
60. Ma, Q.H.; Wang, C.; Zhu, H.H. TaMYB4 cloned from wheat regulates lignin biosynthesis through negatively controlling the transcripts of both cinnamyl alcohol dehydrogenase and cinnamoyl-CoA reductase genes. *Biochimie* **2011**, *93*, 1179–1186. [CrossRef]
61. Craven-Bartle, B.; Pascual, M.B.; Cánovas, F.M.; Avila, C. A Myb transcription factor regulates genes of the phenylalanine pathway in maritime pine. *Plant J.* **2013**, *74*, 755–766. [CrossRef] [PubMed]
62. He, J.; Liu, Y.; Yuan, D.; Duan, M.; Liu, Y.; Shen, Z.; Yang, C.; Qiu, Z.; Liu, D.; Wen, P.; et al. An R2R3 MYB transcription factor confers brown planthopper resistance by regulating the phenylalanine ammonia-lyase pathway in rice. *Proc. Natl. Acad. Sci. USA* **2020**, *117*, 271–277. [CrossRef] [PubMed]
63. Kazan, K.; Manners, J.M. Linking development to defense: Auxin in plant-pathogen interactions. *Trends Plant Sci.* **2009**, *14*, 373–382. [CrossRef] [PubMed]
64. Luo, J.; Zhou, J.J.; Zhang, J.Z. Aux/IAA gene family in plants: Molecular structure, regulation, and function. *Int. J. Mol. Sci.* **2018**, *19*, 259. [CrossRef]
65. Staswick, P.E.; Tiryaki, I.; Rowe, M.L. Jasmonate response locus JAR1 and several related *Arabidopsis* genes encode enzymes of the firefly luciferase superfamily that show activity on jasmonic, salicylic, and indole-3-acetic acids in an assay for adenylation. *Plant Cell* **2002**, *14*, 1405–1415. [CrossRef]
66. Nobuta, K.; Okrent, R.A.; Stoutemyer, M.; Rodibaugh, N.; Kempema, L.; Wildermuth, M.C.; Innes, R.W. The GH3 acyl adenylase family member PBS3 regulates salicylic acid-dependent defense responses in *Arabidopsis*. *Plant Physiol.* **2007**, *144*, 1144–1156. [CrossRef]
67. Zhang, Z.; Li, Q.; Li, Z.; Staswick, P.E.; Wang, M.; Zhu, Y.; He, Z. Dual regulation role of GH3.5 in salicylic acid and auxin signaling during *Arabidopsis-Pseudomonas syringae* interaction. *Plant Physiol.* **2007**, *145*, 450–464. [CrossRef]
68. Jiang, W.; Yin, J.; Zhang, H.; He, Y.; Shuai, S.; Chen, S.; Cao, S.; Li, W.; Ma, D.; Chen, H. Genome-wide identification, characterization analysis and expression profiling of auxin-responsive GH3 family genes in wheat (*Triticum aestivum* L.). *Mol. Biol. Rep.* **2020**, *47*, 3885–3907. [CrossRef]

69. Brown, D.E.; Rashotte, A.M.; Murphy, A.S.; Normanly, J.; Tague, B.W.; Peer, W.A.; Taiz, L.; Muday, G.K. Flavonoids act as negative regulators of auxin transport in vivo in *Arabidopsis*. *Plant Physiol.* **2001**, *126*, 524–535. [CrossRef]
70. Gondor, O.K.; Janda, T.; Soós, V.; Pál, M.; Majláth, I.; Adak, M.K.; Balázs, E.; Szalai, G. Salicylic acid induction of flavonoid biosynthesis pathways in wheat varies by treatment. *Front. Plant Sci.* **2016**, *7*, 1447. [CrossRef]

forests

MDPI

Article

Genome-Wide Identification of miRNAs and Its Downstream Transcriptional Regulatory Network during Seed Maturation in *Tilia tuan*

Xuri Hao [1,2,†], Lei Liu [1,2,†], Peng Liu [1,2], Menglei Wang [1,2] and Yuepeng Song [1,2,3,*]

1 National Engineering Laboratory for Tree Breeding, College of Biological Sciences and Technology, Beijing Forestry University, No. 35, Qinghua East Road, Beijing 100083, China
2 Key Laboratory of Genetics and Breeding in Forest Trees and Ornamental Plants, College of Biological Sciences and Technology, Beijing Forestry University, No. 35, Qinghua East Road, Beijing 100083, China
3 State Key Laboratory of Tree Genetics and Breeding, Northeast Forestry University, No. 51, Hexing Road, Harbin 150040, China
* Correspondence: yuepengsong@bjfu.edu.cn; Tel.: +86-10-62336007; Fax: +86-10-62336164
† These authors contributed equally to this work.

Abstract: Seed maturation not only determines the qualities and yields of seeds, but also affects seed storage and quality preservation. MicroRNAs (miRNAs) are a ubiquitous regulatory factor of gene expression in eukaryotes, which participate in the complex regulatory network of gene expression during seed maturation. However, miRNAs involved in maturation of *Tilia tuan* are still unknown. To reveal the role of miRNAs in *T. tuan*, small RNAs were profiled by high-throughput sequencing during seed maturation at five developmental stages. By predicting the target genes of miRNAs, the expression patterns of miRNAs during seed maturation were analyzed to identify those related to seed maturation. A total of 187 known miRNAs belonging to 42 miRNA families were found at five different seed maturation stages. Based on the analysis of unknown sequences, eight novel miRNAs were identified; 11,775 targets of 195 miRNAs were identified. Large numbers of miRNAs with diverse expression patterns, multiple-targeting and co-targeting of many miRNAs, and a complex regulatory network of miRNA-target genes were identified during seed maturation. These miRNAs and their targets may be involved in fatty acid, ABA, and lignin biosynthesis. Our study provides more information about the miRNA regulatory network and deepens our understanding of the function of miRNAs in *T. tuan*. miRNAs are revealed to be crucial during seed maturation, which provides a basis for further study of the regulatory role of miRNAs during seed maturation.

Keywords: miRNA; *Tilia tuan*; high-throughput sequencing; seed maturation

Citation: Hao, X.; Liu, L.; Liu, P.; Wang, M.; Song, Y. Genome-Wide Identification of miRNAs and Its Downstream Transcriptional Regulatory Network during Seed Maturation in *Tilia tuan*. *Forests* **2022**, *13*, 1750. https://doi.org/10.3390/f13111750

Academic Editor: Claudia Mattioni

Received: 5 October 2022
Accepted: 18 October 2022
Published: 24 October 2022

1. Introduction

The *Tilia tuan Szyszyl.* contains 80 species and is mainly distributed in the northern temperate zone, and intermittently distributed in Europe, Asia, and North America [1]. Because of its beautiful shape, fragrant blossoms, and high capacity to resist hazardous gases, *Tilia tuan* may be employed as a landscape tree species while also serving significant ecological roles. This species also has significant economic value, as it can be used as timber, fiber, a source of honey, and garden ornamental. *T. tuan* seeds are deep dormant, and without germination therapy, they barely germinate in the year after sowing. Dormancy has hampered mating and population growth in this species, and the occurrence of dormancy in *T. tuan* seeds has been acknowledged by numerous experts [2,3]. Many factors contribute to seed dormancy, including seed coat abnormalities, endogenous inhibitors, and physiological post-ripening of seed embryos [4,5]. In particular, *T. tuan* seeds have obvious woody pericarp and are difficult to germinate. However, the molecular mechanisms underlying these traits are still unknown.

MicroRNAs (miRNAs), a class of ~20 to 24 nt non-coding small endogenous RNAs, are important and ubiquitous regulators of gene expression in eukaryotes. Through sequence complementarity, miRNAs bind the mRNA of specific targets to form an RNA-induced silencing complex, which negatively regulates gene expression by initiating mRNA degradation or inhibiting mRNA translation [6,7]. In 1993, a small RNA with negative regulation was first identified while studying the embryonic development of *Caenorhabditis elegans* [8]. In 2002, miRNAs were first discovered in plants [9]. Following that, cloning, sequencing, bioinformatics analysis, and high-throughput sequencing were used to identify a significant number of miRNAs in *Arabidopsis thaliana*, *Oryza sativa*, *Glycine max*, and *Gossypium*, indicating that they play an important role in plant growth and development as well as stress adaptation. Furthermore, studies have revealed that miRNAs are highly conserved and spatiotemporally regulated in plants and participate in several key processes in plant growth and development, including leaf morphogenesis [10,11], flower differentiation and development [12,13], root formation and development [14,15], and the plant transition from the juvenile to the reproductive stage [16]. In addition, they play a significant regulatory role in plant responses to external stresses such as drought stress [17] and salt stress [18,19].

The seed development and maturation program are, to a major extent, regulated by miRNAs, and transcription factors. The involvement of miRNAs in post-transcriptional regulation of seed and fruit development has been documented in apricot [20], rice [21], soybean [22], and *Brassica napus* [23]. Transcription factors play crucial roles in regulating lipid biosynthesis and seed size [24–26]. For example, miR160 negatively regulates auxin response factors involved in *Arabidopsis* seed development [27] and floral organs [28]. The *Sesamum indicum* bHLH transcription factor binds to E- or G-box elements in the FAD2 gene promoter and impacts lipid biosynthesis and accumulation during seed development [29]. MYB89 functions as a negative regulator of seed oil accumulation during maturation in *Arabidopsis* seeds [30]. Accordingly, miRNA-mediated gene expression influences the GA and ABA signal pathways during seed germination in maize [31]. However, although miRNA-mediated regulatory networks controlling seed development have been revealed in model plants, little is known in *T. tuan*.

Above all, it is of great significance for the regulation of seed maturation to reveal the complex miRNA-target gene regulatory network, particularly in terms of fatty acid, ABA, and lignin biosynthesis. In this study, we employed RNA-seq to produce a high-confidence full-length transcriptome dataset of *T. tuan* seed individuals and further used them to identify miRNAs through constructing small RNA libraries in five different seed maturation stages. The related miRNAs present during seed maturation were identified and the target genes were predicted, as well Gene Ontology (GO) and KEGG analyses. Most importantly, specific miRNAs were screened out in fatty acid, ABA, and lignin biosynthesis pathways, with the co-expressed miRNA-target regulatory interactions investigated using transcriptome data. This study provided systematically characterize *T. tuan* seed related miRNAs and the expanded features of putative targets reveal the miRNA inferred regulatory networks during seed maturation, which provides a theoretical basis for further investigation of molecule function of miRNAs during seed maturation in *T. tuan*.

2. Materials and Methods

2.1. Plant Materials

The *T. tuan* seeds were collected from an individual *T. tuan* with good growth and development, at Beijing Forestry University. Five seed samples (Jun. 1, 1 June 2019; Jul. 1, 1 July 2019; Aug. 2, 2 August 2019; Sept. 2, 2 September 2019; and Oct. 2, 2 October 2019) were obtained in *T. tuan* at different seed maturation stages (Figure 1A). Three independent biological replicates of thirty seeds at each maturation stage were collected and immediately frozen in liquid nitrogen, and stored at −80 °C for further analysis.

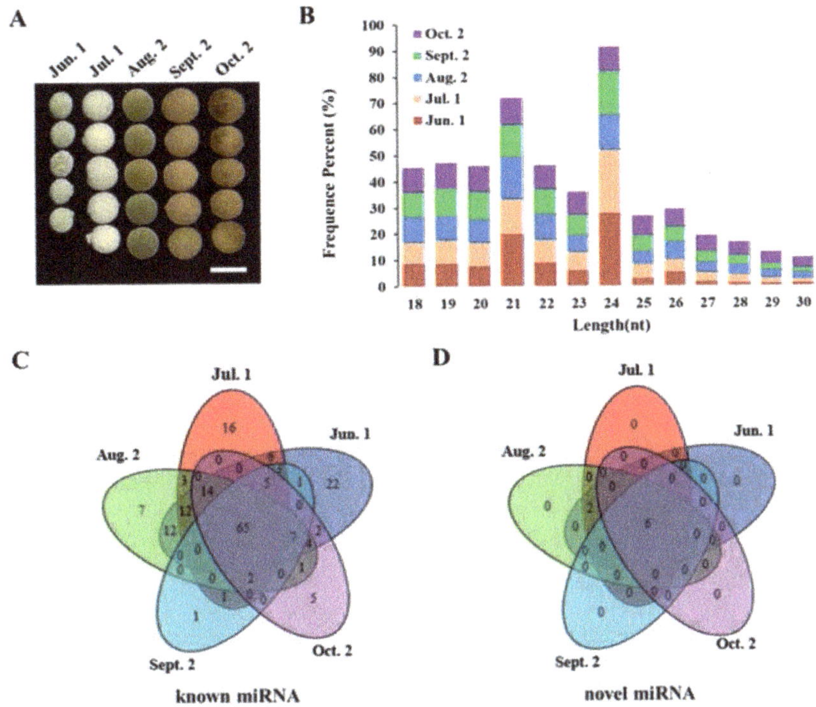

Figure 1. Overview of small RNAs expressed during seed maturation in *T. tuan*. (**A**) Stages of seed maturation used for small RNA sequencing. The images were taken during the sampling period on June 1, July 1, August 2, September 2, and October 2, 2019 (5 time points). Scale bars of 1 cm. (**B**) Size distribution of small RNA at five different seed maturation time stages in *T. tuan*. Jun. 1, Jul. 1, Aug. 2, Sept. 2 and Oct. 2 represent *T. tuan* seed collected at June time point, July time point, August time point, September time point, and October time point 2019, respectively. The abscissa is the length of miRNAs, and the ordinate is the percentage of miRNA at that length. (**C**,**D**) Venn diagram showing known miRNAs and novel miRNAs among five samples.

2.2. Construction of RNA Library and Transcriptome Sequencing

Three biological replicates throughout all five seed maturation stages were used for the transcriptome sequencing. A total amount of 1 μg RNA per sample was used as input material for the RNA sample preparations. Sequencing libraries were generated using NEBNext® Ultra™ RNA Library Prep Kit for Illumina® (New England Biolabs, Ipswich, MA, USA) following manufacturer's recommendations and index codes were added to attribute sequences to each sample. The clustering of the index-coded samples was performed on a cBot Cluster Generation System using TruSeq PE Cluster Kit v3-cBot-HS (Illumia) according to the manufacturer's instructions. After cluster generation, the library preparations were sequenced on an Illumina Novaseq platform and 150 bp paired-end reads were generated.

Raw reads were filtered by removing reads containing adapter, reads containing ploy-N and low-quality reads. The Trinity software [32] with default parameters and a minimum contig length of 150 bp was used for assembly generation. Transcript levels were determined from the short-read data through RSEM [33], with the resulting full-length transcripts used as a reference sequence. The gene level counts were converted into fragments per kilobase of transcript per million mapped reads (FPKM) values.

2.3. Construction of Small RNA Library and High-Throughput Sequencing

Total RNA was extracted by Novogene Company (Beijing, China) for construction of small RNA library and deep sequencing. Detection of total RNA was done using 1% agarose gel to analyze the degree of RNA degradation and contamination, RNA purity was checked using the NanoPhotometer® spectrophotometer (IMPLEN, Westlake Village, CA, USA), RNA concentration was measured using Qubit® RNA Assay Kit in Qubit® 2.0 Flurometer (Life Technologies, Carlsbad, CA, USA), and RNA integrity was assessed using the RNA Nano 6000 Assay Kit of the Agilent Bioanalyzer 2100 system (Agilent Technologies, Santa Clara, CA, USA). After RNA quantification and qualification, a total amount of 3 μg total RNA per sample was used as input material for the small RNA library. Sequencing libraries were generated using NEBNext® Multiplex Small RNA Library Prep Set for Illumina® (New England Biolabs, Ipswich, MA, USA) following manufacturer's recommendations and index codes were added to attribute sequences to each sample. Briefly, RNA bands corresponding to a size range of 16–30 nt were separated and purified from the acrylamide gel.

The small RNA molecules ligated with 5′ and 3′ adaptors were used for reverse transcription and subsequent PCR. After PCR amplification, the target DNA fragments were separated by polyacrylamide gel electrophoresis, and a cDNA library was obtained. At last, library quality was assessed on the Agilent Bioanalyzer 2100 system using DNA High Sensitivity Chips. The clustering of the index-coded samples was performed on a cBot Cluster Generation System using TruSeq SR Cluster Kit v3-cBot-HS (Illumia) according to the manufacturer's instructions. After cluster generation, small RNA samples were sequenced using Illumina HiSeq™ 2500 platform (San Diego, CA, USA) and 50 bp single end reads were generated. The sequencing data obtained in FASTQ files were used for further processing.

2.4. Sequence Data Analysis

Raw reads were obtained from the high-throughput sequencing platform. clean reads were obtained by removing reads containing ploy-N, with 5′ adapter contaminants, without 3′ adapter or the insert tag, containing ploy A or T or G or C and low-quality reads from raw data. At the same time, Q20, Q30, and GC-content of the raw reads were calculated. At the end, all the downstream analyses were performed on sequences in the length range of 18–30 nt. Small RNA derived from rRNAs, tRNAs, small nuclear RNAs (snRNAs), small nucleolar RNAs (snoRNAs), repeat sRNAs, and ta-siRNAs from the databases Rfam14.7 (http://rfam.xfam.org/; accessed on 11 November 2021), GenBank (https://www.ncbi.nlm.nih.gov/genbank/; accessed on 11 November 2021), and Plant Repeat were further identified by Bowtie [34] without mismatches to analyze their expression and distribution on the reference. The high-quality non-redundant set of reads were used for miRNA identification in each sample.

2.5. Identification of Known and Novel miRNAs

The known miRNAs in *T. tuan* were identified using miRBase22.1 (http://www.mirbase.org/; accessed on 15 February 2021) was used as reference by software mirdeep2 [35] and srna-tools-cli (http://srna-tools.cmp.uea.ac.uk/; accessed on 15 February 2021) were used to obtain the potential miRNA and draw the secondary structures. Furthermore, the miRNA counts as well as base bias on the first position of identified miRNA with certain length and on each position of all identified miRNA were obtained, respectively. The characteristics of hairpin structure of miRNA precursor can be used to predict novel miRNAs. The available software miREvo [36] and mirdeep2 [35] were integrated to predict novel miRNAs through exploring the secondary structure, the Dicer cleavage site and the minimum free energy of the small RNA unannotated. Only those miRNAs were detected miRNA* to consider as novel miRNAs. At the same time, the identified miRNA counts as well as base bias on the first position with certain length and on each position of all identified miRNA were obtained, respectively.

2.6. Expression Analysis of miRNA

The expression levels of known and novel miRNAs in each sample were statistically analyzed. miRNA expression levels were estimated by TPM (transcript per million) through the following criteria [37] normalization formula: TPM (transcripts per million reads) = (read count × 1,000,000)/total reads. DESeq2 was used to analyze the differential expression of miRNA [38]. p-value was adjusted using qvalue [39]. miRNAs whose expression levels between any two of the five different seed maturation stages varied significantly ($|\log2$ fold change$| > 1$ and qvalue < 0.01) were assigned as differentially expressed miRNAs by default. Analysis was performed to visualize the expression patterns of miRNAs using the Short Time-series Expression Miner (STEM 1.1) program [40].

2.7. Target Prediction of miRNA and Function Enrichment Analysis

The target genes of known miRNAs and novel miRNAs were predicted by psRobot [41]. Small RNA sequencing results were correlated with the transcriptome data, and target genes corresponding to the differentially expressed key miRNAs were intersected with the differentially expressed genes in the transcriptome. The potential targets of the above key miRNAs were obtained, classified, and functionally annotated. According to the corresponding relationship between miRNAs and their targets, we carried out Gene Ontology (GO) [42] and Kyoto Encyclopedia of Genes and Genomes (KEGG) enrichment analysis [43] on the targets of each group of differentially expressed miRNAs. Finally, Cytoscape (version 3.9.1, Boston, MA, USA) [44] was used to construct a co-expression network.

3. Results

3.1. Small RNA Sequence Statistics

To explore the biological functions of small RNAs during seed maturation in *T. tuan*, five samples were selected during seed maturation (Figure 1A). Small RNA sequencing results showed that raw reads were generated (Table 1). After removing contaminant reads, clean reads were obtained and screened against rRNAs, tRNAs, snRNAs, snoRNAs, and mRNAs in the Rfam and NCBI GenBank databases, resulting in clean reads for further analyses (Supplementary Table S1). They accounted for 92.68%, 94.31%, 96.02%, 87.38%, and 88.77% of the total data, respectively, indicating that the quality of the five small RNA libraries constructed was high (Supplementary Table S2). Length distribution analysis of the sequence shows that most of the fragment lengths are 18–30 nt (Figure 1B). The five libraries shared a similar distribution pattern. The most abundant length of small RNA was 24 nt (~28%), followed by 21 nt (20%), which is the typical length of canonical miRNAs. After analyzing of the small RNA length screening, a total of 4,551,867 (Jun. 1), 11,994,904 (Jul. 1), 7,338,458 (Aug. 2), 8,004,524 (Sept. 2), and 6,323,645 (Oct. 2) total reads of each sample small RNA is obtained (Supplementary Table S3). A total of 1,576,482 (34.63%), 7,945,688 (66.24%), 4,785,383 (65.21%), 6,154,905 (76.89%), 5,191,238 (82.09%) reads that can mapped to the reference sequence in the sample (Table 1). Finally, 2,975,385 (Jun. 1), 4,049,216 (Jul. 1), 2,553,075 (Aug. 2), 1,849,619 (Sept. 2), and 1,132,407 (Oct. 2) unannotated unique small RNA sequences were further analyzed to predict novel miRNAs. To further insight into small RNA function, small RNA after length screening were located on the reference sequence by bowtie, and a comparison of small RNA classification tables is provided in Table 2. In terms of quantity, the proportion of miRNAs was larger than that of other types of non-coding RNAs (snRNAs, snoRNAs, and tRNAs). Among the five libraries, the highest 14,225 reads were annotated as known miRNAs and 28,171 reads were annotated as novel miRNAs, accounting for 0.9% and 0.35% of the total sequences, respectively. The ratios of rRNAs were 34.99%, 37.08%, 29.63%, 35.23%, and 21.50%, all much lower than 60%, indicating that our data is high quality of the sample and reliable.

Table 1. Summary of small RNA sequencing from *T. tuan* seed.

Sample	Jun. 1	Jul. 1	Aug. 2	Sept. 2	Oct. 2
Raw reads	11,780,752	15,366,592	13,279,485	12,378,261	11,573,778
clean reads	10,918,208	14,491,495	12,750,811	10,816,291	10,274,213
Total sRNA	4,551,867	11,994,904	7,338,458	8,004,524	6,323,645
Mapped sRNA	1,576,482	7,945,688	4,785,383	6,154,905	5,191,238

Table 2. Non-coding RNAs among the small RNAs.

Types	Jun. 1	Jun. 1 (Percent)	Jul. 1	Jul. 1 (Percent)	Aug. 2	Aug. 2 (Percent)	Sept. 2	Sept. 2 (Percent)	Oct. 2	Oct. 2 (Percent)
total	1,576,482	100.00%	7,945,688	100.00%	4,785,383	100.00%	6,154,905	100.00%	5,191,238	100.00%
Known miRNA	14,225	0.90%	9212	0.12%	6197	0.13%	6381	0.10%	4432	0.09%
rRNA	551,669	34.99%	2,946,068	37.08%	1,417,839	29.63%	2,168,153	35.23%	1,116,058	21.50%
tRNA	0	0.00%	1	0.00%	0	0.00%	0	0.00%	4	0.00%
snRNA	2162	0.14%	4574	0.06%	2940	0.06%	1507	0.02%	2048	0.04%
snoRNA	2248	0.14%	2374	0.03%	1524	0.03%	1432	0.02%	1271	0.02%
Novel miRNA	24,604	1.56%	28,171	0.35%	23,649	0.49%	9276	0.15%	10,268	0.20%
other	981,574	62.26%	4,955,288	62.36%	3,333,234	69.65%	3,968,156	64.47%	4,057,157	78.15%

3.2. Identification of Known miRNAs during Seed Maturation in *T. tuan*

Comparing to the known miRNAs of 45 selected reference species, 189 unique sequences that showed perfect matches to known miRNAs belonging to 42 miRNA families (with a minimum of two reads) were identified from the five libraries (Figure 1C, Table 3 and Supplementary Table S4). Among them, 23 highly conserved miRNA families were identified in *T. tuan*, indicating that these conserved miRNAs may have fundamental regulatory roles during seed maturation in *T. tuan*. Most members of the miRNA family in *T. tuan* are miR171, which has 21 members, followed by 21, 14, 12, 12, and 11 members from the miR159, miR156, miR395, miR166, and miR167 families, respectively (Supplementary Figure S1). Differential expression of miRNAs was also observed in the five libraries. Among these, 65 miRNAs were expressed in all five libraries, and all of them were conserved. Among those, ttu-miR403 was the most abundant at all five stages, with 130,047, 54,182, 103,165, 46,282, and 12,903 TPM, respectively, followed by ttu-miR394a, ttu-miR156a, and ttu-miR319a; all reads were over 1000 TPM. In addition, the different members in known miRNA families had drastically different expression levels. miR403 (ttu-miR403b) was the most abundant and miR156 (ttu-miR156a) was the second most abundant miRNA in *T. tuan* seeds (Supplementary Table S5). There are a large number of miRNAs in *T. tuan* seeds, indicating that many miRNAs are involved in seed maturation. At the same time, we make a statistical analysis of base bias on the first position of identified miRNA with certain length (Supplementary Figure S2A). In *T. tuan*, the first base in known miRNAs with different lengths varies. For instance, the first nucleic acid is biased to U in 18–21 nt miRNAs, biased toward C in 22–28 nt miRNAs, and biased to U in 29–30 nt miRNAs. Based on statistical analysis of nucleotides at each position of all identified miRNAs, the first and second nucleotides of all known miRNAs in the five libraries were biased towards U, except for those at positions 11 and 21, which were biased towards G; nucleic acids at other sites were biased towards C (Supplementary Figure S3A).

Table 3. Summary of mapped mature and hairpin in known miRNAs.

Types	Total	Jun. 1	Jul. 1	Aug. 2	Sept. 2	Oct. 2
Mapped mature	189	153	127	127	85	105
Mapped hairpin	487	405	372	378	256	304
Mapped uniq sRNA	1935	329	409	418	366	413
Mapped total sRNA	40,447	14,225	9212	6197	6381	4432

3.3. Identification of Novel miRNAs during Seed Maturation in *T. tuan*

In addition to known miRNAs, novel miRNAs were also identified during seed maturation in *T. tuan*. The characteristic hairpin structure of miRNA precursors can be used to predict novel miRNAs. Eight novel miRNAs were identified from 2,458 unique small RNAs (Figure 1D, Table 4 and Supplementary Table S6). These all contained fold-back structure, which has the sequence of the typical hairpin structure (Figure 2A). According to the synthesis mechanism of miRNAs, six miRNA star (*) sequences were detected. The length of these novel miRNAs and miRNAs* varied from 18 to 24 nt (Supplementary Table S6). It is worth noting that no miRNA* was detected for ttu-miR09 and ttu-miR10. There were some differences in the expression of novel miRNAs (Figure 2B). Among them, ttu-miR01 was the most richly expressed specific miRNA in *T. tuan* seeds, which was sequenced in 706,451, 507,841, 721,606, 731,926, and 577,150 TPM of the five libraries, and its expression level was higher than that of some conserved miRNAs. The eight novel miRNAs were divided into two expression patterns. There was synergistic regulation among novel miRNAs, ttu-miR01, ttu-miR04, ttu-miR05, ttu-miR07, ttu-miR08, ttu-miR10 were highly expressed at Jun. 1, Jul. 1, and Aug. 2 stage, and down-regulated at Sept. 2 and Oct. 2 stage. ttu-miR03 was expressed in high abundance only at Sept. 2 and Oct. 2 stage, indicating that novel miRNAs have complementary regulatory functions. Notably, ttu-miR09 were expressed from Jun. 1 to Oct. 2 stage and had two peaks in expression at Jul. 1 and Sept. 2 stage, suggesting that the regulatory mechanism mediated by *T. tuan* specific miRNAs may play a housekeeping functions role during seed maturation. Expression of miRNA* can be grouped into 4 major patterns (Figure 2C and Supplementary Table S7). The majority of miRNA* were preferentially expressed at Jun. 1, Jul. 1, and Aug. 2 stage while were low expressed at Sept. 2 and Oct. 2 stage. ttu-miR03 and ttu-miR05 were strongly expressed only at Jun. 1 stage then decreased to a low level; ttu-miR08 were strongly expressed only at Jul. 1 stage, and ttu-miR07 was down-regulated after Aug. 2 stage; ttu-miR01 and ttu-miR04 were clustered into one group, they only expressed at Jun. 1 and Aug. 2 stage. We found that the expression pattern of novel miRNA* was significantly different from those of novel miRNAs at different stages, indicating that there was no synergistic regulation of miRNA-miRNA*. Further analysis of the distribution of first base of these novel miRNAs showed that the first nucleotide is biased towards U except in 30 nt novel miRNAs, in which the first nucleic acid is biased towards A (Supplementary Figure S2B). The analysis of the distribution of base bias on each position of these novel miRNAs showed that Nucleotides were also biased towards U at positions 1, 3, 4, 14, and 19 of novel miRNAs; towards A at positions 8, 9, 11, 18, and 20; and towards C at the remaining nucleotide sites (Supplementary Figure S3B).

Table 4. Summary of mapped mature and hairpin in novel miRNAs.

Types	Total	Jun. 1	Jul. 1	Aug. 2	Sept. 2	Oct. 2
Mapped mature	8	8	8	8	6	6
Mapped star	6	5	4	3	1	2
Mapped hairpin	8	8	8	8	6	7
Mapped uniq sRNA	2458	307	476	637	411	627
Mapped total sRNA	95,968	24,604	28,171	23,649	9276	10,268

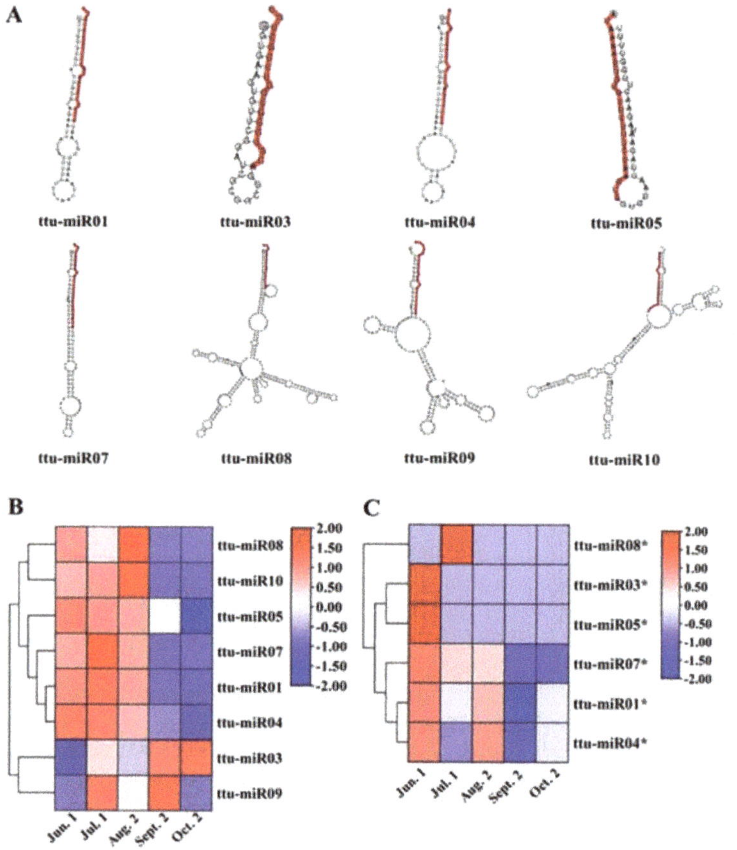

Figure 2. Characterization of novel miRNAs obtained from *T. tuan* seed. (**A**) Examples of stem-loop hairpin secondary structures of predicted novel miRNAs during seed maturation in *T. tuan*. (Segments corresponding to the mature miRNAs are marked in red). Heat map of expressed novel miRNAs (**B**) and novel miRNAs* (**C**) in *T. tuan* seed at five maturational stages.

3.4. Expression Analysis of miRNAs

To understand the expression pattern of differentially expressed miRNAs during *T. tuan* seed maturation and provide clues about their potential function. We first compared the expression patterns of miRNAs at different seed maturation stages; several different expression patterns were observed (Figure 3A and Supplementary Figure S4). To obtain a comprehensive expression profile of known and novel miRNAs, we performed a five-stage time series significance analysis of miRNA expression. Three distinct expression modules were discovered. The red pattern was a monotonously falling mode at Jun. 1 and Jul. 1 stage. The purple pattern shows a significant dynamic expression pattern across time, with a significant decline in Jun. 1 and Jul. 1 stage, followed by a continuous increase at Aug. 2 stage and a decline at Sept. 2 stage (Figure 3B). The green pattern was a monotonously falling mode at Jul. 1, Sept. 2, and Oct. 2 stage. To determine miRNAs that were expressed specifically at the seed maturation stage, hierarchical clustering analysis was performed on the normalized read counts of known miRNAs and novel miRNAs, suggesting that strong stage differential expression of most of the miRNAs (Figure 3C). Differential miRNA clustering analysis was used to determine the clustering pattern of differential miRNA expression under different experimental conditions. To obtain the number of conservatively differentially expressed miRNAs, we depicted in a Venn diagram, which directly showed

62 miRNAs differentially expressed in the whole process of seed maturation (Figure 3D), indicating that these miRNAs participate in regulating seed maturation in *T. tuan*.

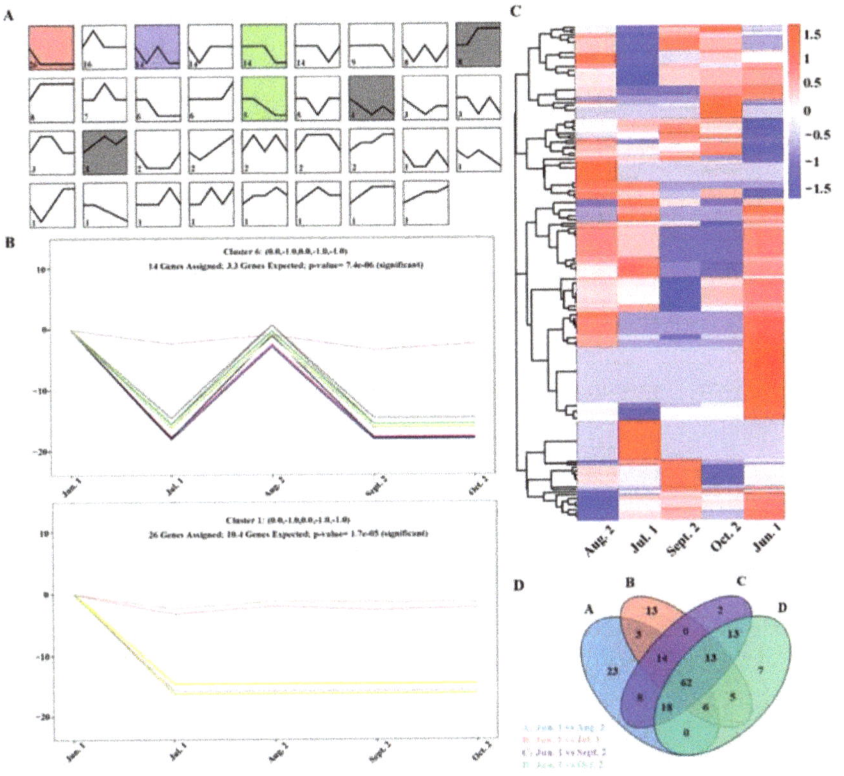

Figure 3. Expressed analysis of miRNAs during seed maturation in *T. tuan*. (**A**) Analysis of miRNAs expression patterns using the Short Time-series Expression Miner (v 1.1, STEM) program. The trend block with color is the trend of significant enrichment, and the different colors are set by the software to distinguish different trends, which have no special significance; the trend block without color is the trend of non-significant enrichment. (**B**) Expression patterns of known miRNAs and novel miRNAs. (**C**) Heat map of expressed known miRNAs and novel miRNAs in *T. tuan* seed at five maturational time stages. (**D**) Venn diagram showing the number of difference miRNA under selection in the four groups. The big circle represents each comparison combination, the sum of the numbers in each big circle represents the total number of difference miRNA of the comparison combination, and the overlapping part of the circle represents the common number of difference miRNA between combinations.

3.5. Target Prediction of miRNAs and Function Enrichment Analysis

To better understand the regulatory functions of miRNAs during seed maturation in *T. tuan*. A total of 11,775 miRNA targets were predicted, and 29,288 miRNA-mRNA pairs, including 11,285 targets for 189 known miRNAs and 757 targets for six novel miRNAs (Figure 4A and Supplementary Table S8). Among them, 268 targets are regulated by both known miRNAs and novel miRNAs. According to the annotation results, the regulatory relationship between miRNAs and targets is not always the same. Some miRNAs regulate multiple targets, and a particular target may be regulated by multiple miRNAs. Among the 11,775 miRNA targets, 4.6% were unknown or had no significant similarity to other genes in the database (Supplementary Table S9). In addition, most of the miRNA targets were plant-specific transcription factors or targets that encode signal transduction pathway components and proteins related to plant metabolism, such as auxin response factors, growth-regulating factors, MYB family transcription factors, and various other proteins (squamosa promoter binding protein) or enzymes (alanyl-tRNA synthetase).

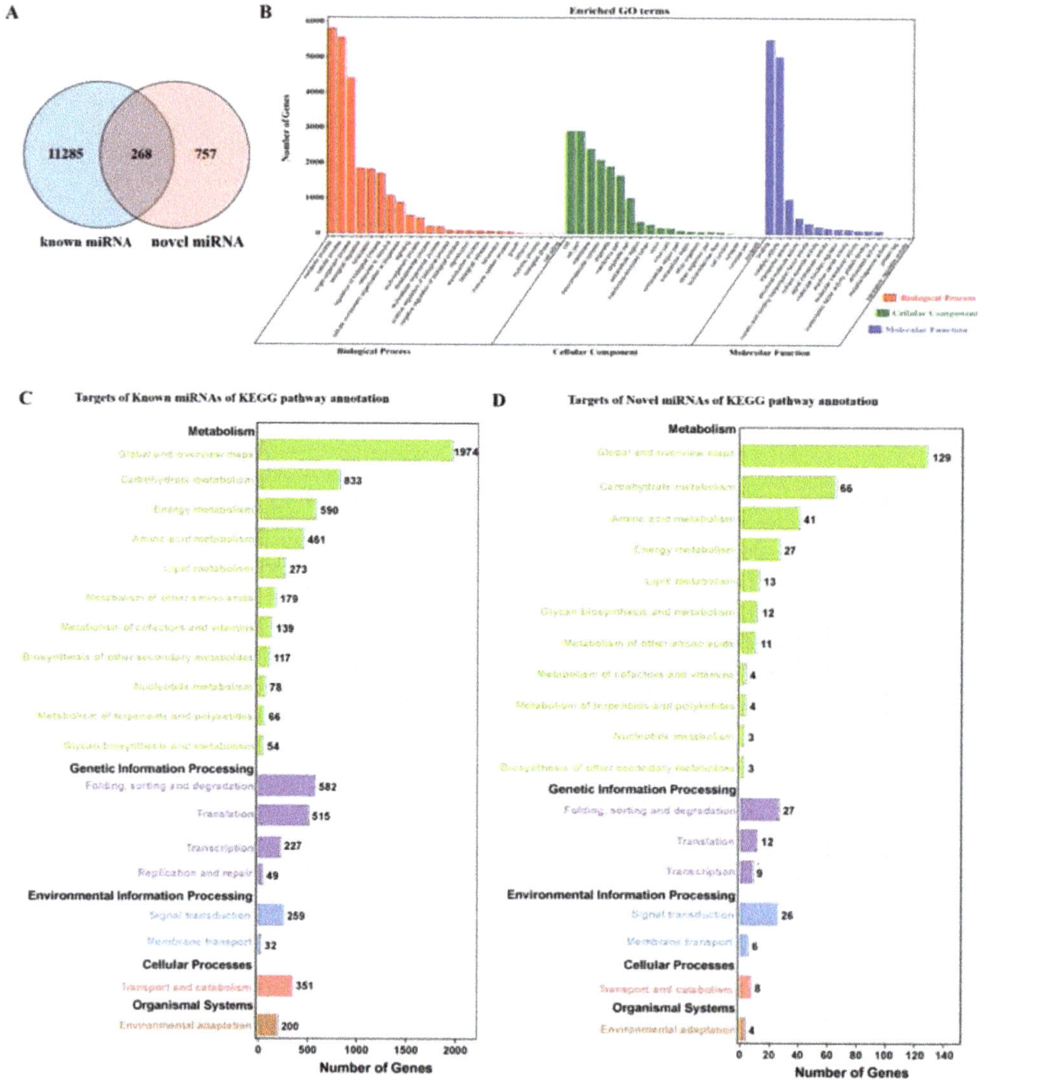

Figure 4. Target prediction of miRNAs and function enrichment analysis. (**A**) Venn diagram showing target prediction of known miRNAs and novel miRNAs. (**B**) Gene ontology (GO) enrichment analysis of the biological functions of targets. Genes were assigned into three main categories: biological processes, cellular components and molecular functions. The y-axis indicates the number of genes in a given category. (**C**) Histogram of cluster of KEGG pathways of known miRNA targets. The results were summarized in five main categories (black words). (**D**) Histogram of cluster of KEGG pathways of novel miRNA targets. The results were summarized in five main categories (black words).

To gain a better understanding of the function of targets in *T. tuan*, we analyzed the function enrichment of gene ontology (GO) and KEGG terms among the genes targeted by miRNAs. GO analysis revealed that the two most abundant types in biological processes were the metabolic process and cellular process; many cellular components were involved, including cell, cell part and membrane; in molecular function, the targets participated in different processes, such as binding, catalytic activity, and transporter activity (Figure 4B). The targets of known miRNAs were annotated by KEGG and matched to 129 different pathways (Figure 4C and Supplementary Figure S5A). Among them, the significantly enriched pathways were MAPK signaling pathway, ABC transporters, Circadian rhythm, and Plant hormone signal transduction. Comparing with targets of known miRNAs, targets of novel miRNAs were annotated by KEGG and matched to 70 different pathways (Figure 4D and Supplementary Figure S5B). The significantly enriched pathways were Plant hormone signal transduction, Glycolysis/Gluconeogenesis, and Glycosylphosphatidylinositol (GPI)-anchor biosynthesis.

3.6. *Transcriptional Regulatory miRNA-mRNA Networks*

miRNAs and their targets are involved in various biochemical metabolic and signal transduction pathways. To explore the biosynthesis and regulatory mechanism of fatty acid, ABA, and lignin biosynthesis during seed maturation in *T. tuan*, we analyzed the expression patterns of related miRNAs and targets (Supplementary Table S10). Fatty acid biosynthesis is an important metabolic process during seed development and maturation, and involves lipid transport and metabolism, biosynthesis of secondary metabolites, and carbohydrate transport and metabolism. A total of 123 pairs of miRNAs and targets are involved in fatty acid and lipid metabolism, including 47 miRNAs and 48 target transcripts. These belong to 16 miRNA families (miR156/157, miR159, miR1511, miR162, miR164, miR166, miR167, miR168, miR171, miR172, miR395, miR396, miR2916, miR5141 and miR8155), and the targets include key enzymes of fatty acid synthesis, such as acetyl-CoA carboxylase1 (ACC1), acyl-carrier-protein (MCAMT), long-chain acyl-CoA synthetase (LACS), stearoyl-ACP desaturase, and fatty acyl-ACP thioesterase B (FATB). These enzymes are important for fatty acid and lipid metabolism and play an important role in the initiation and extension of carbon chains and the Kennedy pathway (Figure 5A and Supplementary Table S11). During seed maturation, ttu-miR8155 expression increased from Jun. 1 to Aug. 2 stage, decreased slightly afterward, and reached a maximal level at Oct. 2 stage. Expression of target KR (PB.22270.1) was highest at Jul. 1 stage and then decreased continuously. Expression of ttu-miR171 was downregulated at seed maturation stage, while target LACS7 (PB.5840.1) was upregulated. ttu-miR2916 and target LACS8 (PB.6192.1) were upregulated. In addition, ttu-miR166 exhibited its highest expression at Jun. 1 stage and then decreased slightly but was not expressed at seed maturation stage. Its target FTM1 (PB.23786.1 and PB.24287.1) increased continuously at Jun. 1, stage and reached its highest level at Sept. 2 stage (Figure 5B,C).

Similarly, ABA plays a significant role in seed development and maturation. A total of 21 pairs of miRNAs and targets are involved in ABA biosynthesis pathways, including 14 miRNAs and 10 target transcripts. These belong to seven miRNA families (miR156, miR157, miR159, miR160, miR171, miR395, and miR397) and the targets include key enzymes of ABA biosynthesis, such as beta-ring hydroxylase (CYP97A3), zeaxanthin epoxidase (ABA1), 9-cis-epoxycarotenoid dioxygenase (NCED5), and xanthoxin dehydrogenase (ABA2). These enzymes play important roles in ABA biosynthesis and in carotenoid pathway starting from β-carotene (Figure 6A and Supplementary Table S11). miR171 is upregulated while its target NCED5 is downregulated at Aug. 2 and Sept. 2 stage. ABA2 is not expressed from Jun. 1 to Aug. 2 stage; however, it is significantly upregulated at Sept. 2 and Oct. 2 stage. The expression levels of miR160 altered dynamically from Jun. 1 to Oct. 2 stage, with the highest level occurring at Oct. 2 stage. ttu-miR162 was only expressed at Aug. 2 stage, while PYL11 (PB.36139.1) showed an opposite trend at Aug. 2 and Sept. 2 stage, with its highest level observed at Sept. 2 stage (Figure 6B, C). Overall, a large number of genes related to ABA biosynthesis were expressed at Sept. 2 stage, and ABA promoted embryo maturation and seed dormancy.

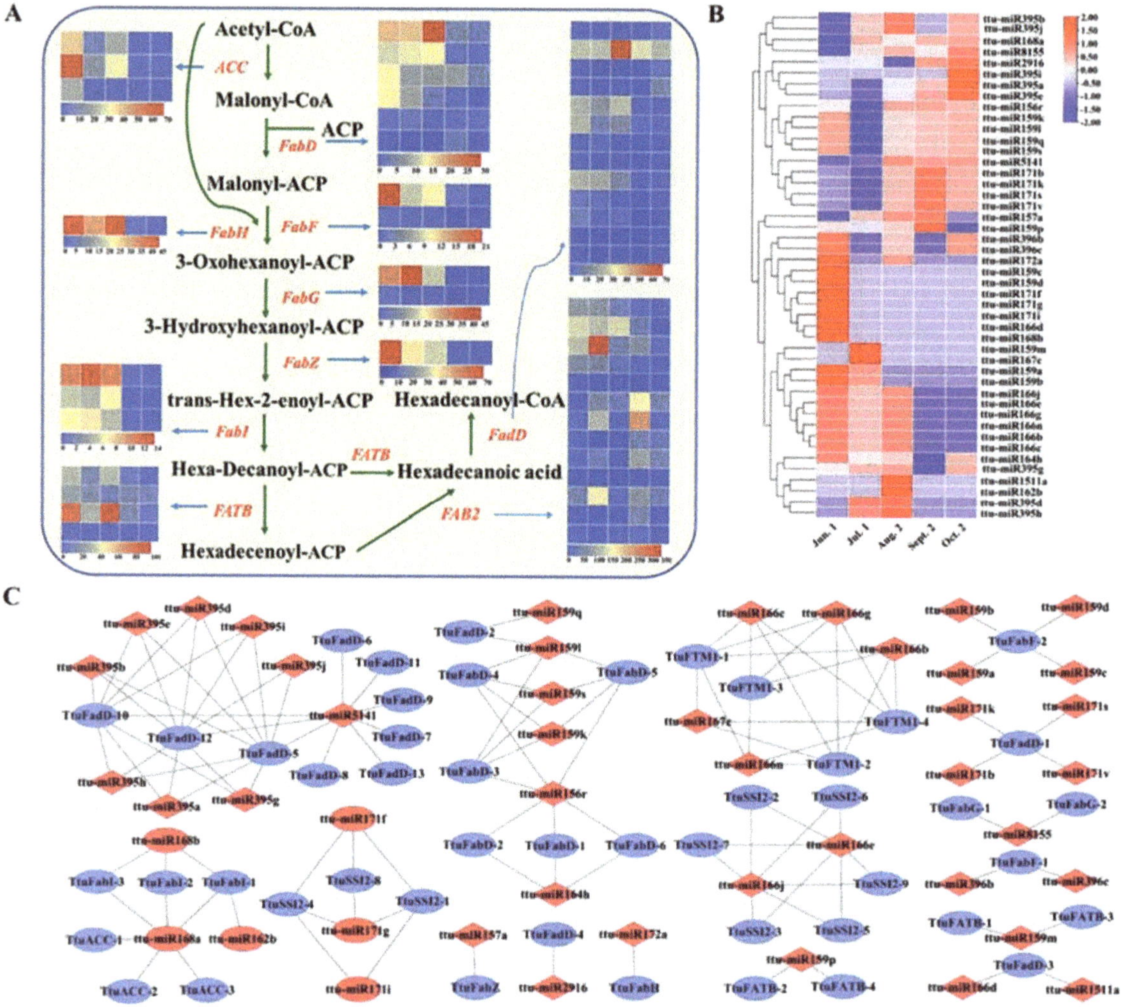

Figure 5. Enrichment analysis of candidate targets in fatty acid biosynthesis pathway. (**A**) Represents targets functions of miRNA involved in fatty acid biosynthesis, Heat maps of gene expression levels (FPKM) at five different seed maturation stages. (**B**) Heat map of expressed miRNAs involved in fatty acid biosynthesis. (**C**) Represents the view of miRNA-mRNA network involved in fatty acid. The red color represents miRNA and the blue color represents targets.

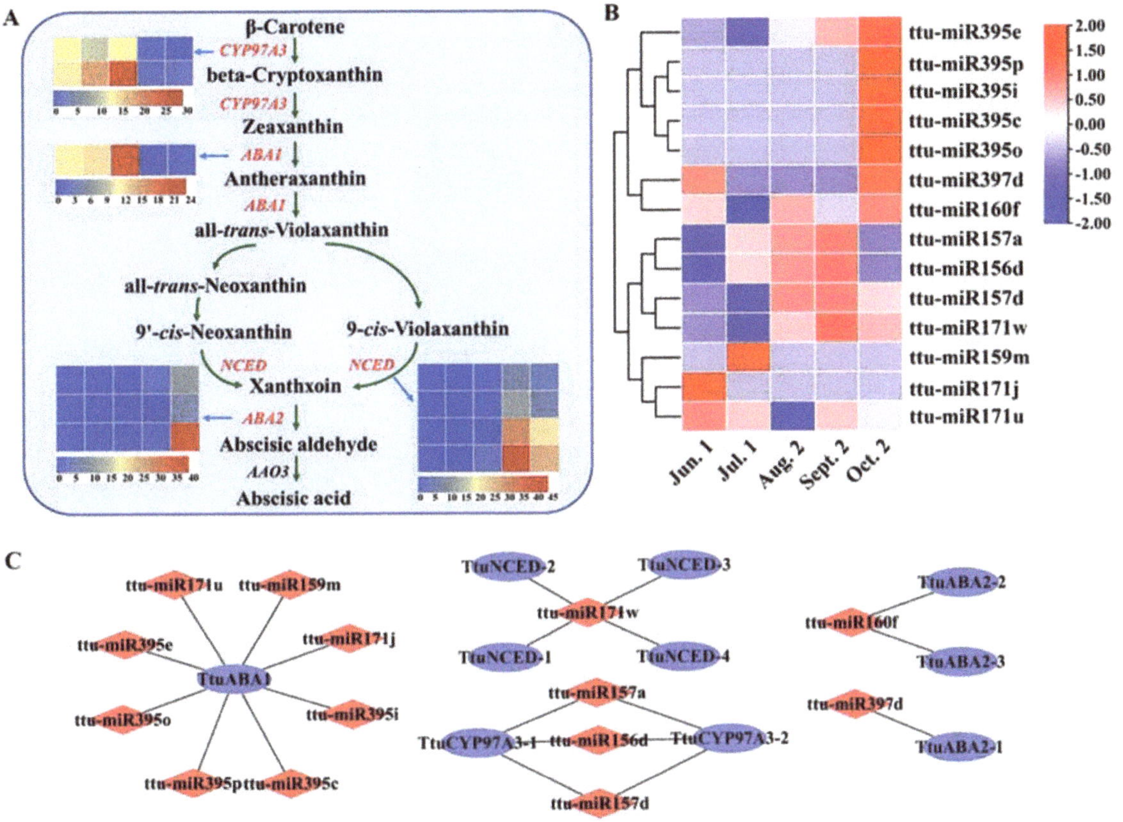

Figure 6. Enrichment analysis of candidate targets in abscisic acid (ABA) pathway. (**A**) Represents miRNA targets functions involved in abscisic acid (ABA) biosynthesis. Heat maps of gene expression levels (FPKM) at five different seed maturation stages. (**B**) Heat map of expressed miRNAs involved in ABA pathway. (**C**) Represents the view of miRNA-mRNA network involved in abscisic acid (ABA) biosynthesis. The red color represents miRNA and the blue color represents targets.

We also found 62 pairs of miRNAs and targets involved in lignin biosynthesis pathways, including 30 miRNAs and 40 target transcripts. These belong to 12 miRNA families (miR156, miR166, miR171, miR172, miR319, miR393, miR396, miR397, miR4995, miR5139, miR1511, and miR8155), among which miR319 was expressed at the highest level, followed by miR5139, miR1511, and miR8155. These targets include some enzymes important for lignin biosynthesis, such as phenylalanine ammonia-lyase (PAL), trans-cinnamate 4-monooxygenase (C4H), 4-coumarate-CoA ligase (4CL), cinnamoyl-CoA reductase (CCR), cinnamyl alcohol dehydrogenase (CAD), and peroxidase (FOD), which all play important roles in lignin biosynthesis and the phenylalanine pathway (Figure 7A and Supplementary Table S11). ttu-miR4995 targets PAL (PB.5109.1, PB.9417.1, PB.10698.1, PB.12095.1, PB.17886.1, PB.24295.1, and PB.31742.1), which is highly expressed from Jun. 1 to Oct. 2 stage, while targets of ttu-miR4995 were downregulated. ttu-miR166 targets C4H (PB.17044.1), and its expression pattern was antagonistic at Jul. 1 to Aug. 2 stage and synergistic in other periods. ttu-miR5139 targets 4CL (PB.11564.1 and PB.11801.1), which is responsible for p-coumaroyl CoA accumulation. Notably, ttu-miR04 targets FOD (PB.26371.1) which is expressed at at Jul. 1 to Aug. 2 stage then decreased to a low level at Sept. 2 and Oct. 2 stage (Figure 7B, C). These results suggest that these differentially expressed miRNAs may be involved in the biosynthesis of fatty acids, ABA, and lignin in *T. tuan* seeds through post-transcriptional regulation.

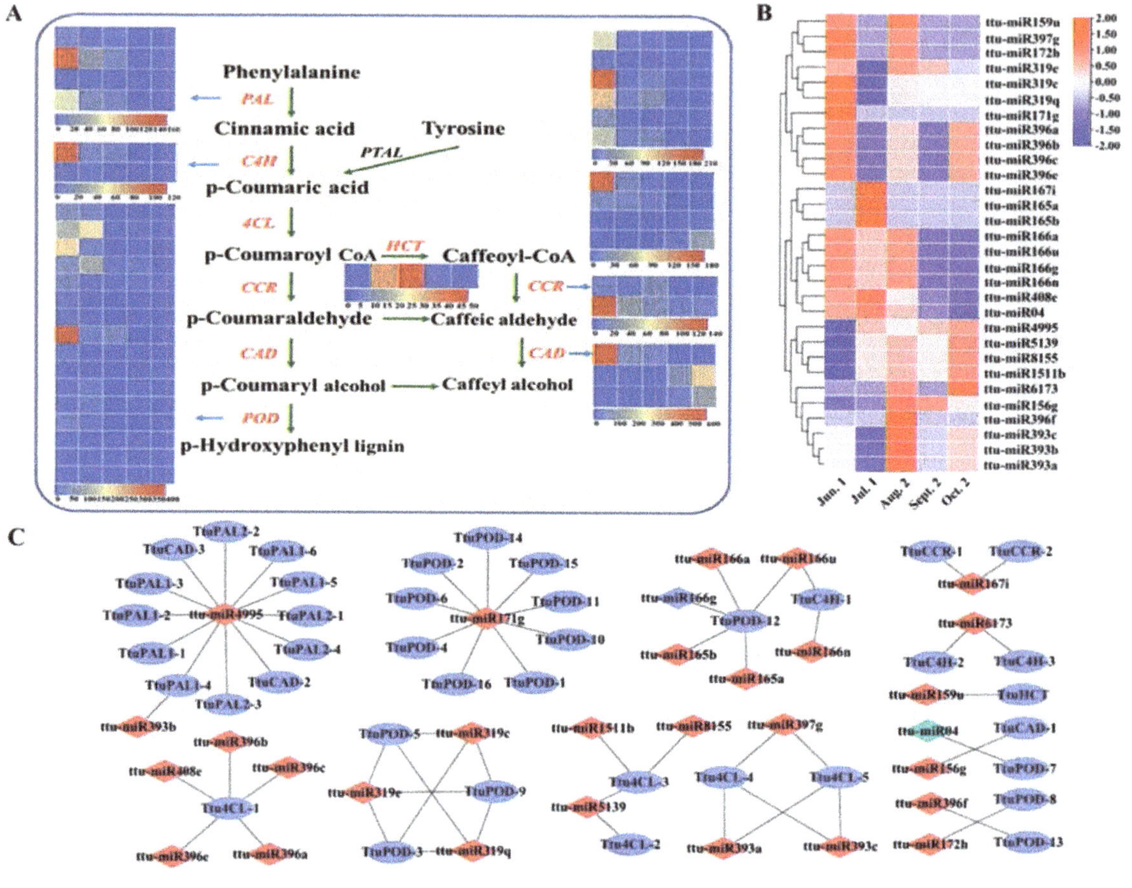

Figure 7. Enrichment analysis of candidate targets in lignin biosynthesis pathway. (**A**) Represents miRNA targets functions involved in lignin biosynthesis. Heat maps of gene expression levels (FPKM) at five different seed maturation stages. (**B**) Heat map of expressed miRNAs involved in lignin biosynthesis pathway. (**C**) Represents the view of miRNA-mRNA network involved in lignin biosynthesis. The red color represents miRNA and the blue color represents targets.

4. Discussion

To identify the potential role of miRNAs in different seed maturation stages in *T. tuan*, we used miRBase to identify hairpin precursor sequences of the newly identified miRNAs, which led to the discovery and annotation of 189 known miRNAs and 8 novel miRNAs hairpin sequences. To date, no miRNAs and corresponding hairpin precursor sequences from *T. tuan* have been deposited in miRBase. These findings have greatly expanded the repertoire of *T. tuan* miRNA genes and provide supporting evidence for newly discovered miRNAs.

miR403 is a member of a miRNA family unique to dicotyledons [45] that plays vital roles in antiviral defense [46,47], stress resistance [48,49], and growth and development of dicotyledons [50]. It may also play a regulatory role in dicotyledons and participate in the regulation of seed maturation. Our data showed that ttu-miR403b was preferentially expressed at Sept. 2 stage, and its expression level was the highest of all known miRNAs, suggesting that miR403 may also play a major regulatory role in *T. tuan* seed maturation. In monocotyledonous plants, miR156 has been reported to be the second most abundant miRNA in rice [51], wheat [52], and barley [53]. Previous studies have shown that it regulates seed dormancy by suppressing the gibberellin pathway through depression of the target gene Ideal Plant Architecture 1 (IPA1) in rice [54]. In *B. napus* seed, the miR156 family is the most abundant family of miRNAs in seeds and mature embryos, most of which are expressed during embryo development. Furthermore, miR156 expression levels increased steadily as the seed maturation [55]. In *Phalaenopsis aphrodite*, miR156 is also highly expressed in different tissue parts such as leaves, roots, flowers, seeds; members of the miR156 family are highly expressed in the seed library [56]. Functional analysis of miR156 has revealed its crucial role during embryogenesis in *Arabidopsis* via regulation of SPL [57] and control of grain size, shape, and quality by *OsSPL16* in rice [58]. miR156 negative regulatory target SPL during seed development in wheat and maize [59,60]. We identified 14 members of the miR156 family in this study. Compared to Jun. 1 stage, expression of ttu-miR156a, ttu-miR156g, ttu-miR156b, and ttu-miR156f was higher at Aug. 2 and Sept. 2 stage, and significantly increased at Sept. 2 stage, confirming that the miR156 module plays an essential role in *T. tuan* seed maturation. These findings imply that miR156 plays a conserved role in seed development and maturation in several plant species.

In this study, these novel miRNAs different expression patterns at five stages. It is worth noting ttu-miR01, ttu-miR04, ttu-miR05, ttu-miR07, ttu-miR08, ttu-miR10 was synergistic regulation among novel miRNAs, and ttu-miR03 plays a complementary regulatory role with them. ttu-miR09 were expressed in both seed development stage and seed maturation stage. GO analysis revealed that the significantly enriched pathways were Plant hormone signal transduction, Glycolysis/Gluconeogenesis, and Glycosylphosphatidylinositol (GPI)-anchor biosynthesis (Figure 4B). The regulatory mechanism mediated by *T. tuan* specific miRNAs may play an important role during seed maturation. In *B. napus* seeds, novel_mir_1706, novel_mir_1407, novel_mir_173, and novel_mir_104 were significantly down-regulated at 21 DAF and 28 DAF, whereas novel_mir_1081, novel_mir_19 and novel_mir_555 were significantly up-regulated in fatty acid biosynthesis during seed development [61]. These results reveal that different novel miRNAs function at different steps via different regulation routes to co-regulate seed development and maturation. Furthermore, we found that the expression pattern of novel miRNA* was significantly different from those of novel miRNA at different stages, and there was no synergistic regulation between miRNA and miRNA*, suggesting that there is a unique molecular mechanism for miRNA* degradation or its role during seed maturation, which is not known at present. According to previous reports, expression pattern of novel miRNA* was also different from novel miRNA in *Arabidopsis*, wheat, and *B. napus* [61–63]. This shows that there is no synergistic regulation between miRNA and miRNA* in wide range of species.

Bioinformatics analysis predicted 11,775 targets for 189 known miRNAs and 757 targets for 6 novel miRNAs. A number of the targets were transcription factors, including growth-regulating factors (GRFs), GRAS family transcription factors, MYB family transcription factors, squamosa promoter binding proteins (SPLs), and auxin response factors (ARFs) (Supplementary Table S9). We determined the biological functions of the genes primarily engaged in regulation using GO enrichment after estimating the relevant targets and their activities. Our analyses revealed that energy metabolism, signal transmission, transcription factors, and gene expression were all involved in seed maturation (Figure 4). we found that some miRNAs were involved in the same pathway (e.g., biochemical metabolic and signal transduction pathways) but targeted different genes. Previous reports have indicated that miR156/157 targets Squamosa-promoter binding proteins (SBPs) or SBP-like proteins (SPLs) [16]. Similarly, ttu-miR5139 targets 4-coumarate-CoA ligase, a key lignin synthetase. In addition, ttu-miR171, ttu-miR2916, ttu-miR1511, and ttu-miR5141 jointly target long-chain acyl-CoA synthetase in the fatty acid biosynthesis pathway. We also discovered a phenomenon in which miRNAs target the same genes. ttu-miR171f targets 9-cis-epoxycarotenoid dioxygenase and ttu-miR4995 targets phenylalanine ammonia-lyase. In

B. napus, miR173, miR400, and miR396 all target pentatricopeptide (PPR) repeat-containing proteins; miR156, miR394, miR319 "co-target" F-box family proteins and miR160, miR167, miR390 and miR156 co-target various ARFs [55]. Our results are consistent with the study on *B. napus* mentioned above and indicate that many miRNAs likely play a role in regulating functionally relevant genes or pathways through multiple-targeting and co-targeting of different miRNAs.

miRNAs and targets are involved in many biochemical metabolic and signal transduction pathways. Plant lipids, in which fatty acids (FAs) are esterified to glycerol, are essential components of the cellular membrane, the major structural and functional barrier of cells and intracellular organelles. Lipids are stored in the form of triacylglycerol (TAG), which is a carbon and energy storage material in seeds [64,65]. In plants, acetyl-CoA carboxylase (ACC), fatty acid synthase (FAS), long-chain acyl-CoA synthetase, and acyl-[acyl-carrier-protein] desaturase are key enzymes for ab initio synthesis of FAs in plastids [66,67]. In a previous study on tree peony seeds, nine miRNAs were involved in fatty acid biosynthesis [68]. In *B. napus* seeds, bna-miR156b, bna-miR156g, bna-miR159, bna-miR395b, bna-miR6029, and 19 novel miRNAs were found to be actively involved in fatty acid biosynthesis using high-throughput sequencing [61]. In this study, ttu-miR166 was highly expressed at Jun. 1 stage and expression gradually decreased with seed maturation; by contrast, its target expression pattern showed a decreasing trend overall. Thus, it is speculated that this miRNA negatively regulates lipids synthesis.

Previous studies have shown that miR166 regulates a variety of developmental processes, such as SAM maintenance; root, stem, leaf, flower, and seed development; and rhizome formation [62,69–72]. However, the exact biological function of miR166 in regulating seed maturation in *T. tuan* remains unclear. The relative expression of ttu-miR171 was highest at Sept. 2 stage and downregulated at Oct. 2 stage. The expression pattern of target LACS2 was the opposite, gradually increasing as a whole. The miR171 family is a highly conserved family with perfect similarity among different species of angiosperm plants. miR171 regulates members of the SCL transcription factor family. In *Arabidopsis*, the targets of miR171 are SCL6 (SCL6-II, SCL6-III, and SCL6-IV), which play an essential role in plant root and leaf development, gibberellin response, phytochrome signaling, lateral organ polarity, meristem formation, vascular development, and stress response [73–76]. We also found that ttu-miR168a regulated ACC synthesis at Jun. 1 stage, ttu-miR8155 regulated FSA synthesis at Jul. 1 stage, and acyl-[acyl-carrier-protein] regulated desaturase synthesis at Aug. 2 and Sept. 2 stage. ttu-miR171 regulates the synthesis of a large number of LACS2 at Oct. 2 stage, indicating these four miRNAs regulate fatty acid biosynthesis in an orderly manner. Further study is needed to elucidate the regulatory mechanisms of miRNAs during seed maturation.

Plant hormones, particularly ABA, are essential regulators of seed dormancy and maturation [77]. Many ABA signal transduction proteins are involved in seed development [78,79]. The inactivation of ABA by 8′-hydroxylase (CYP97A) and the cleavage of carotenoid precursors by 9-cis epoxycarotenoid dioxygenase (NCED) are key steps in regulating metabolism. AtNCED6 and AtNCED9 are necessary for ABA biosynthesis during seed development in *Arabidopsis*. ABA synthesized in the endosperm and embryo participates in the hormone balance that controls seed dormancy and germination [80]. In this study, miR171 targeted CYP97A, miR162 targeted ABA1 and ABA2, and ttu-miR160 targeted NCED, which participates in ABA biosynthesis during seed maturation (Figure 7). CYP97A3 and ABA1 were highly expressed at Aug. 2 stage, while NCED was highly expressed at Sept. 2 stage and ABA2 was highly expressed at Oct. 2 stage. The accumulation of ABA with seed maturation promotes proper embryo growth and maturation, as well as seed shedding. Members of the miR160 family are conserved and play a crucial role in regulating plant morphology [81], enhancing plant resistance [82], regulating flower and embryo development [31], and affecting hormone levels [83]. Until now, functional studies of miR160 and its targets have mainly focused on vegetative and reproductive growth. miR160 can also negatively regulate *AtARF10* and *AtARF16* in *Arabidopsis*, and participate in seed germination and dormancy through the ABA pathway [27,31]. miR162 is involved in a variety of abiotic stress responses in plants. In addition, ABA treatment has been shown to induce miR162 to improve adaptation to drought stress by inhibiting Trehalase precursor 1 (OsTRE1) [84]. In a study on maize, miR162 responded to salt stress, and its accumulation increased 30 min after salt treatment but decreased 5 and 24 h after the treatment [85]. In a study on *Panicum virgatum*, the accumulation of miR162 changed significantly under drought stress [86]. In cotton, miR162 responds to salinity [87]. miR171 was one of the first members of the miRNA family found in plants. It targets Scarecrow-like protein 6, a transcription factor involved in gibberellin signal transduction and gametophyte development in plants; it is also indispensable in plant sex differentiation [88,89]. These results suggest that miR160, miR162, and other key miRNAs affect ABA biosynthesis by inhibiting relevant targets during seed maturation in *T. tuan*. Many studies have found that endogenous inhibitors are a significant cause of seed dormancy, and the most prevalent endogenous inhibitor is ABA.

Lignin is a crucial macromolecular organic matter in plants, which occurs in the thickened secondary cell wall [90]. The poor seed coat permeability caused by lignification is the main reason for the dormancy of *T. tuan* seeds. In the phenylpropanoid biosynthesis pathway, phenylalanine is catalyzed by phenylalanine ammonia-lyase (PAL), trans-cinnamate 4-monooxygenase (C4H), 4-coumarate-CoA ligase (4CL), cinnamyl-alcohol dehydrogenase (CAD), and peroxidase to form p-hydroxyphenyl lignin [91]. Key lignin biosynthesis enzymes are regulated by miRNAs. Previous studies have shown that overexpression of ptr-miR397a in poplar can downregulate the relative expression of 17 laccase genes, resulting in a decrease in lignin content [92]. The overexpression of miR397b in *Arabidopsis* reduces the relative expression of *AtLAC4*, resulting in a decrease in lignin content [93]. In this study, the expression of ttu-miR397 was the highest expressed at Jun. 1 stage, and gradually decreased with seed maturation, while expression of the target 4CL exhibited the opposite. In addition, the levels of ttu-miR4995, ttu-miR5139, ttu-miR1511, and ttu-miR8155 increased gradually with seed maturation, and all reached their highest levels at Oct. 2 stage, while the target PAL decreased gradually overall. Thus, the four miRNAs appear to negatively regulate lignin biosynthesis and have a conserved function. We also found that ttu-miR319 was more highly expressed at Jun. 1 stage compared to other miRNAs, and its expression decreased gradually with seed maturation. A number of studies have revealed that miR319 and its targets play a variety of roles in plant developmental processes, such as leaf morphogenesis, flower development, senescence, and jasmonic acid biosynthesis [11,94–96]. miR319 targets TCP transcription factors (TCP2, TCP3, TCP4, TCP10, and TCP24) to regulate floral formation, and leaf and gametophyte development [97,98]. In tomato, ectopic expression of miR319 downregulates the expression of several TCPs, resulting in larger leaflets and continuous growth of the leaf margin, while decreased miR319 or increased TCP level resulting in a decrease in leaf size [99]. In *Arabidopsis*, the target TCP4 of miR319 can directly bind the promoter of VND7, which regulates lignin biosynthesis, to activate its expression, thereby activating formation of the secondary cell wall and programmed cell death [100]. Thus, miR319 plays a significant role in plant developmental processes and may also regulate seed dormancy. These results reveal that some miRNAs may regulate functional genes directly involved during seed maturation, whereas other miRNAs regulate the seed maturation process by acting on a large number of transcription factors. This thoroughly demonstrates that not only has the seed ripening process been altered (Figure 1), but highly sophisticated metabolic changes have also occurred, and that this process is carried out cooperatively by many regulatory networks (Figures 5–7).

5. Conclusions

miRNAs with diverse expression patterns, multiple-targeting and co-targeting of many miRNAs, and complex relationships between the expression of miRNAs and targets were identified in this study. We identified and characterized the transcriptome of miRNAs in five different seed maturation stages. A total of 189 known miRNAs belonging to 42 miRNA families and 8 novel miRNAs were identified; 62 miRNAs were differentially expressed in five different seed maturation stages. Further joint analysis of transcriptome data at the same stage of seeds showed that there was an antagonistic correlation between the miRNA expression level and the differential expression of target genes. The relative abundance as well as specific temporal and spatial expression patterns of these miRNAs and their targets suggested that miR403, miR156, miR171, miR172, miR396, miR319, and miR397 are major contributors to the network controlling seed maturation through their pivotal roles in plant development. Our results improve our understanding of miRNA-mRNA networks. This work provides new insights into the regulatory mechanisms of miRNAs and targets, offering critical clues to the molecular mechanisms of fatty acid, ABA, and lignin biosynthesis during seed maturation in *T. tuan*. Further research elucidating the molecular mechanism underlying the involvement of these miRNAs in growth and development will be important.

Supplementary Materials: The following supporting information can be downloaded at: https://www.mdpi.com/article/10.3390/f13111750/s1, Figure S1: Number of miRNA family members in *T. tuan* seed; Figure S2: First base composition bias of miRNAs (18–30 nt) in *T. tuan*; Figure S3: Nucleotide bias at each position of miRNAs in *T. tuan*; Figure S4: Several distinct expression patterns of known miRNAs and novel miRNAs in *T. tuan* seed; Figure S5: Kyoto Encyclopedia of Genes and Genomes (KEGG) pathway enrichment analysis of known miRNAs and novel miRNAs targets; Table S1: List of sequencing and mapping of reads on unigenes; Table S2: Summary of small RNA sequencing statistics; Table S3: Summary of small RNA types and quantities; Table S4: The information of known miRNAs in *T. tuan* seed; Table S5: List of known miRNA expression levels in five small RNA libraries; Table S6: The information of novel miRNAs in *T. tuan* seed; Table S7: List of novel miRNA expression levels in five small RNA libraries; Table S8: List of targets prediction results in know miRNAs and novel miRNAs in *T. tuan* seed; Table S9: List of targets annotation in *T. tuan* seed; Table S10: List of miRNA targets expression levels in *T. tuan* seed; Table S11: Key genes expression in crucial pathways in *T. tuan* seed.

Author Contributions: X.H. participated in performing the experiments, data analysis, drafting and revising the manuscript; L.L. participated in performing the experiments, data analysis, drafting the manuscript; P.L. participated in drafting and revised the manuscript; M.W. participated in performing the experiments; Y.S. participated in conceiving the study and revising the manuscript. All authors have read and agreed to the published version of the manuscript.

Funding: This research was supported by: The Opening Project of State Key Laboratory of Tree Genetics and Breeding (Northeast Forestry University) (No. K2019101). The Major Science and Technology Projects of Inner Mongolia Autonomous Region (No. 2021ZD0008), The Project of Youth talent program of Forestry and grass-land Science and Technology Innovation (No. 2020132606) and the 111 Project (No. B20050).

Data Availability Statement: In this section, the sequencing data of this study have been deposited at the National Center for Biotechnology Information Sequence Read Archive (NCBI, SRA, http://www.ncbi.nlm.gov/sra/; accessed on 23 July 2022) under accession number PRJNA861431, PRJNA861958 and PRJNA878938.

Acknowledgments: We are grateful for the sequence information produced by the U.S. Department of Energy Joint Genome Institute (http://www.jgi.doe.gov; accessed on 5 October 2022).

Conflicts of Interest: The authors declare no conflict of interest.

References

1. Tang, Y.; Ren, Z. Geographical distribution of *tilia linn*. *J. Syst. Evol.* **1996**, *34*, 254–264.
2. Yao, W.F.; Shen, Y.B. Effects of gibberellic acid and magnetically treated water on physiological characteristics of *Tilia miqueliana* seeds. *Can. J. Res.* **2018**, *48*, 554–558. [CrossRef]
3. Shi, F.H.; Shen, Y.B.; Shi, J.S. Protection, development and utilization of *Tilia miqueliana* seeds. *J. For. Eng.* **2012**, *26*, 11–14.
4. Wu, Y.; Shen, Y.B. The structural and chemical characteristics of the pericarp are important in *Tilia miqueliana* seed dormancy. *New Forests* **2021**, *52*, 878–888. [CrossRef]

5. Wu, Y.; Shen, Y.B. Dormancy in *Tilia miqueliana* is attributable to permeability barriers and mechanical constraints in the endosperm and seed coat. *Braz. J. Bot.* **2021**, *44*, 725–740. [CrossRef]
6. Llave, C.; Kasschau, K.D.; Rector, M.A.; Carrington, J.C. Endogenous and silencing-associated small RNAs in plants. *Plant Cell* **2002**, *14*, 1605–1619. [CrossRef] [PubMed]
7. Yu, Y.; Jia, T.; Chen, X. The 'how' and 'where' of plant microRNAs. *New Phytol.* **2017**, *216*, 1002–1017. [CrossRef] [PubMed]
8. Lee, R.C.; Feinbaum, R.L.; Ambros, V. The, *C. elegans* heterochronic gene lin-4 encodes small RNAs with antisense complementarity to lin-14. *Cell* **1993**, *75*, 843–854. [CrossRef]
9. Park, W.; Li, J.; Song, R.; Messing, J.; Chen, X. Carpel Factory, a Dicer homolog, and HEN1, a novel protein, act in microRNA metabolism in *Arabidopsis thaliana*. *Curr. Biol.* **2002**, *12*, 1484–1495. [CrossRef]
10. Gao, R.; Wang, Y.; Gruber, M.Y.; Hannoufa, A. miR156/SPL10 modulates lateral root development, branching and leaf morphology in Arabidopsis by silencingAGAMOUS-LIKE 79. Front. *Plant Sci.* **2017**, *8*, 2226. [CrossRef]
11. Palatnik, J.F.; Allen, E.; Wu, X.; Schommer, C.; Schwab, R.; Carrington, J.C.; Weigel, D. Control of leaf morphogenesis by microRNAs. *Nature* **2003**, *425*, 257–263. [CrossRef]
12. Aukerman, M.J.; Sakai, H. Regulation of flowering time and floral organ identity by a microRNA and its APETALA2-like target genes. *Plant Cell* **2003**, *15*, 2730–2741. [CrossRef] [PubMed]
13. Wang, J.W.; Czech, B.; Weigel, D. miR156-regulated SPL transcription factors define an endogenous flowering pathway in *Arabidopsis thaliana*. *Cell* **2009**, *138*, 738–749. [CrossRef]
14. Li, K.; Liu, Z.; Xing, L.; Wei, Y.; Mao, J.; Meng, Y.; Bao, L.; Han, M.; Zhao, C.; Zhang, D. miRNAs associated with auxin signaling, stress response, and cellular activities mediate adventitious root formation in apple rootstocks. *Plant Physiol. Biochem.* **2019**, *139*, 66–81. [CrossRef] [PubMed]
15. Sun, L.; Yang, Y.; Pan, H.; Zhu, J.; Zhu, M.; Xu, T.; Li, Z.; Dong, T. Molecular characterization and target prediction of candidate miRNAs related to abiotic stress responses and/or storage root development in Sweet Potato. *Genes* **2022**, *13*, 110. [CrossRef] [PubMed]
16. Wu, G.; Park, M.Y.; Conway, S.R.; Wang, J.W.; Weigel, D.; Poethig, R.S. The sequential action of miR156 and miR172 regulates developmental timing in *Arabidopsis*. *Cell* **2009**, *138*, 750–759. [CrossRef] [PubMed]
17. Xia, H.; Yu, S.; Kong, D.; Xiong, J.; Ma, X.; Chen, L.; Luo, L. Temporal responses of conserved miRNAs to drought and their associations with drought tolerance and productivity in rice. *BMC Genom.* **2020**, *21*, 232. [CrossRef]
18. Ma, Y.; Xue, H.; Zhang, F.; Jiang, Q.; Yang, S.; Yue, P.; Wang, F.; Zhang, Y.; Li, L.; He, P.; et al. The miR156/SPL module regulates apple salt stress tolerance by activating MdWRKY100 expression. *Plant Biotechnol. J.* **2021**, *19*, 311–323. [CrossRef]
19. Pegler, J.L.; Oultram, J.; Grof, C.; Eamens, A.L. Molecular manipulation of the miR399/PHO2 expression module alters the salt stress response of *Arabidopsis thaliana*. *Plants* **2020**, *10*, 73. [CrossRef]
20. Niu, J.; Wang, J.; An, J.; Liu, L.; Lin, Z.; Wang, R.; Wang, L.; Ma, C.; Shi, L.; Lin, S. Integrated mRNA and miRNA transcriptome reveal a cross-talk between developing response and hormone signaling for the seed kernels of Siberian apricot. *Sci. Rep.* **2016**, *6*, 35675. [CrossRef]
21. Mutum, R.D.; Kumar, S.; Balyan, S.; Kansal, S.; Mathur, S.; Raghuvanshi, S. Identification of novel miRNAs from drought tolerant rice variety Nagina 22. *Sci. Rep.* **2016**, *6*, 30786. [CrossRef] [PubMed]
22. Song, Q.X.; Liu, Y.F.; Hu, X.Y.; Zhang, W.K.; Ma, B.; Chen, S.Y.; Zhang, J.S. Identification of miRNAs and their target genes in developing soybean seeds by deep sequencing. *BMC Plant Biol.* **2011**, *11*, 5. [CrossRef] [PubMed]
23. Wang, Z.; Qiao, Y.; Zhang, J.; Shi, W.; Zhang, J. Genome wide identification of microRNAs involved in fatty acid and lipid metabolism of *Brassica napus* by small RNA and degradome sequencing. *Gene* **2017**, *619*, 61–70. [CrossRef] [PubMed]
24. Khemka, N.; Singh Rajkumar, M.; Garg, R.; Jain, M. Genome-wide profiling of miRNAs during seed development reveals their functional relevance in seed size/weight determination in chickpea. *Plant Direct.* **2021**, *5*, e00299. [CrossRef]
25. Gupta, M.; Bhaskar, P.B.; Sriram, S.; Wang, P.H. Integration of omics approaches to understand oil/protein content during seed development in oilseed crops. *Plant Cell Rep.* **2017**, *36*, 637–652. [CrossRef]
26. Guo, M.; Simmons, C.R. Cell number counts–the fw2.2 and CNR genes and implications for controlling plant fruit and organ size. *Plant Sci.* **2011**, *181*, 1–7. [CrossRef]
27. Liu, P.P.; Montgomery, T.A.; Fahlgren, N.; Kasschau, K.D.; Nonogaki, H.; Carrington, J.C. Repression of AUXIN RESPONSE FACTOR10 by microRNA160 is critical for seed germination and post-germination stages. *Plant J.* **2007**, *52*, 133–146. [CrossRef]
28. Liu, X.; Huang, J.; Wang, Y.; Khanna, K.; Xie, Z.; Owen, H.A.; Zhao, D. The role of *floral organs in carpels*, an Arabidopsis loss-of-function mutation in *MicroRNA160a*, in organogenesis and the mechanism regulating its expression. *Plant J.* **2010**, *62*, 416–428. [CrossRef]
29. Kim, M.J.; Kim, J.K.; Shin, J.S.; Suh, M.C. The SebHLH transcription factor mediates trans-activation of the SeFAD2 gene promoter through binding to E- and G-box elements. *Plant Mol. Biol.* **2007**, *64*, 453–466. [CrossRef]
30. Li, D.; Jin, C.; Duan, S.; Zhu, Y.; Qi, S.; Liu, K.; Gao, C.; Ma, H.; Zhang, M.; Liao, Y.; et al. MYB89 transcription factor represses seed oil accumulation. *Plant Physiol.* **2017**, *173*, 1211–1225. [CrossRef]
31. Liu, J.; Guo, X.; Zhai, T.; Shu, A.; Zhao, L.; Liu, Z.; Zhang, S. Genome-wide identification and characterization of microRNAs responding to ABA and GA in maize embryos during seed germination. *Plant Biol.* **2020**, *22*, 949–957. [CrossRef] [PubMed]

32. Grabherr, M.G.; Haas, B.J.; Yassour, M.; Levin, J.Z.; Thompson, D.A.; Amit, I.; Adiconis, X.; Fan, L.; Raychowdhury, R.; Zeng, Q.; et al. Full-length transcriptome assembly from RNA-Seq data without a reference genome. *Nat. Biotechnol.* **2011**, *29*, 644–652. [CrossRef]
33. Li, B.; Dewey, C.N. RSEM: Accurate transcript quantification from RNA-Seq data with or without a reference genome. *BMC Bioinform.* **2011**, *12*, 323. [CrossRef] [PubMed]
34. Langmead, B.; Trapnell, C.; Pop, M.; Salzberg, S.L. Ultrafast and memory-efficient alignment of short DNA sequences to the human genome. *Genome Biol.* **2009**, *10*, R25. [CrossRef] [PubMed]
35. Friedländer, M.R.; Mackowiak, S.D.; Li, N.; Chen, W.; Rajewsky, N. miRDeep2 accurately identifies known and hundreds of novel microRNA genes in seven animal clades. *Nucleic Acids Res.* **2012**, *40*, 37–52. [CrossRef] [PubMed]
36. Wen, M.; Shen, Y.; Shi, S.; Tang, T. miREvo: An integrative microRNA evolutionary analysis platform for next-generation sequencing experiments. *BMC Bioinform.* **2012**, *13*, 140. [CrossRef]
37. Zhou, L.; Chen, J.; Li, Z.; Li, X.; Hu, X.; Huang, Y.; Zhao, X.; Liang, C.; Wang, Y.; Sun, L.; et al. Integrated profiling of microRNAs and mRNAs: microRNAs located on Xq27.3 associate with clear cell renal cell carcinoma. *PLoS ONE* **2010**, *5*, e15224. [CrossRef]
38. Love, M.I.; Huber, W.; Anders, S. Moderated estimation of fold change and dispersion for RNA-seq data with DESeq2. *Genome Biol.* **2014**, *15*, 550. [CrossRef]
39. Storey, J.D. The positive false discovery rate: A Bayesian interpretation and the q-value. *Ann. Stat.* **2003**, *31*, 2013–2035. [CrossRef]
40. Ernst, J.; Bar-Joseph, Z. STEM: A tool for the analysis of short time series gene expression data. *BMC Bioinform.* **2006**, *7*, 191. [CrossRef]
41. Wu, H.J.; Ma, Y.K.; Tong, C.; Wang, M.; Wang, X.J. Psrobot: A web-based plant small RNA meta-analysis toolbox. *Nucleic Acids Res.* **2012**, *40*, W22–W28. [CrossRef] [PubMed]
42. Young, M.D.; Wakefield, M.J.; Smyth, G.K.; Oshlack, A. Gene ontology analysis for RNA-seq: Accounting for selection bias. *Genome Biol.* **2010**, *11*, R14. [CrossRef] [PubMed]
43. Kanehisa, M.; Araki, M.; Goto, S.; Hattori, M.; Hirakawa, M.; Itoh, M.; Katayama, T.; Kawashima, S.; Okuda, S.; Tokimatsu, T.; et al. KEGG for linking genomes to life and the environment. *Nucleic Acids Res.* **2008**, *36*, D480–D484. [CrossRef] [PubMed]
44. Shannon, P.; Markiel, A.; Ozier, O.; Baliga, N.S.; Wang, J.T.; Ramage, D.; Amin, N.; Schwikowski, B.; Ideker, T. Cytoscape: A software environment for integrated models of biomolecular interaction networks. *Genome Res.* **2003**, *13*, 2498–2504. [CrossRef] [PubMed]
45. Jones-Rhoades, M.W.; Bartel, D.P. Computational identification of plant microRNAs and their targets, including a stress-induced miRNA. *Mol. Cell* **2004**, *14*, 787–799. [CrossRef]
46. Harvey, J.J.; Lewsey, M.G.; Patel, K.; Westwood, J.; Heimstädt, S.; Carr, J.P.; Baulcombe, D.C. An antiviral defense role of AGO2 in plants. *PLoS ONE* **2011**, *6*, e14639. [CrossRef]
47. Bao, D.; Ganbaatar, O.; Cui, X.; Yu, R.; Bao, W.; Falk, B.W.; Wuriyanghan, H. Down-regulation of genes coding for core RNAi components and disease resistance proteins via corresponding microRNAs might be correlated with successful *Soybean mosaic virus* infection in soybean. *Mol. Plant Pathol.* **2018**, *19*, 948–960. [CrossRef]
48. Ebrahimi Khaksefidi, R.; Mirlohi, S.; Khalaji, F.; Fakhari, Z.; Shiran, B.; Fallahi, H.; Rafiei, F.; Budak, H.; Ebrahimie, E. Differential expression of seven conserved microRNAs in response to abiotic stress and their regulatory network in *Helianthus annuus*. *Front. Plant Sci.* **2015**, *6*, 741. [CrossRef]
49. Sun, X.; Xu, L.; Wang, Y.; Yu, R.; Zhu, X.; Luo, X.; Gong, Y.; Wang, R.; Limera, C.; Zhang, K.; et al. Identification of novel and salt-responsive miRNAs to explore miRNA-mediated regulatory network of salt stress response in radish (*Raphanus sativus L.*). *BMC Genom.* **2015**, *16*, 197. [CrossRef]
50. Zhang, C.; Xian, Z.; Huang, W.; Li, Z. Evidence for the biological function of miR403 in tomato development. *Sci. Hortic.* **2015**, *197*, 619–626. [CrossRef]
51. Sunkar, R.; Zhou, X.; Zheng, Y.; Zhang, W.; Zhu, J.K. Identification of novel and candidate miRNAs in rice by high throughput sequencing. *BMC Plant Biol.* **2008**, *8*, 25. [CrossRef] [PubMed]
52. Xin, M.; Wang, Y.; Yao, Y.; Xie, C.; Peng, H.; Ni, Z.; Sun, Q. Diverse set of microRNAs are responsive to powdery mildew infection and heat stress in wheat (*Triticum aestivum* L.). *BMC Plant Biol.* **2010**, *10*, 123. [CrossRef] [PubMed]
53. Schreiber, A.W.; Shi, B.J.; Huang, C.Y.; Langridge, P.; Baumann, U. Discovery of barley miRNAs through deep sequencing of short reads. *BMC Genom.* **2011**, *12*, 129. [CrossRef] [PubMed]
54. Miao, C.; Wang, Z.; Zhang, L.; Yao, J.; Hua, K.; Liu, X.; Shi, H.; Zhu, J.K. The grain yield modulator miR156 regulates seed dormancy through the gibberellin pathway in rice. *Nat. Commun.* **2019**, *10*, 3822. [CrossRef]
55. Huang, D.; Koh, C.; Feurtado, J.A.; Tsang, E.W.; Cutler, A.J. MicroRNAs and their putative targets in *Brassica napus* seed maturation. *BMC Genom.* **2013**, *14*, 140. [CrossRef]
56. Chao, Y.T.; Su, C.L.; Jean, W.H.; Chen, W.C.; Chang, Y.C.; Shih, M.C. Identification and characterization of the microRNA transcriptome of a moth orchid *Phalaenopsis aphrodite*. *Plant Mol. Biol.* **2014**, *84*, 529–548. [CrossRef]
57. Nodine, M.D.; Bartel, D.P. MicroRNAs prevent precocious gene expression and enable pattern formation during plant embryogenesis. *Genes Dev.* **2010**, *24*, 2678–2692. [CrossRef]
58. Wang, S.; Wu, K.; Yuan, Q.; Liu, X.; Liu, Z.; Lin, X.; Zeng, R.; Zhu, H.; Dong, G.; Qian, Q.; et al. Control of grain size, shape and quality by OsSPL16 in rice. *Nat. Genet.* **2012**, *44*, 950–954. [CrossRef] [PubMed]

59. Li, D.; Liu, Z.; Gao, L.; Wang, L.; Gao, M.; Jiao, Z.; Qiao, H.; Yang, J.; Chen, M.; Yao, L.; et al. Genome-wide identification and characterization of microRNAs in developing grains of *Zea mays L. PLoS ONE* **2016**, *11*, e0153168. [CrossRef]
60. Meng, F.; Liu, H.; Wang, K.; Liu, L.; Wang, S.; Zhao, Y.; Yin, J.; Li, Y. Development-associated microRNAs in grains of wheat (*Triticum aestivum* L.). *BMC Plant Biol.* **2013**, *13*, 140. [CrossRef]
61. Wang, J.; Jian, H.; Wang, T.; Wei, L.; Li, J.; Li, C.; Liu, L. Identification of microRNAs actively involved in fatty acid biosynthesis in developing *Brassica napus* Seeds Using High-Throughput Sequencing. *Front. Plant Sci.* **2016**, *7*, 1570. [CrossRef]
62. Tang, X.; Bian, S.; Tang, M.; Lu, Q.; Li, S.; Liu, X.; Tian, G.; Nguyen, V.; Tsang, E.W.; Wang, A.; et al. MicroRNA-mediated repression of the seed maturation program during vegetative development in *Arabidopsis*. *PLoS Genet.* **2012**, *8*, e1003091. [CrossRef] [PubMed]
63. Li, Y.F.; Wei, K.; Wang, M.; Wang, L.; Cui, J.; Zhang, D.; Guo, J.; Zhao, M.; Zheng, Y. Identification and temporal expression analysis of conserved and novel MicroRNAs in the leaves of winter wheat grown in the field. *Front Genet.* **2019**, *10*, 779. [CrossRef] [PubMed]
64. Chapman, K.D.; Ohlrogge, J.B. Compartmentation of triacylglycerol accumulation in plants. *J. Biol. Chem.* **2012**, *287*, 2288–2294. [CrossRef] [PubMed]
65. Bates, P.D.; Stymne, S.; Ohlrogge, J. Biochemical pathways in seed oil synthesis. *Curr. Opin. Plant Biol.* **2013**, *16*, 358–364. [CrossRef] [PubMed]
66. Ohlrogge, J.B.; Jaworski, J.G. Regulation of fatty acid synthesis. *Annu. Rev. Plant Physiol. Plant Mol. Biol* **1997**, *48*, 109–136. [CrossRef]
67. Li-Beisson, Y.; Shorrosh, B.; Beisson, F.; Andersson, M.X.; Arondel, V.; Bates, P.D.; Baud, S.; Bird, D.; Debono, A.; Durrett, T.P.; et al. Acyl-lipid metabolism. *Arab. Book* **2013**, *11*, e0161. [CrossRef] [PubMed]
68. Yin, D.D.; Li, S.S.; Shu, Q.Y.; Gu, Z.Y.; Wu, Q.; Feng, C.Y.; Xu, W.Z.; Wang, L.S. Identification of microRNAs and long non-coding RNAs involved in fatty acid biosynthesis in tree peony seeds. *Gene* **2018**, *666*, 72–82. [CrossRef]
69. Yang, Q.; Liu, S.; Han, X.; Ma, J.; Deng, W.; Wang, X.; Guo, H.; Xia, X. Integrated transcriptome and miRNA analysis uncovers molecular regulators of aerial stem-to-rhizome transition in the medical herb *Gynostemma pentaphyllum*. *BMC Genom.* **2019**, *20*, 865. [CrossRef]
70. Kim, J.; Jung, J.H.; Reyes, J.L.; Kim, Y.S.; Kim, S.Y.; Chung, K.S.; Kim, J.A.; Lee, M.; Lee, Y.; Narry Kim, V.; et al. microRNA-directed cleavage of *ATHB15* mRNA regulates vascular development in Arabidopsis inflorescence stems. *Plant J.* **2005**, *42*, 84–94. [CrossRef] [PubMed]
71. Jung, J.H.; Park, C.M. MIR166/165 genes exhibit dynamic expression patterns in regulating shoot apical meristem and floral development in *Arabidopsis*. *Planta* **2007**, *225*, 1327–1338. [CrossRef]
72. Han, S.Y.; Li, S.G.; Wan, F.; Yang, W.H.; Li, Y.L.; Xu, H.Y. Over-expression of miR166a inhibits cotyledon formation in somatic embryos and promotes lateral root development in seedlings of *Larix leptolepis*. *Plant Cell Tiss. Org.* **2016**, *127*, 461–473. [CrossRef]
73. Llave, C.; Xie, Z.; Kasschau, K.D.; Carrington, J.C. Cleavage of scarecrow-like mRNA targets directed by a class of *Arabidopsis* miRNA. *Science* **2002**, *297*, 2053–2056. [CrossRef] [PubMed]
74. Bolle, C. The role of GRAS proteins in plant signal transduction and development. *Planta* **2004**, *218*, 683–692. [CrossRef] [PubMed]
75. Lee, M.H.; Kim, B.; Song, S.K.; Heo, J.O.; Yu, N.I.; Lee, S.A.; Kim, M.; Kim, D.G.; Sohn, S.O.; Lim, C.E.; et al. Large-scale analysis of the GRAS gene family in *Arabidopsis thaliana*. *Plant Mol. Biol.* **2008**, *67*, 659–670. [CrossRef]
76. Wang, L.; Mai, Y.X.; Zhang, Y.C.; Luo, Q.; Yang, H.Q. MicroRNA171c-targeted SCL6-II, SCL6-III, and SCL6-IV genes regulate shoot branching in *Arabidopsis*. *Mol. Plant* **2010**, *3*, 794–806. [CrossRef] [PubMed]
77. Frey, A.; Godin, B.; Bonnet, M.; Sotta, B.; Marion-Poll, A. Maternal synthesis of abscisic acid controls seed development and yield in *Nicotiana plumbaginifolia*. *Planta* **2004**, *218*, 958–964. [CrossRef] [PubMed]
78. Li, D.; Wang, L.; Liu, X.; Cui, D.; Chen, T.; Zhang, H.; Jiang, C.; Xu, C.; Li, P.; Li, S.; et al. Deep sequencing of maize small RNAs reveals a diverse set of MicroRNA in dry and imbibed seeds. *PLoS ONE* **2013**, *8*, e55107. [CrossRef]
79. Finkelstein, R.; Reeves, W.; Ariizumi, T.; Steber, C. Molecular aspects of seed dormancy. *Annu. Rev. Plant Biol.* **2008**, *59*, 387–415. [CrossRef]
80. Lefebvre, V.; North, H.; Frey, A.; Sotta, B.; Seo, M.; Okamoto, M.; Nambara, E.; Marion-Poll, A. Functional analysis of Arabidopsis NCED6 and NCED9 genes indicates that ABA synthesized in the endosperm is involved in the induction of seed dormancy. *Plant J.* **2006**, *45*, 309–319. [CrossRef]
81. Turner, M.; Nizampatnam, N.R.; Baron, M.; Coppin, S.; Damodaran, S.; Adhikari, S.; Arunachalam, S.P.; Yu, O.; Subramanian, S. Ectopic expression of miR160 results in auxin hypersensitivity, cytokinin hyposensitivity, and inhibition of symbiotic nodule development in soybean. *Plant Physiol.* **2013**, *162*, 2042–2055. [CrossRef] [PubMed]
82. Pinweha, N.; Asvarak, T.; Viboonjun, U.; Narangajavana, J. Involvement of miR160/miR393 and their targets in cassava responses to anthracnose disease. *J. Plant Physiol.* **2015**, *174*, 26–35. [CrossRef] [PubMed]
83. Mallory, A.C.; Bartel, D.P.; Bartel, B. MicroRNA-directed regulation of Arabidopsis AUXIN RESPONSE FACTOR17 is essential for proper development and modulates expression of early auxin response genes. *Plant Cell* **2005**, *17*, 1360–1375. [CrossRef] [PubMed]
84. Tian, C.; Zuo, Z.; Qiu, J.L. Identification and characterization of ABA-responsive MicroRNAs in rice. *J. Genet. Genom.* **2015**, *42*, 393–402. [CrossRef] [PubMed]

85. Ding, D.; Zhang, L.; Wang, H.; Liu, Z.; Zhang, Z.; Zheng, Y. Differential expression of miRNAs in response to salt stress in maize roots. *Ann. Bot.* **2009**, *103*, 29–38. [CrossRef] [PubMed]
86. Sun, G.; Stewart, C.N., Jr.; Xiao, P.; Zhang, B. MicroRNA expression analysis in the cellulosic biofuel crop switchgrass (*Panicum virgatum*) under abiotic stress. *PLoS ONE* **2012**, *7*, e32017. [CrossRef]
87. Salih, H.; Gong, W.; Mkulama, M.; Du, X. Genome-wide characterization, identification, and expression analysis of the WD40 protein family in cotton. *Genome* **2018**, *61*, 539–547. [CrossRef]
88. Chen, J.; Zheng, Y.; Qin, L.; Wang, Y.; Chen, L.; He, Y.; Fei, Z.; Lu, G. Identification of miRNAs and their targets through high-throughput sequencing and degradome analysis in male and female asparagus officinalis. *BMC Plant Bio.* **2016**, *16*, 80. [CrossRef]
89. Ma, Z.; Hu, X.; Cai, W.; Huang, W.; Zhou, X.; Luo, Q.; Yang, H.; Wang, J.; Huang, J. Arabidopsis miR171-targeted scarecrow-like proteins bind to GT cis-elements and mediate gibberellin-regulated chlorophyll biosynthesis under light conditions. *PLoS Genet.* **2014**, *10*, e1004519. [CrossRef]
90. Novaes, E.; Kirst, M.; Chiang, V.; Winter-Sederoff, H.; Sederoff, R. Lignin and biomass: A negative correlation for wood formation and lignin content in trees. *Plant Physiol.* **2010**, *154*, 555–561. [CrossRef]
91. Boerjan, W.; Ralph, J.; Baucher, M. Lignin biosynthesis. *Annu. Rev. Plant Biol.* **2003**, *54*, 519–546. [CrossRef] [PubMed]
92. Lu, S.; Li, Q.; Wei, H.; Chang, M.J.; Tunlaya-Anukit, S.; Kim, H.; Liu, J.; Song, J.; Sun, Y.H.; Yuan, L.; et al. Ptr-miR397a is a negative regulator of laccase genes affecting lignin content in *Populus trichocarpa*. *Proc. Natl. Acad. Sci. USA* **2013**, *110*, 10848–10853. [CrossRef]
93. Wang, C.Y.; Zhang, S.; Yu, Y.; Luo, Y.C.; Liu, Q.; Ju, C.; Zhang, Y.C.; Qu, L.H.; Lucas, W.J.; Wang, X.; et al. MiR397b regulates both lignin content and seed number in *Arabidopsis* via modulating a laccase involved in lignin biosynthesis. *Plant Biotechnol. J.* **2014**, *12*, 1132–1142. [CrossRef] [PubMed]
94. Li, Z.; Li, B.; Shen, W.H.; Huang, H.; Dong, A. TCP transcription factors interact with AS2 in the repression of class-I KNOX genes in *Arabidopsis thaliana*. *Plant J.* **2012**, *71*, 99–107. [CrossRef]
95. Schommer, C.; Palatnik, J.F.; Aggarwal, P.; Chételat, A.; Cubas, P.; Farmer, E.E.; Nath, U.; Weigel, D. Control of jasmonate biosynthesis and senescence by miR319 targets. *PLoS Biol.* **2008**, *6*, e230. [CrossRef] [PubMed]
96. Nag, A.; King, S.; Jack, T. miR319a targeting of TCP4 is critical for petal growth and development in *Arabidopsis*. *Proc. Natl. Acad. Sci. USA* **2009**, *106*, 22534–22539. [CrossRef] [PubMed]
97. Luo, M.; Zhang, Z.M.; Gao, J.; Zeng, X.; Pan, G.T. The role of miR319 in plant development regulation. *Hereditas* **2011**, *33*, 1203–1211. [CrossRef] [PubMed]
98. Schommer, C.; Bresso, E.G.; Spinelli, S.V.; Palatnik, J.F. Role of microRNA miR319 in plant development. *Signal. Commun. Plants* **2012**, *15*, 29–47. [CrossRef]
99. Ori, N.; Cohen, A.R.; Etzioni, A.; Brand, A.; Yanai, O.; Shleizer, S.; Menda, N.; Amsellem, Z.; Efroni, I.; Pekker, I.; et al. Regulation of LANCEOLATE by miR319 is required for compound-leaf development in tomato. *Nat. Genet.* **2007**, *39*, 787–791. [CrossRef]
100. Sun, X.; Wang, C.; Xiang, N.; Li, X.; Yang, S.; Du, J.; Yang, Y.; Yang, Y. Activation of secondary cell wall biosynthesis by miR319-targeted TCP4 transcription factor. *Plant Biotechnol. J.* **2017**, *15*, 1284–1294. [CrossRef]

MDPI
St. Alban-Anlage 66
4052 Basel
Switzerland
Tel. +41 61 683 77 34
Fax +41 61 302 89 18
www.mdpi.com

Forests Editorial Office
E-mail: forests@mdpi.com
www.mdpi.com/journal/forests

www.ingramcontent.com/pod-product-compliance
Lightning Source LLC
Chambersburg PA
CBHW061607210926
51850CB00009B/2238
* 9 7 8 3 0 3 6 5 5 9 4 7 6 *